Non-Being

Non-Being

New Essays on the Metaphysics of Non-Existence

Edited by

SARA BERNSTEIN AND
TYRON GOLDSCHMIDT

OXFORD
UNIVERSITY PRESS

OXFORD
UNIVERSITY PRESS

Great Clarendon Street, Oxford, OX2 6DP,
United Kingdom

Oxford University Press is a department of the University of Oxford.
It furthers the University's objective of excellence in research, scholarship,
and education by publishing worldwide. Oxford is a registered trade mark of
Oxford University Press in the UK and in certain other countries

First Edition published in 2021

Impression: 2

Published in the United States of America by Oxford University Press
198 Madison Avenue, New York, NY 10016, United States of America

British Library Cataloguing in Publication Data
Data available

Library of Congress Control Number: 2020950669

ISBN 978–0–19–884622–2

DOI: 10.1093/oso/9780198846222.001.0001

Printed and bound in the UK by
TJ Books Limited

Dedicated to Bianca and David Bernstein, and Hannah and Michal Goldschmidt

Contents

List of Contributors

Arif Ahmed

Fatema Amijee

Sara Bernstein

Filippo Casati

Roberto Casati

Eddy Keming Chen

Bryan Frances

Naoya Fujikawa

Tyron Goldschmidt

John A. Keller

Lorraine Juliano Keller

Samuel Lebens

Graham Priest

Jacob Ross

Daniel Rubio

Carolina Sartorio

Aaron Segal

Roy Sorensen

Koji Tanaka

Achille C. Varzi

Craig Warmke

Introduction

Sara Bernstein and Tyron Goldschmidt

We are surrounded by things that exist, like chairs, tables, phones, and people. But we are also surrounded by things that don't exist, like holes, shadows, omissions, and negative properties. We read stories of non-existent unicorns and magical creatures. We reason about scenarios that don't exist, from the small ("what if I'd have studied an hour longer?") to the large ("what if World War II hadn't occurred?"). We refer to non-existents ("that paper doesn't exist yet"). And we hold people morally responsible for things that they don't do ("you should have rescued the rabbit!").

Non-existence is ubiquitous, yet mysterious. This volume of new essays covers some of the trickiest questions about non-being and non-existence—from *Could there have been nothing at all?* to *What are holes?*—alongside answers from diverse philosophical traditions. The essays explore analytic, continental, Buddhist, and Jewish philosophical perspectives, and range from metaphysics to ethics, from philosophy of science to philosophy of language, and beyond. While each essay stands alone, they are organized in the following natural groupings.

The first four essays are about fundamental questions of non-being:

Chapter 1 by Sara Bernstein argues that there are different *modes* of non-being, drawing from the contemporary debate about modes of being. She defends *ontological pluralism about non-being*, the view that there are multiple kinds of non-being, and shows how the view applies to various metaphysical problems—about time, absences and fictional objects.

Chapter 2 by Graham Priest argues that nothingness is fundamental to reality. Drawing on work by Heidegger and Nishida, Priest contends that everything (the totality of all objects) and nothing (the absence of all objects) can each be defined as a certain mereological sum. The absence turns out to be a contradictory object, and this contradictory object is the ground of all reality.

Chapter 3 by Roy Sorensen aims to answer an old riddle of Thales: what is older, day or night? Drawing on early insights about the stability of night and day—as well as Lewis Carroll— Sorensen argues that night is older than day and older than the Earth itself.

Sara Bernstein and Tyron Goldschmidt, *Introduction* In: *Non-Being: New Essays on the Metaphysics of Non-Existence.* Edited by: Sara Bernstein and Tyron Goldschmidt, Oxford University Press (2021).
DOI: 10.1093/oso/9780198846222.001.0001

Chapter 4 by Fatema Amijee argues that some negative existential facts are fundamental. She argues that totality facts, facts such that their instances exhaust the relevant domain, are fundamental, and that the usual reasons for rejecting negative facts at the fundamental level do not apply to totality facts.

The next four essays concern sparse ontologies, including the idea that nothing exists:

Chapter 5 by Filippo Casati and Naoya Fujikawa respond to Markus Gabriel's view that the world does not exist. They summarize and formalize Gabriel's argument, show how it does not succeed, and engage with Graham Priest's contribution to this volume along the way.

Chapter 6 by Koji Tanaka explores a Buddhist view that denies the existence of all truths and facts, and how Buddhists have supported this doctrine. He clarifies the meaning of the doctrine, objections against it, and how Buddhists can engage with the objections.

Chapter 7 by Bryan Frances argues for a novel view of how ordinary objects reduce to pluralities of pluralities. The predicate 'is a tree' fails to apply to reality in the familiar way, as 'is an electron' does: 'is a tree' is true of reality because, roughly, there are "tree-unified" pluralities of pluralities of tiny bits that make up a tree. But in a sense 'is a tree' fails to apply to any object, singular or plural.

Chapter 8 by Eddy Keming Chen argues that there is nothing much in time or space. Drawing from work on time's arrow and quantum mechanics, he depicts a fundamental cosmic void, makes sense of appearances to the contrary, and answers philosophical and scientific objections along the way.

The next two chapters concern the influence of negative entities:

Chapter 9 by Roberto Casati and Achille Varzi argues that holes are influential immaterial objects. They explore how the US presidential election of 2000 was ultimately decided by criteria for identifying holes—not their material surroundings, which everyone could detect, but the holes themselves.

Chapter 10 by Aaron Segal argues that it's possible for something to be brought into existence by something that is non-actual. He distinguishes his argument from arguments for causation by omission, and connects the topic to Jewish mystical traditions.

The next two chapters are on non-being and modality:

Chapter 11 by Tyron Goldschmidt and Sam Lebens argues that various modal metaphysics rule out the possibility of there being nothing at all. They conclude that the most prominent pictures of the nature of possibility entail the existence of something, and thus might answer the question of why there is something rather than nothing.

Chapter 12 by Craig Warmke explores the debate over merely possible objects, clarifies the distinction between actualism and possibilism, and reconciles actualism with the reality of possibilities and non-existents. Focusing on late work by Derek Parfit, Warmke proposes and defends an "ostrich actualism" that permits even actualists to quantify over mere possibilities.

The next two chapters focus on language and thought:

Chapter 13 by Lorraine Juliano-Keller and John Keller treats the case of nonsense that appears to make sense. They argue for the existence of what Gareth Evans termed 'illusions of thought', and reply to several arguments, with a focus on those of Herman Cappelen.

Chapter 14 by Arif Ahmed is about the meaning and importance of our counterfactual thoughts. Pursuing a Quinean assumption, he explores why we think and care about what might have existed but does not, even while there are no non-existent things.

The final three chapters focus on the intersection of non-being with broadly normative topics:

Chapter 15 by Jacob Ross clarifies the traditional moral distinction between actions and omissions. He levels various objections against counterfactual and causal ways of drawing the distinction, and proposes instead an explanatory view that avoids the objections while capturing our moral judgments about cases.

Chapter 16 by Carolina Sartorio continues on the topic of acts and omissions, and explores whether and how questions about non-existence and ethics get entangled. Focusing on responsibility for omissions, she shows how metaphysics matters morally in some cases, but not others.

Chapter 17 by Daniel Rubio defends Epicurus's famous argument that death cannot harm us because we no longer exist after we die. Focusing on the deprivationist account of the harm of death, Rubio contends that death is not especially harmful in the ways that are often suggested.

The essays bear on each other in ways not captured by their order, and they also bear on a range of other important philosophical topics not within the direct scope of the volume, including causation, action theory, moral responsibility, and logic, to name just a few. Questions about non-existence and non-being are of interest in themselves, and are connected to myriad philosophical debates. We have made much ado about nothing, and we hope that the breadth and depth of the volume will appeal to a wide audience.

The editors owe thanks to many people for aiding in the creation of this volume, including Yael Goldschmidt, Kris McDaniel, Peter Momtchiloff, and Daniel Nolan. We also wish to thank Mack Sullivan for compiling the index. Finally, thanks to the Thomas J. and Robert T. Rolfs Professorship for its continued research support.

1

Ontological Pluralism about Non-Being

Sara Bernstein

Neither square circles nor manned lunar stations exist. But might they fail to exist in different ways? A common assumption is "no": everything that fails to exist, fails to exist in exactly the same way. Non-being doesn't have joints or structure, the thinking goes—it is just a vast, undifferentiated nothingness. Even proponents of *ontological pluralism*, the view that there are multiple ways of being, do not entertain the possibility of multiple ways of non-being.

This paper is dedicated to the latter idea. I argue that *ontological pluralism about non-being*, roughly, the view that there are multiple ways of non-being, is both more plausible and more defensible than it first seems, and it has many useful applications across a wide variety of metaphysical and explanatory problems.[1]

Here is the plan. In section 1, I lay out ontological pluralism about non-being in detail, drawing on principles of ontological pluralism about being. I address whether and how the two pluralisms interact: some pluralists about non-being are monists about being, and vice-versa. I discuss logical quantification strategies for pluralists about non-being. In section 2, I examine precedent for pluralism about non-being in the history of philosophy. In section 3, I discuss several applications of pluralism about non-being. I suggest that the view has explanatory power across a variety of domains, and that the view can account for differences between non-existent past and future times, between omissions and absences, and between different kinds of fictional objects.

1. Ontological Pluralism

Ontological pluralism, the view that there are multiple fundamental ways of being, has enjoyed a resurgence of popularity in recent years. According to the ontological pluralist, entities can exist differently than each other: a number, for example, exists in a different way than a chair. According to the ontological pluralist, there are several fundamental different ways, modes, or kinds of being: some things exist in different ways than other things. These types of being are fundamental and

[1] Ontological pluralism about non-being holds that there are fundamental differences in types of non-being, not just differences in the characteristics of non-existents.

Sara Bernstein, *Ontological Pluralism about Non-Being* In: *Non-Being: New Essays on the Metaphysics of Non-Existence.* Edited by: Sara Bernstein and Tyron Goldschmidt, Oxford University Press (2021). © Sara Bernstein.
DOI: 10.1093/oso/9780198846222.003.0001

irreducible to each other. For some ontological pluralists, there is no univocal category, *being*, to which all things belong. Rather, there is $being_1$, $being_2$, etc.[2] For other ontological pluralists, there is a univocal category of being that is less fundamental than types of being. I will remain neutral on these different pluralist strands.

Ontological pluralism suggests a connection between something's existence and its essence: there is a relationship between what kind of being something has and the particular sort of thing that it is. A number can $exist_1$, for example, but cannot $exist_2$: a number can never be a chair, no matter how much it changes. Specifically, there is a relationship between a thing's *strict* essence—what it is to be that thing—and the kind of being that it has. If what it is to be a chair is to have four spatially extended legs and a seat, for example, then being a chair implies that the chair is a *concretum*. For the pluralist, questions about an entity's being and its essence overlap heavily.[3]

If there are multiple ways of being, then taking an exhaustive inventory of reality requires more than listing what there is. As Cameron (2018) puts it, ontological pluralism means that there is more structure in the world than we thought there was: an extra dimension of existential sorting for which we must account. Drawing on the Quinean connection between existence and existential quantification, contemporary friends of ontological pluralism like Turner (2010; forthcoming) and McDaniel (2009) take seriously the idea that any theory that accurately describes reality makes use of more than one singular first-order existential quantifier in order to represent this extra structure. For some pluralists, these multiple restricted quantifiers are more "natural" than the singular unrestricted existential quantifier—they describe reality in a more accurate and finer-grained way.

Suppose that a pluralist takes there to be a fundamental difference between *abstracta* and *concreta*. When she says that there are numbers and there are chairs, she means that there are_1 numbers and there are_2 chairs. Both existential quantifiers, $\exists_1$ and $\exists_2$, carve nature at the joints: the existential quantifiers $\exists_1$ and $\exists_2$ are more fundamental than $\exists$.[4] If one is taking an inventory of everything that there is, the pluralist's "is" is ambiguous between $\exists_1$ and $\exists_2$, and the items in being must be sorted into either category. The pluralist's inventory is finer-grained than the list that falls in the domain of the single first-order existential quantifier, since it includes everything that there either is_1 or is_2.

[2] Canonical forms of ontological pluralism take there to be two equally fundamental ways of being, but there might be more than two.

[3] See McDaniel (2017, chapter 9) for a historically-rooted discussion of the relationship essence and existence.

[4] There is some debate about whether the pluralist should recognize a generic quantifier that ranges over all of being, with more fundamental restrictions, or simply deny that there is a generic quantifier. I do not take a stand on this issue here, but see Rettler (forthcoming) for an interesting take. See Simmons (forthcoming) for a detailed look at whether the pluralist can accept a generic notion of being.

The pluralist about being is motivated by a desire to account for multiple ranges of existents that exhibit very different features from each other. A pluralist might believe that numbers exist differently than chairs, that God exists differently than humans, or that *abstracta* exist differently than *concreta*, to name a few examples. McDaniel (2017) and Spencer (2012) point to three overlapping main categories of argument for ontological pluralism: theological, phenomenological, and ontological. Theological motivations for pluralism involve the ability to explain God's different mode of existence from other non-God things. God is so different from other things, the thinking goes, that she must exist differently than everything else. The phenomenological strategy uses the apparent experiential differences between, for example, perceiving a number and perceiving a chair as evidence of multiple ways of being. *Abstracta* and *concreta* are given so differently in experience that different sorts of being are the best explanation. The ontological strategy proceeds from the idea that different sorts of entities behave differently, and ontological pluralism is the best explanation for these fundamental differences.

Now consider that there are many sorts of non-existents: omissions, holes, shadows, possibilia, impossibilia, and fictions, to name a few examples. Plausibly, there are some differences within and between these sorts of non-existents. The pluralist about non-being shares some basic motivations with the pluralist about being: she can best explain ontological, phenomenological, and theological phenomena by positing multiple forms of non-being. The ontologically motivated pluralist might take the difference between impossible and possible non-existent objects, or the difference between non-existent past and future times, to be best modeled by a joint in non-being. Another pluralist might seek to explain phenomenological differences between thoughts about non-existent numbers versus thoughts about non-existent people. And pluralism about non-being opens up a heretofore underexplored option in theological space: a theist can believe that God doesn't always exist, but can plausibly come into being and go out of being. It would be natural for her to hold that God's non-being is different than run-of-the-mill non-being had by mere objects and persons: it's a special, divine sort of non-being. (In section 2 below, I discuss some historical precedent for this view.)

With these motivations in hand, we are in a position to investigate non-being. Call *ontological pluralism about non-being* the view that there are several fundamental different ways, modes, or kinds of non-being. Non-being has structure beyond the list of what does not exist: things that fail to exist, fail to exist differently than each other. If one is a certain kind of pluralist about non-being for *concreta* and *abstracta*, for example, non-existent chairs and numbers do not share a univocal property of non-being. If we wish to speak of both, we must say that the chair has non-being$_1$, and the number has non-being$_2$. Non-being is not a univocal property: speaking of something's non-being is ambiguous between non-being$_1$ and non-being$_2$.

The pluralist about non-being might or might not embrace the same attitude towards being: she can believe in ways of non-being and being, or just one or the other.[5] Call a *bilateral pluralist* one who believes in multiple ways of being and non-being, and a *unilateral pluralist* one who believes in just one or the other. Such a unilateral pluralist could hold, for example, that a square circle and a non-existent chair have different ways of non-being, but that all existents exist the same way. The bilateral pluralist need not believe that the joints in non-being mirror those in being: she might accept differences in non-existence between impossible and possible objects, but differences in existence between *abstracta* and *concreta*.[6] Call bilateral pluralists who believe in different joints in being and non-being *asymmetric* pluralists, and those who believe in equivalent joints in being and non-being *symmetric* pluralists.

The pluralist about non-being stipulates that there is a sort of structure in non-being. Though different kinds of pluralists might stipulate different kinds of structure, a common view of structure is a "pegboard" model, thus described by Turner:

> Ontological structure is the sort of structure we could adequately represent with a pegboard and rubber bands. The pegs represent things, and the rubber bands represent ways these things are and are interrelated. (Turner 2011: 2)

The non-being pluralist accepts a "multiple pegboards" picture, according to which there are two different kinds of propertied and related items in non-being. As there can be relations across kinds of being (I, a concretum, can think of a number, an abstractum), there can be relations across kinds of non-being (Sherlock Holmes is such that he does not eat square circles).

Just as the ontologist of being has principles for discerning how many things exist, so too the ontologist of non-being can ask how many things don't exist. The latter takes the task of creating an ontological inventory one step further: she asks how many entities fail to exist in more specific ways. The pluralist about non-being is as much an ontologist as that of being, since she seeks a sorted inventory of everything that fails to exist.

Believing in ways of being transforms questions about existence into questions about multiple forms of existence. McDaniel, for example, suggests that ontological pluralism splits the question of why there is something rather than nothing into multiple questions:

[5] Plausibly, the Stoics had this view. See Caston (1999) for a subtle interpretation of the Stoics on non-being and non-existence.

[6] Both symmetric and asymmetric pluralists may be what Caplan (2011) calls superpluralists, roughly, those who believe in different ways of being an ontological pluralist.

> If there are modes of being, that is, different ways to be, then either in addition to or instead of the question "why is there something, rather than nothing?", we should pursue, for each mode of being, the question of why there is, in that way, something rather than nothing. (McDaniel 2013: 277)

Similarly, the friend of ways of non-being splits the something-rather-than-nothing question into multiple finer-grained questions. The unilateral pluralist turns that question into: "why is there something rather than nothing_1 or nothing_2?" The bilateral pluralist would ask: "why is there_1 something_1 or there_2 something_2 rather than nothing_1 or nothing_2?"

Denying that something exists is different than conveying that it has a specific sort of non-being. The former involves straightforward negative existential quantification, whereas the latter requires stipulation of an entity that has a specific kind of non-being. Supposing I am a unilateral pluralist about non-being, when I say "There is no Tyrannosaurus Rex with pink feathers in South Bend, Indiana", I do not necessarily mean that there is a Tyrannosaurus Rex with pink feathers that has non-being_1. Rather, I intend to convey that there just isn't anything that corresponds to that description. Note the difference between this sort of statement and one that is intended to convey that a non-existent object is in some sense "out there" in liminal reality, as in "There is a Greek god of war."

This juncture is where one might turn to existential quantification in order to sort things out. One option follows Parsons (1980), Jacquette (1996), Zalta (1988), and Priest (2005) in positing different notations for "there is" ($\exists$) and "there exists" (E!). Depending on one's system, one can either have a special quantifier, or an existence predicate for only things that exist. Here I focus on the predicate strategy. On this scheme, the logical form for "There is an x such that x doesn't exist" is $\exists x(\varphi x \,\&\, \neg E!x)$. "There is a square circle but it doesn't exist", for example, becomes $\exists x(SCx \,\&\, \neg E!x)$. Now, one might be tempted to hold that the logical form for a unilateral non-being pluralist's claim is $\exists x(\varphi x \,\&\, \neg E!_1 x)$, or "There is an x such that x doesn't exist_1". The specific claim about the square circle becomes $\neg\exists x(SCx \,\&\, \neg E!_1 x)$, or "There is a square circle that doesn't exist_1". The problem with this logical form is that it is better interpreted as a claim made by a pluralist about being rather than a pluralist about non-being: it denies a particular positive way of being to the square circle, but does not postulate a specific way of non-being.

With a bit of tweaking, however, the dual notation strategy can be easily adopted by the friend of non-being. As above, let $\exists$ denote ontologically neutral "there is" and E! denote ontologically committed "there exists". Subscripts denote ways of being. Distinguish between two claims that a pluralist about non-being may wish to make: (i) there are no square circles, and (ii) square circles have non-being_1. The former denies that there is anything in being or non-being meeting the description "square circle"; the latter accords a spot in non-being_1 to a square circle. The first claim can be represented with $\exists x(SCx)$, to be interpreted as "There are no square circles." The second, substantive claim about non-being can be represented with $(\exists_1 x)(SCx \ \& \ \neg E!x)$, or "There is$_1$ a square circle, and anything that exists is not it." (A more perspicuous, less introduction-to-logic-y translation is "There is$_1$ a square circle, and it does not exist.")

Here is one way to understand the latter claim. Assuming that there is an ontologically neutral sense in which the square circle is "out there", that leaves two options with respect to heavy-duty ontological commitment to the square circle: either the square circle has non-being, or it has existence. A square circle can't have existence. But it can have non-being. By utilizing both the neutral quantifier and the committed existence predicate, the friend of non-being can hold that square circles have a specific kind of non-being without having existence. What is distinctive for the pluralist is that the subscripted notation "$\exists_1 x$" specifies a particular mode of non-being—a way of being "out there"—for the square circle, while "$\neg E!x$" denies the existence of the square circle.

Another option for representing assertions of pluralistic non-being is to imbue logical negations themselves with ontological import. Let $\neg_1$ mean "there is not$_1$" and $\neg_2$ mean "there is not$_2$." For the pluralist about non-being, $\neg_1\exists$ and $\neg_2\exists$ carve non-being closer to the joints than $\neg\exists$. Note that these notations are different than $\neg\exists_1$ and $\neg\exists_2$: the former represent ways of non-being, whereas the latter represent negations of ways of being. Suppose that a pluralist believes in a fundamental difference between possible and impossible non-existents. If she wants to hold that a square circle has non-being$_1$, she would represent such a claim as $\neg_1\exists x(SCx)$, or "There is not$_1$ a square circle." This claim is substantively different than "The square circle doesn't exist$_1$", which only denies a certain form of positive being. The notation with the restricted logical negation explicitly reserves a spot for the chair in the inventory of non-being$_1$. The friend of this strategy incurs a few extra explanatory burdens: she must explain what subscripted negation is. She must also reckon with the meaning of the subscripted negation in contexts with less ontological importance. For example, she should explain what it means to be not$_1$ hungry or not$_2$ red. Nonetheless, it is an option worth exploring.

Now, a natural objection to ontological pluralism about non-being is that it overly reifies non-existence. The thought is that being has a kind of *oomph* that distinguishes it from non-being. The pretheoretic concept of non-being is that it is a hazy, unstructured nothingness—it does not include natural joints and structure. While being enjoys rich structure and complexity, non-being is just a label under which non-existent things fall. Being is ontologically thick, the thinking goes, while non-being is thin and formless.

A closely related objection holds that pluralism about non-being reifies specific non-existents. Consider the atheist who says: "Look. When I say that God does not exist, I mean that she *really does not exist.* I do not mean that there is an omniscient, all-powerful being sitting around in non-being, with all of the details, properties, and contours of an existent, but inhering in a different ontological category. I mean that there isn't *anything* like that, in any sense." If the things that have non-being have substance, the worry goes, they become very being-like. We should be able to deny that things exist, full stop.

The pluralist has several lines of response to these lines of thinking. In reply to the objector who worries about reifying non-existents with too much specificity, she can hold that not every description corresponds to an item in non-being. Consider the description "being such that one is a golden dragon if each member of the Beatles wears a red hat on a Tuesday". Even if nothing of that description exists, one need not accept that this description correspond exactly to an item in non-being: plenitudinous descriptions do not necessarily equate to plenitudinous items in non-being.

Accepting reified non-existents can also be theoretically useful. Suppose that a theist and an atheist disagree on the existence of God on Cartesian grounds. The theist thinks that God must exist because existence is more perfect than non-existence. The atheist thinks that God doesn't exist because non-existence isn't necessarily better than existence. Here, the atheist would be well-served by a reified non-existent, God, about whose nature she can argue. Utilizing straightforward negative existential quantification is less useful than granting God a kind of non-being, but arguing about her nature.

2. Historical Precedent for Pluralism about Non-Being

The pluralist follows Meinong (1904) in accepting the idea that things can have a kind of being without having existence. Meinong famously distinguishes between objects that *exist* (you, your iPhone, the Eiffel Tower), things that *subsist* (the number twelve, the proposition that snow is white), and impossible things that neither exist nor subsist (a round square, the proof that 2+2 = 5).[7] Pluralism about non-being captures some of the spirit of Meinongianism insofar as some non-existent things have what others take to be the hallmarks of being: properties, relations, and classification under distinct ontological categories. Subsistence is an ontologically rich form of non-being rather than a hazy nothingness without structure.

There are many available Meinongian positions in logical space available to the pluralist about non-being. One option is to hew very closely to the letter of Meinong's theory, while another option is to abandon the letter and remain close to the spirit. Consider the unilateral pluralist who believes in one way of being, but two ways of non-being: one for impossible things and one for merely

[7] Here I follow Reicher (2019) in taking this to be a plausible interpretation of Meinong, though Meinong interpretation is a controversial matter.

non-existent things. This sort of pluralist shares a tripartite ontology of being and non-being with Meinong, as the major ontological joints fall in very similar, and possibly identical, places. Other pluralists might embrace the spirit of Meinongianism but fall farther from the original view. For example, some pluralists about non-being might take the division in non-existent things to lie between, say, God and non-God things rather than possible and impossible things. The symmetric pluralist postulates joints in being in addition to those in non-being. How many joints there are, and where they fall, determine whether a pluralist is Meinongian or merely neo-Meinongian. Either way, accepting the substantivity of non-being has a strong whiff of Meinongianism.

In addition to Meinong's friendliness to substantive non-being, there is scattered historical precedent for accepting different ways of non-being. Here I will discuss a few instances, though I expect that there are more if one searches for them.

Following Moran and Guiu (2019), I interpret John Scotus Eriugena as positing five modes of being and correlative modes of non-being. There are things accessible to senses (and things that are not), orders of created natures (and their differences), actual things (and potential non-things), things perceived by the intellect alone (and those that are not), and those infused with divine grace (and those that are not.) The joints in non-being mirror those in being. Arguably, Eriugena also makes use of a distinctive form of non-being to make sense of God's self-creation. He holds that God is beyond being and non-being, but gradually self-creates from "divine darkness" into light. Such "divine darkness" is a special kind of non-being from which being stems, and is different than ordinary non-existence.[8]

Simone Weil (1952: xxi) makes similar use of a special form of non-being to make sense of an "absent god". According to Weil, God "withdrew" from full existence in order to make room for the universe. Persons, too, are created from the space which God has deserted: a distinct form of non-being from whence being arises.

Theological motivations were not the only underpinnings of historical pluralism about non-being. The Stoics posit a status, subsistence, that characterizes some non-existent objects, including time, place, void, and expressibles. Following Long and Sedley (1987: 162–165), I understand the Stoics as positing that what it is to be something is to be an object of thought and discourse. But certain objects like centaurs, while being proper objects of thought and discourse, do not even subsist. They are "mere somethings" that do not exist. (Long and Sedley also raise the possibility that the Stoics are committed to a third category of non-existent, not-somethings, but see Caston (1999) for objections to this objection.)

[8] Bosley and Tweedale (2006: 573) also support this reading.

Essentially, there are non-existent "mere somethings" that are different than other subsistent non-existents. It is clear that the Stoics were friendly to different ways of thinking about non-being, on Long and Sedley's interpretation.

Sartre (1956) affirms the reality of nothingness ("le néant"), and distinguishes between at least two sorts of non-beings. There is a concrete kind of nothingness as represented by an absence—for example, a friend failing to show up for a meal—and a more abstract kind of nothingness exemplified by square circles. Absences are brought about by human consciousness insofar as they are products of expectations. Sartre's view draws on his admiration of Heidegger's work on nothingness, in which he infamously claimed "The nothing itself nothings." Nozick took up the task of ontologically interpreting Heidegger's claim:

> Imagine this force as a vacuum force, sucking things into non-existence or keeping them there. If this force acts upon itself, it sucks nothingness into nothingness, producing something or, perhaps, everything, every possibility. If we introduced the verb "to nothing" to denote what this nothingness force does to things as it makes or keeps them nonexistent, then (we would say) the nothingness nothings itself. (Nozick 1981: 123)

While Nozick's approach doesn't stipulate pluralism about non-being or push us towards such a view, such a conception of non-being takes it seriously as having distinctive behavior. Viewing non-being as a kind of force or actor is a foundation for the idea that different non-existents behave differently.[9]

3. What Ontological Pluralism about Non-Being Can Do

Ontological pluralism about non-being can be applied to a number of issues in metaphysics. There are a few points to which I will attend before enumerating them. First, one might wish to deploy degrees of non-being rather than ways of non-being for some of these issues. Here I do not focus on this view, but it is worth mentioning the possibility. Second, it should be obvious that one would not want to hold all of these pluralisms about non-being at once; this discussion is simply intended to be a case study of various applications. Finally, the list is not exhaustive: there are likely many more applications of ways of non-being than I discuss in this section.

[9] See Skow (2010) for an analysis of Nozick's claim informed by contemporary physics.

3.1 Presentist Ontological Pluralism about Non-Present Events and Objects

Presentists about time believe that only the present events and objects exist. They are to be contrasted with *eternalists*, who believe that all events and objects exist, and *growing block* theorists, who hold that past and present events and objects exist. For growing block theorists, existence distinguishes future events from past and present ones. For both presentists and eternalists, there are no ontological differences between past and future events: they don't exist for presentists, and they do exist for eternalists.

One explanatory burden for ontologies of time is to account for the apparent differences between the past and the future. For example, the past seems fixed and unchangeable in a way that the future is not. Humans often prefer pain to be in their past and pleasure to be in their future. And the direction of causation seems to run from the past to the future.

Presentists have a unique explanatory possibility, however. The presentist can accept a certain kind of pluralism about non-being, according to which the past and the future are fundamentally different kinds of non-being. Presentist pluralism about non-present times challenges the dominant assumption in the presentist literature that the two kinds of unreality are the same kind.[10] Past and future events have different kinds of non-being, and they do not share a univocal property of non-being. Consider a past and future event: your birth and your lunch one month from now. The pluralist presentist can hold that the birth has past non-existence and the lunch has future non-existence. The present moment is the ontological cleavage between the two fundamental ways of non-being.[11] Events do not fail to exist *simpliciter*; they fail to exist in more specific ways.

Different ways of non-being can help explain phenomenological differences between experiences of the past and the future: we remember one, but not the other. The past and the future differ in the way they are given to us in experience. The view also supports ontological differences between past and future—for example, the fixity of the past and the openness of the future.[12]

According to some essentialist interpretations of ontological pluralism, something that has one kind of being can never have the other kind of being. To use an earlier example, a chair can never be a number. The presentist friend of pluralism should deny the equivalent view about non-being, since moments that have one

[10] Prior (1972: 245) hints at this view, presumably unintentionally, in writing that "The present simply is the real considered in relation to two particular species of unreality, namely the past and the future."

[11] McDaniel (2017: 81–6) proposes that pluralism be applied to ontological differences between the past and the present.

[12] In this vein, Cameron (2011), a rare contemporary friend of pluralism about non-being, argues that the view can help reconcile presentism with truthmaker theory.

kind of non-being will eventually have the other kind of non-being: future moments will become past moments.

3.2 Omissions versus Absences

Intuitively, there are differences between *omissions*, roughly, events that are close to occurring but do not occur, and *absences*, roughly, things that are not close to occurring and do not occur. I caused my plant's death by omitting to water it; I very well could have watered it. I also did not go shopping with Abraham Lincoln last night, leaving me to wonder whether he would have liked the shoes that I eventually picked out. But I could not have gone shopping with Abraham Lincoln: such an event was not even close to occurring. A puzzle for causation theorists is how to distinguish between omissions and absences: both do not exist, but one seems intuitively different from the other. Omissions cause things to happen; mere absences do not, or at least do not exert the same kind of causal power.

It might be initially tempting to distinguish between absences and omissions on the basis of their possibility: absences are not causally efficacious because they are impossible events, but omissions are causally efficacious because they are possible. It is impossible to go shopping with Abraham Lincoln, after all, while it is possible to set an alarm clock.

But drawing the line between omissions and absences on the basis of possibility is wrong, for several reasons. First, some omissions are impossible. Suppose that the assistant professor fails to prove that 2+2=5, and is thus denied tenure. In Bernstein (2016), I argue for the position that such omissions are causally efficacious. Suppose that one accepts a simple counterfactual account of causation, according to which c is a cause of e if e would not have occurred had c not occurred. Then many omissive causal statements come out as true, including ones involving impossible omissions. The counterpossible "If she hadn't failed to prove that 2+2=5, she would have been awarded tenure" is true and non-vacuous. Such causal counterpossibles also furnish correct predictions and explanations. In some contexts, impossible events are closer to actuality than possible ones.

Another reason not to draw the absence/omission distinction in terms of possibility is that many absences are intuitively possible, but causally inefficacious. There is no actual-size replica of the city of Paris in the empty fields between Indianapolis and Chicago, but such a thing is possible. It's not even close to occurring: it's simply not there. Without a particular causal or predictive context, this absence doesn't cause anything to happen, even though it is possible. Impossibility and possibility do not correctly carve the absence/omission distinction.

The ontological pluralist about non-being has a ready solution, however: she can hold that absences and omissions have different ways of non-being. Here's how it would work. In the case of my failing to water the plant, there are at least two non-beings: the omission of my watering the plant, and the absence of my watering the plant.

Supposing that absences have non-being$_1$ and omissions have non-being$_2$, they are fundamentally ontologically distinctive. One virtue of this view is that one need not identify a particular non-event as an absence *or* an omission, since both non-beings correspond to a particular non-event. There is an absence with non-being$_1$ of the plant- watering, and an omission with non-being$_2$ of the plant-watering. One is non-causal and the other is causal. Context makes one or the other salient.

A virtue of the view is that it helps with the *problem of profligate omissions*. The problem is as follows. Suppose that one accepts a simple counterfactual account of causation, according to which *c* is a cause of *e* if *e* would not have occurred had *c* not occurred. And suppose that one accepts that omissions can be causes. Then, for any particular omission that is a cause, there will also be countless other counterfactual dependence-generating non-occurrences. For example, the counterfactual "Had I not failed to water the plant, the plant would not have died" is intuitively true, but so is "Had Barack Obama not failed to water the plant, the plant would not have died." Many more non-occurrences count as causes than are intuitively so.

The pluralist about non-being, however, has a ready explanation for this problem.

She can hold that there are a select few omissions, non-beings with causal efficacy, which have one way of non-being. And she can hold that there are profligate absences, non-beings without causal efficacy, which have another way of non-being. This pluralist accepts a plentitude of non-beings that are absences, but only a select few non-beings that are omissions. That way, the pluralist can account for the countless non-occurrences that are happening at any given time without ascribing them all causal efficacy.

For the proponent of this solution, multiple relevant distinctions will be hyperintensional. There is a *hyperintensional* distinction when two necessarily extensionally equivalent expressions are not intersubstitutable *salva veritate*—that is, when changing out the positions of necessary equivalents changes the truth value of a sentence. Some impossible omissive statements are hyperintensional: every world at which the circle fails to be a square is also a world in which two plus three fails to equal six. But these are different omissions. Omissions and absences might also be hyperintensional: every world where the mathematician couldn't have proved that 2+2=5 is also a world where she failed to prove that 2+2=5. But, intuitively, the absence is different than the omission. Pluralism about non-being does justice to these differences between negative entities relevant to causation and causal explanation.

3.3 The Ontology of Fictions

Another area where positing ways of non-being is useful is in accounting for the ontology of fictional objects. Fictional objects are those objects posited by works of fiction, like Captain Yossarian in Joseph Heller's *Catch-22*, the nameless narrator in Margaret Atwood's *The Handmaid's Tale*, and Issa Dee in HBO's *Insecure*. On the one hand, such objects do not intuitively exist in the "full" sense that you and I exist—we cannot physically interact with them, change them, or bump into them in the supermarket. On the other hand, fictional objects seem to exist in some other, more robust sense than fully non-existent objects.

Ways of non-being can account for this difference: the pluralist about non-being can hold that fictional objects have one kind of non-being and other non-existent objects have another kind of non-being. This fundamental ontological distinction respects the intuitive difference between fictional objects and simply non-existent objects, while doing justice to the idea that they don't exist the way that you and I exist.

Another place that pluralism about non-being can be of use is in distinguishing between impossible and possible fictions. Impossible fictions are fictions that describe impossible entities or scenarios. Such scenarios are particularly common in fiction involving time travel. Pluralism accounts for such differences by positing different kinds of non-being for impossible and possible fictional entities: impossible mathematical entities, like the proof of the inconsistency of mathematics in Ted Chiang's "Division by Zero", have different non-being than Yossarian.

Pluralism can also be of service in accounting for *nested* fictions, or fictional entities within fictional entities. The HBO television show *Insecure* features several secondary shows-within-the-show. "Due North" is a show-within-the-show set in the pre-Civil War South with its own actors and well-developed fictional narrative. The third season of *Insecure* includes "Kev'yn", a comedy series-within-the-show. And the fourth season features "Looking for LaToya", a fictional true crime show-within-the-show. In each case, the nested show is a distinct fictional entity from *Insecure*, with its own plot and characters. The characters in *Insecure* think about and discuss each nested show, but like us, they do not physically interact with fictions.

One reason it is important to distinguish between nested and primary fictions is that we want a way of justifying statements of the form "According to the fiction, ____." Truth-according-to-a-fiction is often seen as different than truth *simpliciter*: it is true according to the fiction that Sherlock Holmes smokes a pipe, but false that Sherlock Holmes smokes a literal pipe. Determining truth-according-to-a-fiction is a fairly easy task in cases in which the claim in question is explicitly stated in the fiction. For example, Issa Dee, the protagonist of *Insecure*, lives in Inglewood, so "According to the fiction, Issa Dee lives in Inglewood" is true because it is explicitly displayed in the fiction. But in cases of nested fictions, it is not necessarily the case that something true according to the primary fiction is true according to the secondary fiction, and vice versa. In *Kev'yn*, for example, Kev'yn and Yolonda stage a protest. It is true according to *Kev'yn* that they stage a protest, but it is not necessarily true according to *Insecure*. Similarly, it is not necessarily true according to *Kev'yn* that Issa Dee lives in Inglewood.

Pluralism about non-being can account for nested fictions by positing distinct kinds of non-being for "primary" fictional entities, like those in *Insecure*, and "secondary" nested fictional entities, like those in *Kev'yn*. The characters and entities in *Insecure* have one sort of non-being, and the characters in each nested fiction have another. This way, truths-according-to-*Insecure* and truths-according-to-*Kev'yn* are grounded in different kinds of non-existence. "Kev'yn and Yolonda staged a protest" is true according to *Kev'yn*, and "Issa Dee lives in Inglewood" is true according to *Insecure*. The difference in truth conditions is grounded in an ontological joint in non-being.

4. Conclusion

The preceding discussion has suggested that *ontological pluralism about non-being*, the view that there are multiple ways, kinds, or modes of non-being, is worthy of serious philosophical consideration. The view has not enjoyed the same attention as pluralism about being, but it is a natural complement to it. The view also has promising explanatory power for a range of theological, metaphysical, and phenomenological *explananda*, and deserves extensive further investigation. One need not think that non-being is, well, *nothing*: it might have explanatory and metaphysical structure unto itself.

Acknowledgments

Thanks to Kris McDaniel, Daniel Nolan, Michael Rea, Brad Rettler, Byron Simmons, Mack Sullivan, and Peter van Inwagen for helpful feedback on this paper. Thanks also to audiences at Syracuse University, the University of St Andrews, Trinity College Dublin, and University College Cork.

References

"Alexius Meinong (1904). Über Gegenstandstheorie", in *Untersuchungen zur Gegenstandstheorie und Psychologie* pp. 1–51, (Leipzig: Barth).

Bernstein, Sara (2016). "Omission Impossible", *Philosophical Studies* 173 (10): 2575–89.

Bosley, Richard and Martin Tweedale (2006). *Basic Issues in Medieval Philosophy: Selected Readings Presenting Interactive Discourse Among the Major Figures*, 2nd edn (Peterborough: Broadview Press).

Cameron, Ross (2011). "Truthmaking for Presentists", in *Oxford Studies in Metaphysics*, vol. **6**, ed. Karen Bennett and Dean W. Zimmerman (Oxford: Oxford University Press), pp. 55–100.

Cameron, Ross (2018). "Critical Study of Kris McDaniel's The Fragmentation of Being", *Res Philosophica* **95** (4): 785–95.

Caplan, Ben (2011). "Ontological Superpluralism", *Philosophical Perspectives* **25** (1): 79–114.

Caston, Victor (1999). "Something and Nothing: The Stoics on Concepts and Universals", *Oxford Studies in Ancient Philosophy* **17**: 145–213.

Jacquette, Dale (1996). *Meinongian Logic: The Semantics of Existence and Nonexistence* (New York: De Gruyter).

Long, A. A. and D. N. Sedley (1987). *The Hellenistic Philosophers, Volume 1: Greek and Latin Texts with Notes and Bibliography* (Cambridge: Cambridge University Press).

McDaniel, Kris (2009). "Ways of Being", in *Metametaphysics: New Essays on the Foundations of Ontology*, ed. David Chalmers, David Manley, and Ryan Wasserman (Oxford: Oxford University Press), pp. 290–319.

McDaniel, Kris (2013). "Ontological Pluralism and the Question of Why There is Something Rather than Nothing", in *The Philosophy of Existence: Why Is There Something Rather Than Nothing?*, ed. Tyron Goldschmidt (New York: Routledge), pp. 290–320.

McDaniel, Kris (2017). *The Fragmentation of Being* (Oxford: Oxford University Press).

Moran, Dermot and Adrian Guiu (2019). "John Scottus Eriugena", *Stanford Encyclopedia of Philosophy*, ed. Edward N. Zalta (Winter 2019 edn).

Nozick, Robert (1981). *Philosophical Explanations* (Cambridge, MA: Harvard University Press).

Parsons, Terence (1980). *Nonexistent Objects* (New Haven, CT: Yale University Press).

Priest, Graham (2005). *Towards Non-Being: The Logic and Metaphysics of Intentionality* (Oxford: Oxford University Press).

Prior, A. N. (1972). "The Notion of the Present", in *The Study of Time: Proceedings of the First Conference of the International Society for the Study of Time*, ed. J. T. Fraser, F. C. Haber, and G. H. Müller (Berlin: Springer).

Reicher, Maria (2019). "Nonexistent Objects", *Stanford Encyclopedia of Philosophy* (Winter 2019 edn), ed. Edward N. Zalta.

Rettler, Bradley (forthcoming). "Ways of thinking about ways of being", *Analysis*.

Sartre, Jean Paul (1956). Being and Nothingness: An Essay on Phenomenological Ontology, trans. H. E. Barnes (New York: Philosophical Library) [French orig. *L'Être et le Néant: essai d'ontologie phénoménologique*, 1943].

Simmons, Byron (forthcoming). "Ontological Pluralism and the Generic Conception of Being", *Erkenntnis*.

Skow, Bradford (2010). "The Dynamics of Non-Being", *Philosophers' Imprint* **10** (1): 1–14.

Spencer, Joshua (2012). "Ways of Being", *Philosophy Compass* 7 (12): 910–18.

Turner, Jason (2010). "Ontological Pluralism", *Journal of Philosophy* **107** (1): 5–34.

Turner, Jason (2011). "Ontological Nihilism", in *Oxford Studies in Metaphysics*, vol. 6, ed. Karen Bennett and Dean W. Zimmerman (Oxford: Oxford University Press), pp. 3–54.

Turner, Jason (forthcoming [2021]). "Ontological pluralism", in *The Routledge Handbook of Metametaphysics* (London: Routledge), pp. 184–96.

Weil, Simone (1952). *Gravity and Grace*, trans. Emma Crawford and Mario von der Ruhr (London: Routledge & Kegan Paul) [French orig. *La pesanteur et la grâce* (Paris: Librarie PLON, 1947)].

Zalta, Edward N. (1988). *Intensional Logic and the Metaphysics of Intentionality* (Cambridge, MA: MIT Press).

2
Nothingness and the Ground of Reality
Heidegger and Nishida

Graham Priest

1. Introduction

Nothingness is a strange object. So is the ground of reality if it has one. In this essay, I will argue that reality does indeed have a ground (in a sense that I will make clear), and that this is, in fact, nothingness.[1]

In the first part of this paper I will explain what I mean by nothingness being the ground of reality, and argue for the view. In the rest of the paper, I will look at two philosophers whom I take to be on my side about the matter, Heidegger and Nishida. An interlude along the way provides some background on Zen Buddhism necessary to understand Nishida. An appendix discusses a connection between Heidegger and Zen.

2. Nothingness

To the substantial philosophical issue, then.

First, note that the word 'nothing' can be used as a quantifier, but it also has a perfectly good use as a noun phrase, meaning *nothingness*. (Hegel and Heidegger wrote about nothing, but said quite different things about *it*.) In what follows, to avoid any confusion, when I wish to use 'nothing' as a noun phrase I will boldface it, thus: **nothing**.

Nothing is the absence of all things. It is, as it were, what remains after everything has been removed; and by 'everything', here, I mean absolutely *everything*, *all* things.

It follows that **nothing** is ineffable. To talk about something requires one to predicate something of it. One can predicate nothing of **nothing** simply because there is nothing there of which to predicate it. One might also put the point this way. To predicate P of something, a, requires a to be an object. (I do not say

[1] I endorsed the view, in effect, in Priest (2014a: 11.9, 13.11). Here I want to look more closely at things.

Graham Priest, *Nothingness and the Ground of Reality: Heidegger and Nishida* In: *Non-Being: New Essays on the Metaphysics of Non-Existence.* Edited by: Sara Bernstein and Tyron Goldschmidt, Oxford University Press (2021). © Graham Priest.
DOI: 10.1093/oso/9780198846222.003.0002

existent object.) The very syntax *Pa* tells you this. But **nothing** is not an object: it is the result of *removing* all objects.

Of course, we are in paradoxical territory here. **Nothing** *is* an object (as well). After all, one can refer to it by the name '**nothing**'. Consequently, it is effable, as well. Thus one can say, as I did, that **nothing** is what remains after all objects have been removed. I have discussed the paradoxical territory elsewhere, and I will not go into it further here.[2] It is **nothing** as the ground of reality which will be my concern in what follows.

3. The Ground of Reality

Ontological dependence, or as it is often called nowadays, *grounding*, has been the subject of much discussion in the recent literature on analytic metaphysics. In truth, the notion of ontological dependence has always played an important role in metaphysics, East and West.[3] However, the recent literature has forced it and its properties onto centre stage.

There is much that should be said if the notion—or notions; arguably there is more than one—of ontological dependence is to be sorted out.[4] However, we can ignore most of the details here, though let me make a few comments. Many argue that the notion is not definable in terms of something more basic. If so, so be it. However, I think it is natural to understand dependence—at least in the sense that will be operative here—as follows. *A*'s being the case depends on *B*'s being the case just if (if *B* were not be the case *A* would not be the case). That is, $\neg B > \neg A$, where $>$ is the counterfactual conditional.[5] (And since dependence is factual, one had better conjoin *A* and *B*.[6])

Now, turning to the subject at hand: some things depend for being what they are on other things. Thus, being the shadow of a tree (s) depends for being what it is on the tree (t) being a tree. If t were not a tree, s would not be the shadow of a

[2] Priest (2014a: 2.4, 6.13), Priest (2014b), Priest (ms). [3] See Bliss and Priest (2017).

[4] For some of this, see Bliss and Trogdon (2014), and Tahko and Lowe (2015).

[5] How to understand such conditionals is somewhat moot. But see Priest (2008: ch. 5), and Priest (2018a).

[6] There are some standard objections to a counterfactual analysis of dependence. This is not the place to discuss them in detail, but let me just note the following. It is often claimed that counterfactual conditionals with necessarily false antecedents are vacuously true, so the analysis does not give the right results. However, it is perfectly straightforward to give an analysis of such counterfactuals according to which this is not the case. (See Berto et al. (2018). See, further, Wigglesworth (2013), and Wrigglesworth (2015), from whom I take the idea that one may use impossible worlds in an analysis of ontological dependence.) Next, it may be claimed that counterfactuals have the wrong structural properties. Dependence is transitive and anti-reflexive. Counterfactual conditionals are not transitive but are reflexive. The properties of dependence are contentious, but if one subscribes to those cited, one can take the counterfactual to be merely a sufficient condition for dependence; a necessary and sufficient condition is being in the transitive closure of the counterfactual relation. And one can make dependence anti-reflexive simply by defining it as $(\neg B > \neg A) \wedge \neg(\neg A > \neg B)$.

tree. The dependence does not go the other way. If s ceased to be the shadow of a tree (say, if the sun went in), t would still be a tree.

Similarly, being the set $s = \{0, 1, 2\}$ depends for being what it is on containing the number 0. If 0 were not a member of s, s would not be $\{0, 1, 2\}$. Again, the dependence does not go the other way. If s were not $\{0, 1, 2\}$, 0 could still be a member of it.

Next, some things depend for being what they are on being *distinct* from something else. Thus, being the spouse (s) of a person (p) depends on s being distinct from p. If s were the same (person) as p, s would not be the spouse of p. The dependence does not go in the other direction. If s is not the spouse of p, it does not follow that s is p.

Similarly, being a hill (h) depends for being what it is on being distinct from its surrounding plane (p). If h were the same (height) as p, it would not be a hill. Again, the dependence does not go the other way. If h is not a hill, it does not follow that it is p. It might be a ravine.

Now, *being something* can be said in many ways. However, there is a most fundamental one, namely being an object. It is fundamental, in that being anything at all presupposes being an object. Something cannot have *any* property unless it is an object. Let us consider this most fundamental sense of being something.

Something (g) being an object depends on its being distinct from **nothing**. If g were the same (in ontological status) as **nothing**, it would not be an object, since **nothing** is not an object. The dependence does not go the other way. If g were not an object, it would not follow that it is identical with **nothing**. There may non-objects other than **nothing**.[7]

Indeed, one may say that what it is to be an object is to "stand out" against the background of nothingness, in just the way that a hill is what it is because it stands out against the background of the surrounding plain. Recall that *exist* comes etymologically from the Latin *ex* (out) *sistere* (made to stand), and so means literally something like *made to stand out*.[8] One could picture it thus:

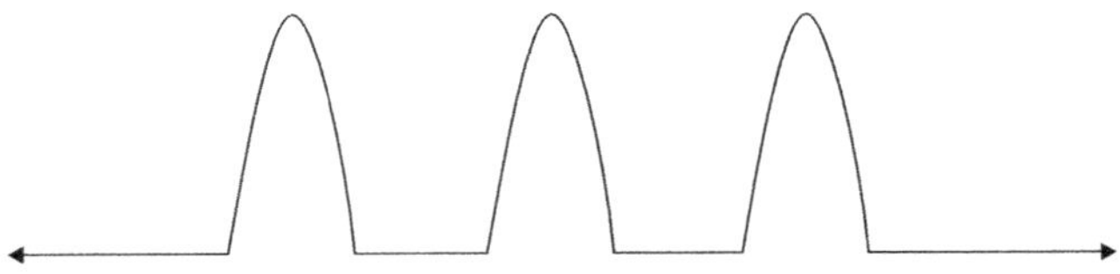

[7] Thus, see Priest (2014a), esp. Part 1. As Priest (2014a: 180) notes, though, there is a different dependence in the other direction. For something to be **nothing** depends on its not being g: if it were g, it would be an object, and so not **nothing**.

[8] True, I do not take being an object to be the same as being existent an object; but many people do.

The peaks might represent hills standing out against the surrounding ground; or they might represent objects standing out against the background of **nothing**.

Hence, **nothing** is the ground of reality, in the sense that it is the ground of every object, reality being composed of objects. One should recall, however, that we are in a dialetheic situation. **Nothing** *is* an object; so **nothing** being an object depends on its not being **nothing**. Indeed **nothing** ≠ **nothing**.

4. Heidegger

So much for **nothing** being the ground of reality, in the sense of being the ground of each being. Let us now turn to two philosophers who have been here before us. The first is the German philosopher Martin Heidegger (1889–1976).

In 1927 Heidegger published *Sein und Zeit*. At the beginning of this he asks: what is being; that is, what is it to be?[9] And immediately he tells us (giving no reason) that, whatever it is, it is not a being. (There is an 'ontological difference' between being and beings.) The question is not answered in *Sein und Zeit*. We are told that to answer it, we must first understand the kind of thing that can ask the question: *Dasein*, people. The book gets no further than addressing that question. The *Seinsfrage* was, however, to drive Heidegger's philosophical inquiries for the rest of his life.

In 1928, *Sein und Zeit* won Heidegger the chair of philosophy at the University of Freiburg, which had just become vacant due to the retirement of his teacher, Edmund Husserl. And in 1929 Heidegger gave his inaugural lecture, 'Was ist Metaphysik?'. The lecture is a discussion on *Das Nichts*. This is often translated into English as *the nothing*. This is just a poor translation. German puts a definite article before abstract nouns, where English (mostly) does not. A better translation is simply *nothing* (used as a noun phrase)—**nothing**.

And what does Heidegger say about **nothing**? First he tells us what it is (agreeing with how I have explained it):[10]

[T]he nothing is the complete negation of the totality of beings.

That is, **nothing** is what remains after all objects are removed.

He also notes that **nothing** is ineffable, for the same reasons that I noted (pp. 98–9):

> What is the nothing? Our very first approach to the question has something unusual about it. In our asking we posit the nothing in advance as something that

[9] Note that, for Heidegger, *to be* does not mean *to exist*. It just means *to be an object*. Thus, "everything we talk about, mean, and are related to is in being one way or another" (Heidegger, trans. Stambaugh 1996: 5). And "[w]hen we say something 'is' and 'is *such and so*', then that something is, in such an utterance, represented as an entity" (Heidegger, trans. Fried and Polt 2000: 93).

[10] Heidegger, trans. Krell (1977: 100). In quotations from Heidegger in this section, page numbers refer to this edition unless otherwise noted.

> 'is' such and such; we posit it as a being. But that is exactly what it is distinguished from. Interrogating the nothing—asking what, and how it, the nothing, is—turns what is interrogated into its opposite. The question deprives itself of its own object.
>
> Accordingly, every answer to this question is impossible from the start. For it necessarily assumes the form: the nothing "is" this and that. With regard to the nothing question and answer alike are inherently absurd.

This, of course, thrusts us straight into the paradox of ineffability that I noted.

However, of more importance for the present is what Heidegger says about the relationship between **nothing** and objects. He says (p. 105):

> The nothing is neither an object nor any being at all. The nothing comes forward neither for itself nor next to beings, to which it would, as it were, adhere. For human existence the nothing makes possible the openedness of beings as such. The nothing does not merely serve as the counterconcept of beings; rather it originally belongs to their essential unfoldings as such. In the Being of beings the nihilation of the nothing occurs.

In other words, **nothing** provides a "space in which objects appear". That is, standing out against it is what makes it possible for something to be an object.

Heidegger also thinks that one can experience **nothing** in a mood he calls 'anxiety'. I will return to that matter in the appendix to this paper. For the present, we need merely note the following, where he makes the same point again (p. 105):

> In the clear night of the nothing of anxiety the original openness of beings as such arises: they are beings—and not nothing. But this 'and not nothing' we add in our talk is not some kind of appended clarification. Rather it makes possible in advance the revelation of beings in general. The essence of the originally nihilating nothing lies in this, that it brings Dasein for the first time before beings as such.

For Heidegger, then, **nothing** is the ground of all objects, that is, of reality.

Why does he hold this view? He does not explain at length; but an answer is provided by his view concerning the relationship between being and **nothing**. He says (p. 110):

> "Pure Being and pure Nothing are the same." This proposition of Hegel's (*Science of Logic*, vol. I, *Werke* III, 74) is correct. Being and the nothing do belong together, not because both—from the point of view of the Hegelean concept of thought—agree in their indeterminacy and immediacy, but rather because Being itself is essentially finite and reveals itself only in the transcendence of Dasein which is held out into the nothing.

In other words, he thinks that being and **nothing** are the same.[11] But Heidegger holds that being is what makes beings be. Thus, when asking the *Seinsfrage* at the beginning of *Sein und Zeit*, Heidegger says:[12]

> What is *asked about* in the question to be elaborated is being, that which determines beings as beings, that in terms of which beings have always been understood, no matter how they are discussed.

Being is what determines beings as beings. If being were not, no being would be a being. So something's being a being depends on being. And if being is **nothing**, the same goes for **nothing**.

Indeed, commenting on the paradox of ineffability of being, Heidegger says:[13]

> If we painstakingly attend to the language in which we articulate what the principle of reason [*Satz vom Grund*] says as a principle of being, then it becomes clear we speak of being in an odd manner that is, in truth, inadmissible. We say: being and ground/reason [*Grund*] 'are' the same. Being 'is' the abyss [*Abgrund*]. When we say something 'is' and 'is such and so: then that something is, in such an utterance, represented as a being. Only a being 'is'; the 'is' itself—being—'is' not.

Here, Heidegger clearly states that being is the ground (*Grund*) of objects. And since being is **nothing**, it is equally an abyss (*Abgrund*), over which, one might say, objects "hover".

Heidegger's views on **nothing** being the ground of reality are, then, in agreement with those I explained and defended in the first part of this essay.

5. Interlude on Zen

In the next section we will turn to the second of the philosophers I wish to discuss, Nishida. It is virtually impossible to understand his thinking, however, unless one knows the Buddhist, and specifically Zen, philosophical tradition on which he is drawing. So in this section I want to provide the appropriate background.[14] I will say slightly more than is necessary to understand Nishida on the matter to hand because it will become relevant when I talk of Heidegger and Zen in the appendix to this essay.

[11] This is an aspect of Heidegger's view with which I do not concur. (See Priest (2014: 4.6).) However, this is of no relevance here.

[12] Heidegger, trans. Stambaugh (1996: 4f.). Italics original.

[13] Heidegger, trans. Lilly (1991: 51f.).

[14] For a longer account of the following, see Priest (2014c), and Priest (2018b), esp. chs. 4, 7, and 9.

Let us start with Indian Buddhism. In all schools of Buddhism—of which there are many—there is a standard distinction between conventional reality (*saṃvṛti satya*) and ultimate reality (*paramārtha satya*).[15] How each term of this pair is understood varies from school to school; but, roughly, conventional reality is the world with which we are familiar, our *Lebeswelt*; while ultimate reality is the world as it is is understood by, or appears to, one who is enlightened. Naturally, the latter is, in some sense, more profound or accurate. Indeed, the Sanskrit *saṃvṛti* means 'conventional'; but it also means *concealing* or *obscuring*. Conventional reality occludes the ultimate, blocking the path to enlightenment.

The Buddhism that went into China, and thence Japan, was Mahāyāna Buddhism. So let us focus on the Mahāyāna account in more detail. The earliest Mahāyāna school of Buddhism was Madhyamaka, traditionally taken to be founded by Nāgārjuna (*fl.* 1st or 2nd cent. CE). According to this, the objects of conventional reality are empty (*śūnya*). What this means is that each thing is dependent for being what it is on other things, notably, its parts, its causes (and maybe effects), and our concepts. In Madhyamaka, ultimate reality is often referred to by the epithet emptiness (*śūnyatā*). Exactly what this is, is more contentious—though it is clear that it, too, is empty; but Nāgārjuna himself appears to suggest that it is ineffable. Ultimate reality is 'without distinction... and free from conceptual construction'.[16] Since to describe is to apply concepts, it cannot be described.

The other, and later, school of Indian Mahāyāna is Yogācāra, traditionally taken to be founded by the half-brothers Asaṅga and Vasubandhu (*fl.* 4th or 5th cent. CE). Yogācāra is standardly interpreted as a form of idealism. Thus, in Yogācāra, objects of conventional reality are empty, as for Madhyamaka; but they have no external reality: they are all "in the mind".

Yogācāra philosophy backs up this view with a sophisticated analysis of consciousness. At the most superficial level, there is ordinary thinking. In particular, it is intentional. That is, it comprises thoughts that are directed towards objects (as in, I am seeing/feeling/thinking of a tree). The objects may appear to be outside the mind, though, in fact, they are not. There is a deeper level of consciousness, however: the storehouse consciousness (*ālaya vijñāna*). In some ways, this is like the unconscious in modern Western thought. In particular, it is the goings-on in this which produce what happens at the higher levels, and in particular the (illusory) objects of intentional states. It is therefore the ultimate reality of such objects. This reality is just as ineffable as it is in Madhyamaka. (Concepts deliver only conventional reality.) In particular, there are no distinctions present in the *ālaya*: no *thises* rather than *thats*. Most notably, the duality between subject and

[15] The Sanskrit word *satya* can mean both *truth* and *reality*. The former is the more usual translation; but in many contexts, including the present one, the latter is more appropriate.

[16] As the dedicatory verses of the *Mūlamadhyamakakārikā* put it. (Garfield 1995: 2)

object, characteristic of the higher levels of consciousness, is itself absent. The *ālaya* itself is pure, though pre-enlightenment its form is impure, poisoned by 'karmic seeds'—the traces of previous actions.

When Buddhism goes into China, it meets the native philosophy of Daoism. And a particular interpretation of this was to have a significant impact on the development of Chinese Buddhism. According to this, behind the flux of our experienced world—the myriad things—there is a principle, *dao* (道) of which these are the manifestations. The *dao*, generating all objects, is not itself an object. Hence it is ineffable. As the opening verses of the *Daodejing* put it, 'the dao that can be talked about is not the true dao'.[17] It is therefore common to see it described as **nothing** (無, Chin: *wu*; Jap. *mu*) as opposed to the beings (有, *you*) which are its manifestations.

The similarity between the Indian Buddhist conventional/ultimate distinction and the Daoist 無/有 distinction is clear enough. And in the development of the distinctively Chinese forms of Buddhism, the two distinctions are identified. In texts of Chinese Buddhism one finds ultimate reality referred to as both 空 (Chin: *kong*, Jap: *kū*, emptiness) and 無, depending on whether it is its emptiness or its ineffability that is at issue.

Moreover, with a bit of help from certain *tathāgatagarbha* (如来藏, womb of Buddhahood) sūtras (which we need not go into here), the notion of the *ālaya* undergoes a striking development. It becomes one's Buddha nature (佛性, Chin: *foxing*). That is, it is the part of a person which is already enlightened. This enlightenment is cloaked by its impurity. Put bluntly, people are already enlightened: they just don't realise it.

Which brings us at last to Zen (禪, Chin: *Chan*).[18] Zen is one of the distinctly Chinese forms of Buddhism. In all forms of Buddhism, experiencing ultimate reality though meditative practices, and hence getting rid of the unhappy consequences of misunderstanding the nature of reality, is of great importance; but it is absolutely central to Zen. This is achieved in the experience of *satori* (悟, Chin: *wu*), a direct experience which, due to the nature of 無 cannot be described. For the same reason, enlightening people cannot be done by teaching with words. There must be a 'direct transmission'. All the teacher can do is help the student to have the experience. Meditation is important in this, and Zen developed a number of distinctive forms of meditation. But it also developed many other techniques such as *kōan* (公案, Chin: *gong an*) practice and shock tactics, which we need not go into here. The training can be long and disciplined, but according to many schools of Chan, the experience of *satori*, when it comes, is sudden and dramatic. If the appropriate preparation has been made, it can be triggered by a blow, or by something mundane, such as the sound of a tile falling, or the sight of the rising moon.

[17] Kwok, Palmer, and Ramsay (1993).

[18] For more on Chan, see Hershock (2019).

Does Zen Buddhism take ultimate reality, 無, to be the ground of reality? Yes, though one has to be slightly careful here. Objects of conventional reality depend on ultimate reality for their being. In all Mahāyāna Buddhisms, Zen included, the objects of conventional reality are conceptual constructions. If there were no ultimate reality for us to apply concepts to, there could be no conventional objects. The objects of reality therefore depend on ultimate reality. However, it would be wrong to suppose that ultimate reality is an ultimate ground, that is, a groundless ground. For, following Nāgārjuna, ultimate reality is as empty as everything else. Hence, it depends on something. What this is might be a somewhat debatable point; but the natural answer, at least in Chinese Buddhisms, is that it depends on the objects of conventional reality. One cannot have the manifestations of something without the thing of which these are manifestations. But conversely, one cannot have something whose nature it is to manifest itself in a certain way without those manifestations. Given this, the dependence between ultimate reality and conventional reality is reciprocal. So the relation of ontological dependence is not anti-symmetric.[19]

6. Nishida

With this background we can now turn to the second of the philosophers who hold **nothing** to be the ground of reality. This is Nishida Kitarō (1870–1945). Nishida was the founder of the Kyoto School of Philosophy, and arguably the most influential Japanese philosopher of the twentieth century.[20]

Nishida is a difficult philosopher: he was constantly reworking his ideas because of his dissatisfaction with them. Roughly speaking, his thought falls into three phases. In the first of these, he was concerned with an analysis of pure experience. In the second, he developed his theory of *basho* (場所). In the third he turned his thought to the sociopolitical consequences of his metaphysical views. It is the second of these periods which will concern us.

Nishida's style of expression is also not easy to follow. He does not present his ideas systematically. His thought appears to jump around, and it is not at all clear how (or whether) all the pieces fit together. For that reason, I am not sure that I have entirely understood Nishida's theory of *basho*.[21] It is probably more complex than I shall describe. However, I think I have it roughly right, and as to what he says about **nothing** I'm pretty sure that I have it exactly right.[22]

[19] For further discussions of ontological dependence in a Buddhist context, see Priest (2018c).

[20] For a general account of Nishida and his thought, see Maraldo (2015). For the Kyoto School, see Davis (2019).

[21] And for the same reason, I shall generally not quote Nishida. Pellucid explanations are not Nishida's forte. The picture has to be rather painfully put together from what he says in many places.

[22] For a discussion of the intricies of Nishida's account, see Warago (2005).

Let us start with the notion of *basho*. One might translate this is *place* or *topos*. A *basho* could be a physical place, but it general it is much more abstract than this, as we will see. The *basho* are also arranged in a hierarchy. **Nothing**, it will turn out, is the most fundamental of these. We will get there in due course, but let us start simply.

Consider a physical object, such as the moon, *m*. This satisfies the condition *is a sphere*, *S*. This, or at least its extension, is a *basho* of *m* in which *m* finds itself. We may depict matters thus:

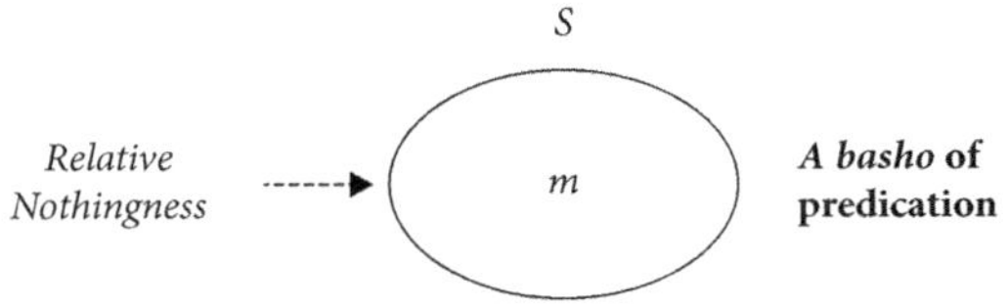

This *basho*, and each of the *basho* we shall meet till further notice, is a *relative nothingness* (相対無, *sōtai mu*). It is a nothingness because it is not itself present in the *basho*. However, this nothingness is relative to that *basho*, because it can occur in other *basho*.

In particular, that the moon is a sphere is a judgment, *Sm*. Hence this *basho* finds itself in a larger *basho*: the *basho* of judgment, thus:

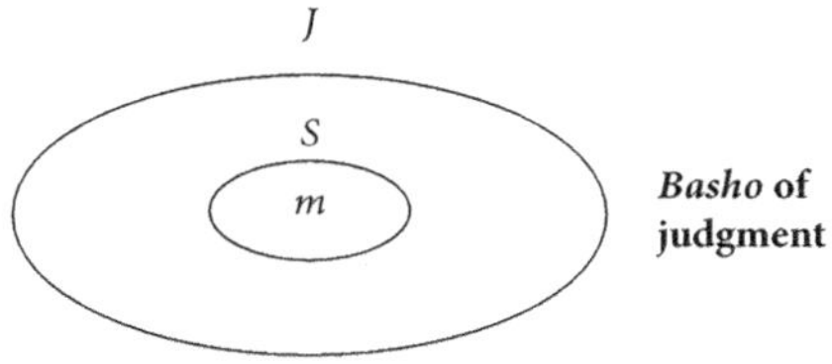

Note that the *basho* are cumulative. Everything in the first is in the second, but the second contains things not in the first, not only other judgments, but *S* itself.

To appreciate the next level, we need to understand something of Nishida's views on consciousness. He distinguishes two kinds: *consciousness that is conscious of* and *consciousness that is conscious*. We might call the first of these *intentional consciousness*, and the second *consciousness simpliciter*. The next level of *basho* is that of intentional consciousness. Let us write this as C_i.Then C_i may be depicted as follows (I leave out the contents of *J* to avoid clutter):

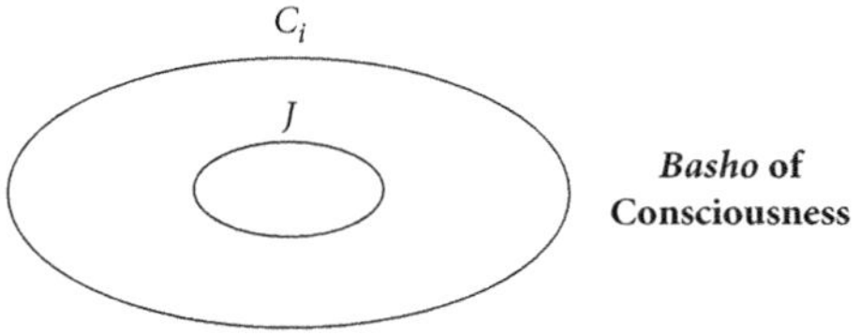

The contents of this *basho* are the things we would standardly think of as the contents of consciousness. This includes judgments, J, but it will also include other mental states, such as desires, emotions, etc.

I note that in English, the word 'judgment' is ambiguous. It can mean an act ('her judgment was made very fast') or a content ('her judgement was true'). Arguably, the confusion caused by this ambiguity bedevilled Western philosophy until it was cleared up by Frege and Husserl. It is clear from the way that J is formed that this contains judgments in the sense of contents. The *basho* of consciousness has them as mental states of activities. Does this imply a confusion on the part of Nishida? Yes, I'm afraid that it does.

This brings us to the final and most fundamental level of *basho*, which is the level of consciousness *simpliciter*. Let us write this as C_s. This is as follows:

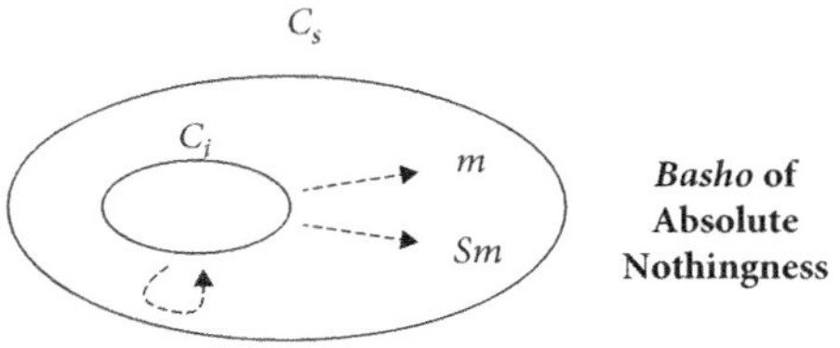

This *basho* has C_i as part of its contents. One may think of this as the *subject* of intentional states. The dotted arrows go to the *objects* of such states. These can be judgements such as Sm or objects such as m. Strictly speaking, these are within C_i itself, but I have moved them outside in the diagram to avoid clutter and make subject/object duality clearer. Note that one of the object poles of the subject/object distinction is the subject itself. In fact, Nishida thinks that any intentional state involves awareness of the subject itself. Does the fact that all other objects of intentional states are within consciousness itself imply a sort of idealism? Yes, I think it does. This is partly a result of running together judgments as acts and judgments as contents.[23] But it is also in line with the Yogācāra idealism that fed into Zen.

The *basho* C_s is that of absolute nothingness, (絶対無, *zettai mu*). It is a nothingness like all the other *basho*, since it does not occur within the *basho*. But it is absolute because there is no greater *basho* for it to occur within. It is, as Nishida sometimes puts it, a predicate which can never be a subject. The contents of the *basho* have a subject/object duality, but the *basho* itself does not. Indeed, *zettai* means something like *free from duality*. One may think here of the *ālaya*, and the Buddha nature into which this morphed in Chinese Buddhism.

[23] And could be avoided by having different *basho* for things in the world and their mental representations.

And finally, *zettai mu* is what till now I have called **nothing**. It is what remains, as it were, after all objects—indeed, all objects including the special object which is the subject of intentional states—have been removed. It is also ineffable. If one could predicate anything of it, it would perforce be in a larger *basho*, because of the way that predication works, as we saw right at the start.

Moreover, *zettai mu* is the ground of all objects. It is what objects appear *within*, and so what determines objects as objects. Without a place for them to be located, there could be no objects at all. Nishida puts it thus:[24]

> [T]he ultimate universal has the sense of being the noematic plane of the self-consiousness of absolute nothingness. Our entire life is reflected here. In this way, objective determination receives its deepest, most profound foundation.

And again:[25]

> When the self-consciousness of absolute nothingness determines itself, its noematic plane is the topos of the final universal that determines all that exists, and in its noetic direction we find the flow of infinite life.

For Nishida, too, then, **nothing** is the ground of all reality.

7. Conclusion

In the first part of the paper I argued that **nothing** is indeed the ground of reality, in the sense that **nothing** is what objects "stand out against". Without it, there could be no objects, just as there could be no hills if there were no surrounding plain.

In the later parts of the paper we have looked at two important philosophers who subscribe to this view—though each puts a distinctive spin on it in terms of larger projects—being in Heidegger's case, and consciousness in Nishida's.

As I have indicated, and as both Heidegger and Nishida were aware, this matter ties into further issues concerning ineffability and paradox. However, these will have to wait for another occasion.

8. Appendix on Heidegger and Zen

In this appendix, I want to take up the matter of the similarity between Heidegger's and Nishida's views on **nothing**. The similarity is indeed striking. For both, **nothing** plays an important role in their thinking; for both, **nothing** is ineffable; and for both, **nothing** is, in the sense we have seen, the space in which

[24] Warago (2005: 199). [25] Warago (2005: 207).

objects appear, the ground of reality. Perhaps the similarity is not surprising. It is of course well known for the same idea to occur to different people independently.

However, the confluence of views is made even closer, given Heidegger's views on the phenomenology of the experience of *das Nichts*, compared with the Zen experience of 無. If one knows something of Zen thought, it is impossible to read Heidegger's essay 'Was ist Metaphysik?' without being struck by the similarities, which appear to come from nowhere. Let us examine the matter.

Heidegger says that in a mood he calls *anxiety* one comes face to face with **nothing**:[26]

> Does such an attunement, in which man is brought before the nothing itself, occur in human existence?
>
> This can and does occur, although rarely enough and only for a moment, in the mood of anxiety.

Compare: in Buddhism our *Lebenswelt* is that of conventional reality, though **nothing** can be experienced in moments of *satori*. Next (p. 102):

> But just when moods of this sort [which have an object] bring us face to face with beings as a whole they conceal us from the nothing we are seeking.

Recall that one meaning of *saṃvṛti* is *concealing* or *obscuring*.

In both Zen and Heidegger's thought, **nothing** is experienced when the objects of conventional reality drop away, and we are left face to face with their background. Thus Heidegger (p. 104):

> This nothing reveals itself in Anxiety—but not as a being... [T]he nothing makes itself known with beings and in beings expressly as a slipping away of the whole.

Anxiety, then, is not an intentional state, directed towards some object or other. Indeed, not only is it objects which slip away, but the subject too (p. 103):

> We "hover" in anxiety. More precisely, anxiety leaves us hanging because it induces the slipping away of beings as a whole. This implies that we ourselves—we men who are in being—in the midst of beings slip away from ourselves. At bottom therefore it is not as though 'you' or 'I' feel ill at ease; rather, it is this way for some 'one'. In this unsettling experience of this hovering where there is nothing to hold on to, pure Dasein is all that is still there.

[26] Krell (1977: 102). Page references to Heidegger in this section are to this text.

In other words, the subject/object duality disappears—as in Zen—and all there is is just a "something happening". Heidegger calls this *Dasein*. A Zen Buddhist might call it Buddha nature.

Moreover, our awareness of **nothing** is, in a certain sense, always present (pp. 106–7):

> But now a suspicion we have been suppressing too long must find expression. If Dasein can relate itself to beings only by holding itself out into the nothing and can exist only thus, and if the nothing is disclosed only in anxiety; then must we not not hover in anxiety constantly in order to be able to exist at all? And have we not ourselves confessed that the original anxiety is rare? But above all else, we all do exist and relate ourselves to beings which we may or may not be—without this anxiety. Is this not an arbitrary invention and the nothing attributed to it a flight of fancy?
>
> Yet what does it mean that this original anxiety occurs only in rare moments? Nothing else than that the nothing is at first and for the most part distorted with respect to its originality. How, then? In this way: we originally lose ourselves altogether among beings in a certain way. The more we turn ourselves towards beings in our preoccupations the less we let beings slip away as such and the more we turn away from the nothing. Just as surely do we hasten into the public superfices of existence.
>
> And yet this constant if ambiguous turning away from the nothing accords, within certain limits, with the most proper significance of the nothing. In its nihilation the nothing directs us precisely towards beings. The nothing nihilates incessantly without our really knowing this occurrence in the manner of everyday knowledge.

In other words (p. 108):

> This implies that the original anxiety in existence is usually repressed. Anxiety is there. It is only sleeping. Its breath quivers perpetually through Dasein, only slightly in those who are jittery, imperceptibly in the 'Oh, yes'...and most assuredly in those who are basically daring. But those daring ones are sustained by that on which they expand themselves—in order to thus preserve a final greatness of existence.

That is, in Buddhist terms, we are already enlightened, though this is hidden from us.

Moreover, when 'the daring' do experience **nothing** this may happen quite suddenly and unexpectedly (p. 108):

> Original anxiety can awaken in existence at any moment. It needs no unusual event to rouse it. Its sway is as thoroughgoing as its possible occasions are trivial. It is always ready, though it only seldom springs, and we are snatched away and left hanging.

Or in Buddhist terms, *satori* can be sudden, and triggered by quite mundane events.

There remains the point that Heidegger calls this mood in question 'anxiety', which implies an unpleasant experience—which one would not associate with an experience which is supposed to lead to liberation. But things are not so straightforward. From the Heideggerian side, he says things about the experience which are hardly unpleasant. Anxiety is not to be confused with fear (p. 102, Heidegger's ellipses):

> Much to the contrary, a peculiar calm pervades it. Anxiety is indeed anxiety in the face of..., but not in the face of this or that thing.

And again (p. 108):

> The anxiety of those who are daring cannot be opposed to joy or even the comfort of tranquilized bustle. It stands—outside all such opposition—in secret alliance with the cheerfulness and gentleness of creative longing.

On the other hand, Heidegger's use of the word 'anxiety' is not capricious (p. 103):

> In anxiety, we say 'one feels ill at ease'.... The receding of beings that closes in on us in anxiety oppresses us... We can get no hold on things. In the slipping away of beings only this "no hold on things" comes over us and remains.

No doubt the slipping away of the familiar world can be a disconcerting experience.

But one should also note, from the side of Zen, that some Zen thinkers have referred to the initial state of awakening as the *Great Death*. The term was coined by Zhaozhou (Jap.: Jōshu),[27] and taken up by Dōgen, and Hakuin. According to some accounts, the Great Death is likened by Hakuin to leaping from a high cliff into a void. Jumping off a cliff can certainly be an anxiety-generating experience—at least until one realises that there is no ground to hit.

The similarities between Heidegger and Zen on the phenomenology of the experience of **nothing** are, then, manifest and clear. Of course, this, again could just be coincidence. But this is not so plausible if there is another explanation; and one is suggested by the following. Tanabe Hajime was assistant professor to Nishida: indeed, he became Nishida's successor in his chair of philosophy at Kyoto University. Tanabe, in fact, studied with Heidegger in the early 1920s.[28] It is very plausible that Heidegger learned of the Zen ideas from him, and applied them to his own ideas concerning being.

[27] Cleary and Cleary (2005), Case 41.

[28] See Davis (2019: 3.1) and Heisig (2001: 108).

References

Berto, Francesco, Rohan French, Graham Priest, and David Ripley (2018). "Williamson on Counterpossibles", *Journal of Philosophical Logic* **47**: 693–713.

Bliss, Ricki and Kelly Trogdon, (2014). "Metaphysical Grounding", *Stanford Encyclopedia of Philosophy*, ed. Edward N. Zalta (Winter 2016 edn).

Bliss, Ricki and Graham Priest (2017). "Metaphysical Grounding, East and West", in *Buddhist Philosophy: A Comparative Approach*, ed. S. Emmanuel (Hoboken, NJ: Wiley-Blackwell), pp. 63–85.

Cleary, Thomas and J. C. Cleary (trans.) (2005). *The Blue Cliff Record* (Boston, MA: Shambala).

Davis, Bret W. (2019). "The Kyoto School", *Stanford Encyclopedia of Philosophy*, ed. Edward N. Zalta (Summer 2019 edn).

Garfield. Jay L. (trans.) (1995). *The Fundamental Wisdom of the Middle Way: Nāgārjuna's Mūlamadhyamakakārikā* (New York: Oxford University Press).

Heidegger, Martin (1977). *Basic Writings*, ed. David Farrell Krell (New York: Harper and Row).

Heidegger, Martin (1991). *The Principle of Reason*, trans. Reginald Lilly (Bloomington, IN: Indiana University Press).

Heidegger, Martin (1996). *Being and Time*, trans. Joan Stambaugh (Albany, NY: State University of New York Press).

Heidegger, Martin (2000). *Introduction to Metaphysics*, trans. Gregory Fried and Richard Polt (New Haven, CT: Yale University Press).

Heisig, James W. (2001). *Philosophers of Nothingness: An Essay on the Kyoto School* (Honolulu, HI: University of Hawai'i Press).

Hershock, Peter (2019). "Chan Buddhism", *Stanford Encyclopedia of Philosophy*, ed. Edward N. Zalta (Spring 2019 edn).

Kwok, Man-Ho, Martin Palmer, and Jay Ramsay (trans.) (1993). *Tao Te Ching* (Shaftesbury: Element Books).

Maraldo, John C. (2015). "Nishida Kitarō", *Stanford Encyclopedia of Philosophy*, ed. Edward N. Zalta (Winter 2019 edn).

Priest, Graham (2008). *Introduction to Non-Classical Logic: From If to Is* (Cambridge: Cambridge University Press).

Priest, Graham (2014a). *One: Being an Investigation into the Unity of Reality and of its Parts, including the Singular Object which is Nothingness* (Oxford: Oxford University Press).

Priest, Graham (2014b). "Much Ado about Nothing", *Australasian Journal of Logic* **11** (2): no. 4, https://doi.org/10.26686/ajl.v11i2.2144.

Priest, Graham (2014c). "Speaking of the Ineffable…", in *Nothingness in Asian Philosophy*, ed. JeeLoo Liu and Douglas L. Berger (New York: Routledge), pp. 91–103.

Priest, Graham (2018a). "Some New Thoughts on Conditionals", *Topoi* **37**: 369–77.

Priest, Graham (2018b). *The Fifth Corner of Four* (Oxford: Oxford University Press).

Priest, Graham (2018c). "Buddhist Dependence", in *Reality and its Structure: Essays in Fundamentality*, ed. Ricki Bliss and Graham Priest (Oxford: Oxford University Press), pp. 126–39.

Priest, Graham (ms), "*Everything and Nothing*", manuscript.

Tahko, Tuomas E. and E. Jonathan Lowe (2015). "Ontological Dependence", *Stanford Encyclopedia of Philosophy*, ed. Edward N. Zalta (Fall 2020 edn).

Wargo, Robert J. J. (2005). *The Logic of Nothingness: A Study of Nishida Kitarō* (Honolulu, HI: University of Hawai'i Press).

Wigglesworth, John. "*Metaphysical Dependence and Set Theory*", PhD dissertation, City University of New York, 2013.

Wigglesworth, John (2015). "Set-Theoretic Dependence", *Australasian Journal of Logic* **12** (3): no. 3, https://doi.org/10.26686/ajl.v12i3.2131.

3
Thales' Riddle of the Night

Roy Sorensen

Which is older: day or night? Thales of Miletus is said to have answered: "Night is the older by one day" (Diogenes Laertius I. 34–6). My aim is to vindicate the answer attributed to Thales. The eerie truth is that Night, as the shadow of Earth, is the oldest thing on Earth.

1. Deciphering the Riddle

The riddler, I conjecture, identifies the night as Earth's shadow. If the sun is a stable light source, in relative motion around the Earth, then the Earth's shadow has the same antiquity as the shadow caster. And given that the absence of light is the default state, night precedes the first day—regardless of whether there is a first night.

"*When* does the night begin?" entwines "*Where* does the night begin?" Yet the spatial origin of night was only studied a thousand years later. When asked in concert, the two questions present an opportunity to view astronomy in reverse perspective. Light is submerged in the background as darkness comes to the fore. 'When and where does the night begin?' yields a photographic negative—astronomy as recounted from the dark side.

Thales was a legendary astronomer. According to Herodotus (*c*.485–*c*.420 BC),

> On one occasion [the Medes and the Lydeans] had an unexpected battle in the dark, an event which occurred after five years of indecisive warfare: the two armies had already engaged and the fight was in progress, when day was suddenly turned into night. This change from daylight to darkness had been foretold to the Ionians by Thales of Miletus, who fixed the date for it within the limits of the year in which it did, in fact, take place (Herodotus 1920: 1.74).

Reportedly, the battle ceased. Cyaxares, king of the Medes, accepted mediation. One of the two mediators was Nebuchadnezzar (best known for his destruction of Jerusalem in 584 BC).

Roy Sorensen, *Thales' Riddle of the Night* In: *Non-Being: New Essays on the Metaphysics of Non-Existence.*
Edited by: Sara Bernstein and Tyron Goldschmidt, Oxford University Press (2021). © Roy Sorensen.
DOI: 10.1093/oso/9780198846222.003.0003

Historians of astronomy agree that this is almost certainly a tall tale about Thales. There are credible reports that Thales worked as a military engineer for the son of King Alyattes of Lydia. Possibly, Thales witnessed the eclipse near the past or future battleground. Possibly possibly, this coarser *observation* was mis-recalled as a fine-grained *prediction*.

After Isaac Newton, astronomers predicted eclipses with an exactitude that makes Herodotus' tale seem anachronistically plausible to post-Newtonian laymen. Twentieth-century astronomers have pinpointed Herodotus' total eclipse as occurring about an hour before sunset on May 28, 585 BC (Stephenson 1997, 281)! Ancient eclipses insert archipelagoes of precision into a sea of vagueness.

Regardless of whether Thales addressed the riddle of day and night, the answer attributed to him was regarded as insightful. I ignore interpretations that make the riddle nonsensical. This excludes the reading in which 'day' and 'night' are units of time. For then 'Night is older than the day' commits the category mistake of assigning ages to units. *Names* of units have ages. *Uses* of units have ages. But not the units themselves.

Units do have relative magnitudes. On Earth, a year is always longer than a day. (The reverse holds on Mercury and Venus.) Of any particular Earth day, we may ask whether that day is longer than the following night (relative to our location). If the day and night riddle were posed in this sense, the answer attributed to Thales would be an obvious untruth.

A more charitable interpretation emerges from the fact that units are based on standard objects or phenomena that are datable. We cannot ask for the age of the foot unit. But we can ask how long there have been human feet. The unit term 'night' is based on a primal sense of 'night' that denotes the darkness that follows each sunset and that yields to light with each sunrise. We can ask how long that darkness has lasted.

2. The Unity of the Night

You cast a distinct shadow after each new sunrise because your shadow does not persist through the night. However, the night persists after sunrise. The night is the Earth's shadow (Figure 3.1).

Since the Sun is an extended light source, the Earth has both an umbra and a penumbra. The umbra is a region of total light blockage and the penumbra is a region of partial light blockage. The apex of the umbral cone averages 108 Earth diameters behind the Earth (varying about 10% with the position of the Earth). The night is the umbra. So, the night is a cone that extends 108 Earth diameters into space, occasionally landing on the Moon. (The generalization holds regardless of the size of the ball; the shadow of the Sputnik satellite was 108 Sputnik diameters.)

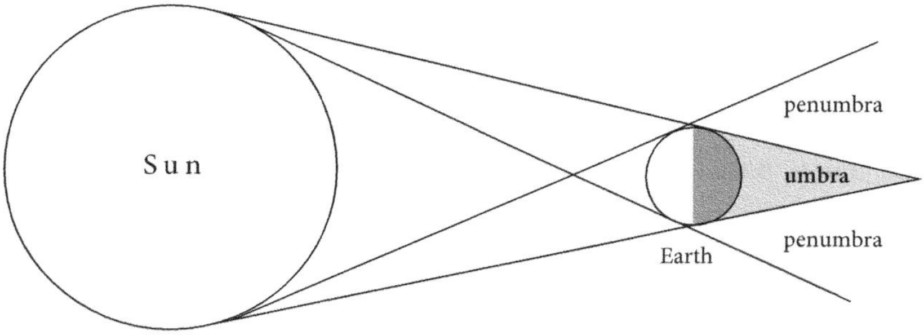

Figure 3.1 The Earth's shadow.

From a perspective on Earth, the night appears shapeless and vague. Generally, we describe a dark region as a shadow only when we can view its boundaries. Therefore, we do not spontaneously classify the night as a shadow. Consequently, earth-dwellers do not ask for the night's shape, height, or speed.

Too bad for earth-dwellers. All of these unasked questions are well-defined. Indeed, the questions are naturally posed from an extra-terrestrial perspective.

Perspective shifts became popular after the Earth lost its status as the center of the universe. To dislodge geocentricism, young Johannes Kepler imagined a trip to the Moon. For lunar astronomers, the movement of the Earth is as obvious as the Moon's movement is obvious to terrestrial astronomers. This thought experiment eventually became the basis of Kepler's posthumously published novel "The Dream"—which makes the first work of science fiction contemporaneous with the origin of science.

Protagoras was an egalitarian about perspectives. According to his slogan "Man is the measure", any perspective is as valid as any other.

Kepler's aristocratic hero Plato opposed Protagoras' egalitarianism. Plato may be subtly ranking perspectives in the dialogue "Theaetetus". The interlocutors discuss the tale of a Thracian maidservant finding Thales in a well. She infers the other-worldly philosopher has been so distracted by the heavens that he has absent-mindedly fallen into the hole! This is commonly interpreted as a concession that philosophers are impractical. But ancient astronomers deliberately climbed down wells (and other tubes) to focus vision on a circumscribed portion of the sky. Friedemann Buddensiek (2014) argues that the intended lesson of Plato's story is that young Theaetetus (and readers) must transcend the maidservant's coarse perspective. She is a Thracian after all, as notorious for stupidity as the Cretans for mendacity. Contrary to Protagoras, some awkward-looking perspectives are *superior*. To control for illusions induced by an object's proximity to the horizon, view them upside down between your legs. We ought to adopt the celestial perspective that allows us to ask and answer questions that may seem excessively precise.

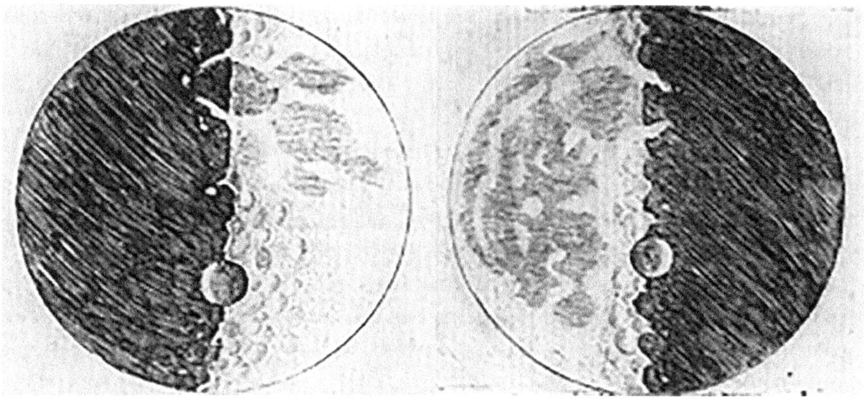

Figure 3.2 Galileo's drawing of the lunar night.

The boundaries of Earth's shadow are not entirely invisible to all earthlings. Olympian observers with a long line of sight see it as a dark, bluish region of sky opposite the direction of the rising Sun or the setting Sun; the Earth's shadow is being cast on the atmosphere, clouds, and mountains. Surprisingly, the edge of night can be photographed during the day (Lynch and Livingston 2001: 38).

At high altitudes, the night's edge can be seen beneath the observer. The explanation is readily gleaned from the white spots of Galileo's drawing of the lunar night (see Figure 3.2).

Galileo correctly construed the spots as the tops of lunar mountains. Since the night has the shape of a cone, an observer on a mountain can be *above* the night.

A new night is akin to the new moon. There is a new moon when the Moon becomes visible again. 'New moon' does not denote a heavenly body that replaces an earlier moon.

The analogy between the new moon and the new night is likely to have occurred to Thales. The new moon was culturally salient to the Greeks (as it was in all agrarian and maritime societies). Thales made an intensive study of the lunar cycle. One purpose of this study was calendar reform. There is also evidence that Thales based his prediction of the solar eclipse on the correlation with lunar eclipses (O'Grady 2002). (The correlation only predicts a solar eclipse somewhere, not its location.)

There is a new night when the night becomes visible again. The night ends when the darkness is no longer visible. From a terrestrial perspective, the rising Sun seems to destroy the darkness. But at dawn we are merely moving out of a stable shadow. The shadow of the Earth is the shadow of a spinning sphere.

Does the night spin? Uranus has a spinning night because it was knocked sideways. If there were a mountain at its equator, the rotation of its shadow would be discernible by the rotating protrubence. Since Uranus is round, the night's spin is invisible.

A planet with no tilt imparts no spin to its shadow. Instead the shadow scrolls (Sorensen 2008, 94–6). Each slice of the planet transfers responsibility for a shadow slice to a neighboring slice of the planet. Since the Earth has a tilt of 23 degrees, its shadow mostly scrolls but also spins.

Different regions of the sphere rotate through the shadow. Any region exiting the shadow will re-enter the same shadow. If residents of the sphere picture the Sun as being extinguished (as a burning ship sinking in water), then they will infer that they are entering a fresh shadow after each sunset.

Heraclitus (born about 535 BC) claimed that a new Sun forms each morning. He explained the night in terms of the demise of the main light source rather than by occlusion. If Heraclitus' hypothesis of successive Suns were correct, each night would be destroyed at dawn.

The Greeks presupposed that the earth is at rest. The puzzle was how this stability could be achieved without support. To avoid an unsupported supporter, Xenophanes contends that Earth is infinitely deep (Aristotle, *de Caelo*, ii. 13; 294 a 21).

Thales' solution is water. Everything is made of water. One form of water, such as ice, can float in another form, liquid water. Water is both the universal physical support and the universal constitutor.

Aristotle criticizes the hypothesis that the earth floats on water. After all, a rock dropped into a lake sinks – *like a rock!*

Except when the rock is pumice. Thales lived in a region in which such stones are well known (Heyd 2014). Aggregates of these stones occasionally formed floating islands.

Thales was also familiar with stalagmites that accrete from dripping water. The Greeks regarded caves as sanctuaries from surface distractions (Ustinova 2009). Principles revealed from the depths explain the superficial realm. Plato's Allegory of the Cave is an unusual reversal of the theme that caves reveal rather than distort reality. For most Greeks, the wise escape *to* a cave rather than escape *from* a cave.

Only with Nicolaus Copernicus (1473–1543) was there a good framework for the hypothesis that the Earth is always falling. He modeled the Earth as a satellite of the Sun. This implies that the Earth and its shadow race at tremendous speed.

Johannes Kepler later showed that the Earth has an elliptical orbit. Consequently, its shadow waxes and wanes. The shape of the night also undergoes slight, continuous change because the Earth is not a perfectly round, spinning ball. Despite this flux, the Earth has only had a single shadow.

Thales was ignorant about the age of heavenly bodies. If the Sun is identical to Apollo, then the Sun is immortal because all gods are immortal. Thales pioneers reductive explanations. This policy of impersonalization opens the possibility of the Sun dying out. Thales seems to have explained the Sun's fire as an effect of evaporation (O'Grady 2002, 63). Several observations suggest a role for water in fire. In Thales' region there are hot springs. There are methane emissions in the stagnant backwaters of the Meander River. These spontaneously ignite. Thales was also familiar with the heat radiating from fermenting manure and lawn cuttings. In any case, the belief that the sun's fire is the product of evaporation suggests that the Sun will stop shining after drying. Thales' pupil Anaximander may have been elaborating on this idea when he spoke of our well-defined world returning to the infinite, undifferentiated state from which it originated.

Astronomers now know that the Sun has been shining for 4.6 billion years. They estimate that the Sun began with only 70% of its current brightness. It has been very gradually brightening. In another 1.5 billion years the last snow will have fallen on Earth. And 3.5 billion years after that, the aging Sun will become a red giant. The Sun will bloat to about the present orbit of the Earth—thereby engulfing Mercury and Venus. The Earth will survive (as a burnt-out cinder) because the Sun will have lost mass, which will widen the orbits of the planets. After the Sun shrinks into a white dwarf, it will cast a modest amount of light. As the Sun wastes away and becomes a black dwarf, its light wanes. The night will wither into the darkness.

Shadows are holes in the light. They are destroyed when their host is destroyed. But other forms of shadow destruction are spurious. Black lava fields make shadows *appear* to disappear. But laying a white cloth on the ground reveals that the absence of light of was merely mistaken as the black lava. Since black surfaces are visible by virtue of the light they absorb, the surface became causally idle and so invisible. Instead of the shadows becoming invisible, patches of the lava field become invisible. All you see are shadows, not the black lava that is no longer absorbing light.

3. Threats to the Night

Most shadows are short-lived. But we can exclude all of the ways the night might have been interrupted since the formation of the Earth. Shadows can be washed out by the intrusion of *alternative* light sources. For instance, if a supernova occurred close to Earth, the intense light would have erased the night. But an event of that magnitude would have left traces in meteorites. Scientists possess meteorites as old as the Earth. None bear a mark of a local supernova.

Of course, some starlight does penetrate the night. But shadows tolerate light pollution.

Why doesn't the normal operation of the stars preclude the Earth from casting a shadow? In 1826, Wilhelm Olbers noted that there is a star anywhere you point into the night sky. Since the sky is therefore covered with light sources, why does the sky darken after sunset?

True, light dims with the square of the distance. But the number of stars also doubles with the square of the distance, so the two effects cancel.

If there were an obstacle blocking the starlight, then the laws of thermodynamics imply that this shadow caster would heat up and eventually become luminous (Harrison 1987: 99).

Further solutions to Olbers' paradox have been canvassed. They involve implausible assumptions about the size of the universe and the distribution of the stars. For instance, if the stars were lined up in rows, most of the light sources would be blocked. But the stars are distributed haphazardly.

The stars at the center of the Whirlpool Galaxy are clustered so thickly that all of their planets are nightless. But an absence of night is also predicted under the assumption that there are infinitely many stars distributed randomly in an infinite universe, or almost any non-contrived pattern.

Astronomers after Olbers eventually regarded the night as a powerful clue about the fundamental structure of the universe. Contemporary astronomers believe the night to be a legacy of the Big Bang. As space expands, starlight takes longer to reach an observer. Eventually, the space will expand at a rate faster than the speed of light. Each portion of the universe will become isolated from other distant portions. Eventually, the Big Bang will erase evidence of itself (Krauss 2012: 117). Future astronomers, from all parts of the universe, will converge on the nineteenth-century view that only their own galaxy exists.

There is no record of Thales or any other Greek noticing Olbers' paradox. This oversight is remarkable because the dark sky is predicted by their assumption that the universe is finite. In *The Paradox of Olbers' Paradox* Stanley L. Jaki wonders why Aristotle should resort to esoteric proofs of the universe's finitude when he could have drawn the conclusion from a premise that was plain as night. And why should Newtonian astronomers have later strayed from this readily confirmed hypothesis to the readily *dis*confirmed hypothesis that the universe is infinite?

I have noted that supernovas pose a threat to the Earth's shadow that ancients could not have anticipated. The ancients could understand how eclipses pose a threat. The Moon and Sun share the same *apparent* diameter. Perhaps Thales measured the *objective* diameters of the Sun and Moon. If sunlight were entirely blocked by the Moon, then the Earth's shadow would be consumed by the Moon's shadow. However, the Moon's umbra only falls on a narrow slice of the Earth's surface. At worst, the Earth's shadow was diminished somewhat during these events, not destroyed.

The fate of the Moon's shadow differs. During a total lunar eclipse, the Moon falls completely into Earth's umbra. The lunar night is destroyed by Earth's night.

Although all direct sunlight is blocked during a lunar eclipse, some light is refracted through the Earth's atmosphere and lands on the Moon. Thus, the Moon often looks reddish during a lunar eclipse. (An astronaut on the Moon would see a bright red ring around the Earth—thereby witnessing all the sunrises and sunsets simultaneously!) The exact color of the Moon depends on how cloudy and dusty our atmosphere is. Some may hope that this indirect light is enough to sustain the Moon's shadow.

But there is no hope that the Moon's shadow can *always* be nursed along by earthshine. For this indirect light sometimes dwindles to an insignificant level. Volcanic dust from Mount Pinatubo rendered the Moon nearly invisible during the total lunar eclipse of December 1992. Cataclysmic meteors and asteroid collisions have occasionally made Earth's atmosphere almost completely dark. The Moon's shadow could not have survived during those circumstances (even if one thought it survives during a regular lunar eclipse). Unlike the Earth, the Moon has had a succession of shadows. Its present night is therefore much younger than the Earth's night.

The only other way that objects can lose their shadows is by becoming transparent. But the Earth has never lost its opacity. Therefore, the shadow of the Earth is *at least* as old as the Earth.

4. The Night is Older than the Earth

As hypothesized by Immanuel Kant in his *Universal Natural History and Theory of the Heavens* (1755), the solar system coalesced from a cloud of dust. Some of this dust began to gently combine. As the collections enlarged, so did the role of gravity. Pebbles were drawn to rocks, rocks were drawn to boulders, boulders to proto-planets. Since the distinction between 'protoplanet' and 'planet' is vague, there is vagueness as to when the Earth formed. One might expect 'the Earth's shadow' to inherit this vagueness. However, the Earth's shadow pre-dates the Earth. A protoplanet cast a shadow which later became the shadow of the Earth.

The age of a thing can exceed the period for which its description became applicable. The Agia Sofia Mosque was built in 548 as a Byzantine church, before Mohammed's birth in 570. The building was converted to a mosque in 1453. Just as the Agia Sofia mosque manages to be older than Islam, the Earth's shadow manages to be older than the Earth.

The Sun was shining throughout the formation of the protoplanet Earth. The mere fact that light shines on an object is not enough for it to constitute the day of that object. Night and day require a large scale. The light that falls upon a meteor does not qualify as the day of the meteor and the shadow of the meteor does not constitute the night of the meteor. Since the Earth developed from the accretion of rocks, the shadow that became the Earth's shadow existed before there was any day.

In the case of black holes, there is no difference in light to sustain a day/night distinction. All light is uniformly absorbed. In 2019 astronomers used the Event Horizon Telescope to photograph a black hole at the center of galaxy Messier 87. The lightless object is made visible by contrast from surrounding light (emanating from the accretion disk and ambient stars).

5. Night without Day

The Greeks may have pictured 'Which is older: day or night?' as akin to the riddle 'Which came first, the chicken or the egg?' If one answers 'The day is older', the riddler has the rejoinder that each day is preceded by a night. If one answers 'The night is older', the riddler will reply with the principle that each night is preceded by a day.

Many feel the connection between night and day is more intimate than that between chicken and egg. If there is day, there must *simultaneously* be night somewhere else. This reciprocal relationship is voiced in William Shakespeare's *A Midsummer Night's Dream*:

> O grim-looked night, O night with hue so black,
> O night which ever art when day is not!

The symmetry between day and night prompts many to respond to 'Which is older, day or night?' with 'They are equally old'.

However, other people—perhaps including Thales—are inclined to credit the night with greater antiquity because they deem the absence of light as the natural starting state. This is reinforced by creation stories that assume a dark prelude.

Most explanations of the "birth" of the universe are modeled on human reproduction. Hesiod's (*c.*700 BC) *Theogony* personifies night as the goddess Nyx who came forth from Chaos (associated with nothingness or as a disorderly mix of matter). According to Hesiod, Chaos is the first god, making Nyx one of the oldest gods. Hesiod claims Night in turn bore Eris (Discord or Strife), the Moirai (Fates), Hypnos (Sleep), Nemesis (Retribution), Thanatos (Death), and the three Hesperides (goddesses of the sunset). These divine children were born fatherless. Nyx also had children by the god Erebus: Aether (Air) and Hemera (Day). Thus, Hesiod characterizes Night as the mother of Day. Although each is very old and immortal, Night must always be older than Day. Hesiod also describes where Nyx lives:

> There also stands the gloomy house of Night;
> ghastly clouds shroud it in darkness.
> Before it Atlas stands erect and on his head
> and unwearyingly arms firmly supports the broad sky,
> where Night and Day cross a bronze threshold
> and then come close and greet each other. (*Theogony*, 744 ff.)

Night and Day always go out by turns. As frightful as Nyx was, prayers were addressed to her from those who craved the cover of darkness.

The priority of darkness is also observed by creation stories that forego sexual imagery. These are modeled on performative speech acts, such as marriage, that summon new things into existence. To secure uptake, the speaker exploits conditions that are apt to be incorporated into the content of the ceremony. For instance, a storyteller uses the blinders of background darkness to funnel attention. Characters are stipulated into existence with "Once upon a time". In Genesis, the earth is at first "without form and void; and darkness was upon the face of the waters." God then declares "Let there be light". If night is darkness and day is light, then night precedes day. Since 'day' can also be used as a unit, one can coax out an interpretation in which night is older than day by one day.

Philosophers improve the logical credentials of creation myths. Double negation suggests, not implausibly, that two negatives can make a positive. With this pair of forceps, John Scotus Eriugena pulls out a model of self-creation in which God negates "divine darkness", a kind of non-being, to create light (Bernstein, this volume Chapter 1).

Jewish mystics hint that this transformation from nothingness proceeds from infinity. The Kabbalist Azriel of Gerona (*c.*1160–*c.*1238) discourages speculation as to how the process arises. Finite minds cannot understand the infinite! Undeterred, Aaron Segal (this volume Chapter 10) conjectures that infinite causal overdetermination can generate something from nothing via a "before effect". If a beginningless sequence of gods stand ready to create the world given that no predecessor god has already done the deed, creation occurs despite no god being the creator.

'Exist' derives from ex' (out) and 'sistere' (made to stand). Thus, Latin etymology suggests that nothing is the background for each object. Graham Priest (this volume, Chapter 2) foregrounds this background: nothing is the ground for all objects. Objects stand to nothing as hills to their plane.

Hope of establishing the priority of the night through darkness is dimmed by extra conditions that are needed to make darkness qualify as night. There is permanent darkness at the floor of the Shackleton crater near the Moon's south pole. This dark region is only a local shadow, not a miniature night.

Any planet that does not receive light lacks a day. Merely facing a light source is not good enough. The nearest star could be too far away to sustain a difference between day and night. Even if it is near enough, another heavenly body could perpetually block the light. These dayless planets are also nightless. They have plenty of darkness but no night.

All of the planets in our solar system have nights even though the difference between night and day decreases with distance from the sun. Pluto's day is 1,000 times fainter than an Earth day (but that is still 100 times brighter than the light from a full moon). The dimness makes Pluto well named. 'Pluto' is the Roman name for Hades, god of the underworld.

What is the minimum amount of light necessary to sustain the distinction between night and day? This question resembles 'What is the minimum contrast needed for there to be a shadow?' Our standards for what qualifies as a shadow shifts with our ability to detect light differences. Faint differences suffice for sensitive equipment. On July 15, 2015, the New Horizon spacecraft photographed the night side of Charon (Pluto's largest moon). For astronomers, the difference between day and night on Charon is as well defined as the difference between day and night on Earth.

Since shadows, by definition, require light sources, we naturally expect day when there is night. However, there are special conditions under which night can linger without the day. Suppose the Sun stops shining. Since the speed of light is finite, there will be a moment when the Earth's shadow still exists even though the sunward side of Earth is as dark as the night side of the Earth. For this fraction of a second, there will be night without day.

Darkening the atmosphere can also stop the day. Debris from volcanoes, meteors, and comets has periodically stopped sunlight from reaching the surface of the Earth. During those periods there have been nights without days.

The Earth still casts a shadow when the atmosphere is darkened. One might object that the shadow is then the shadow of the atmosphere rather than the shadow of the Earth. But this is analogous to saying that a woman in a burqa fails to cast a shadow. The burqa covers the woman's entire body so none of her skin blocks the light. Yet the woman is still blocking the light because the burqa would not have the contours it has without her. The shadow tracks the movements of the woman.

Like the night, the day is surprisingly stable. At sunrise, we encounter a new day only in the sense we encounter a new moon. What is really happening is that the old day has become newly visible. The day, as a natural phenomenon, lasts for millions of years. This persistence sets the stage for the circumnavigator's paradox.

6. Where Does the Day Begin?

Lewis Carroll (1850: 31–3) imagines a ring of land girdling the globe. The earthbound ring is populated with English speakers. A speedy traveler departs London Tuesday at 9 a.m. He races just fast enough to orient the Sun in the same position in the sky. As the traveler speeds beneath the stilled sun, he verifies the time with the locals by asking 'What time is it?'. Their constant answer: *9 a.m.* Indeed, *9 a.m.* is the response when, 24 hours later, the circumnavigator returns to London. But the Londoners also report the day as *Wednesday* rather than Tuesday. The traveler wonders: *Where did Wednesday begin?*

There is an echo of this riddle in *Through the Looking Glass* when Alice is traveling with the Red Queen. The pair must run faster and faster just to keep in place. When they finally pause Alice is puzzled:

> "Well, in *our* country," said Alice, still panting a little, "you'd generally get to somewhere else—if you ran very fast for a long time, as we've been doing."
> "A slow sort of country!" said the Queen. "Now, *here*, you see, it takes all the running *you* can do, to keep in the same place. If you want to get somewhere else, you must run at least twice as fast as that!" (Carroll 1897: 50)

Whereas Carroll's circumnavigator keeps stationary with respect to the Sun, the Red Queen keeps stationary with respect to a moving Earth.

The speedy circumnavigator subverts the urgency of having only one day left: "Do you think that I count the days? There is only one day left, always starting over: it is given to us at dawn and taken away from us at dusk." (Jean-Paul Sartre *The Devil and the Good Lord,* Act 10, scene 2) The speedy circumnavigator's riposte: *All life on Earth is for a single day.*

"God is light, and in him is no darkness at all" (1 John 1:5). Augustine reads such passages literally. God *is* light. In John Milton's *Paradise Lost* (1667), Satan races around the globe to flee the light:

The space of seven continued nights he rode
With darkness; thrice the equinoctial line
He circled, four times crossed the car of Night
From pole to pole, traversing each colure (VIII. 63–6).

Jason George's 2018 Belgian television series "Into the Night" provides a physically realistic rationale for nocturnal circumnavigation. Ionizing gamma rays from the sun has made daylight fatal to all organisms. To survive, passengers in a jet must keep flying into the night.

I will not join the list of commentators who have attempted to make geographical sense out of Milton's four lines. But I will return to Carroll's riddle of where Wednesday originated: Wednesday has not yet begun *for the circumnavigator.* The Sunday, Monday, Tuesday, ... , Saturday cycle counts the number of encounters with the day. This makes the day vary with one's location. But the system also has the side-effect of making the day of the week vary with the history of the observer. A circumnavigator will have a different number of encounters with the day than a stationary observer.

This biographical solution to the circumnavigator's paradox exposes a shortcoming in the calendar system of Lewis Carroll's era. If the population contains many circumnavigators, schedulers will have to know traveler histories. Great Britain was the leading maritime nation and so harbored many circumnavigators. Lewis Carroll must have realized that the high speed of his traveler merely makes the observer relativity salient. Circumnavigation at any speed alters the number of encounters with the day (adding or subtracting depending on whether the circumnavigators travel east or west).

The circumnavigator's paradox was detected *a priori* by a Syrian prince who became forgotten because of the exorbitant tax his name imposes on memory: Isma'il ibn 'Ali ibn Mahmud ibn Muhammad ibn Taqi ad-Din 'Umar ibn Shahanshah ibn Ayyub al Malik al Mu'ayyad 'Imad ad-Din Abu 'l-Fida (1273–1331). The paradox was then re-discovered *a posteriori* by Ferdinand Magellan's skeletal crew. They inadvertently sailed around the world in 1522 (minus Magellan, who died along the way). The man in charge of keeping the ship's log relates the incident:

> On Wednesday, the ninth of July, we arrived at one these islands named Santiago, where we immediately sent the boat ashore to obtain provisions. . . . And we charged our men in the boat that, when they were ashore, they should ask what day it was. They were answered that to the Portuguese it was Thursday, at which they were much amazed, for to us it was Wednesday, and we knew not how we had fallen into error. For every day I, being always in health, had written down each day without any intermission. But, as we were told since, there had been no mistake, for we had always made our voyage westward and had returned to the same place of departure as the Sun, wherefore the long voyage had brought the gain of twenty-four hours, as is clearly seen. (Pigafetta 1969, I, 147–8)

After a period of befuddlement, commentators realized that dating discrepancies are inevitable—unless everybody stays put or everybody moves together. Since most people are stationary, the least disruptive solution is to relativize the days of the week to a fixed point on the Earth. But which? As shipping increased, the custom was to relativize the day of the week to the point of departure or arrival. But since ships have many such points, memory was again over-taxed. Scheduling confusions caused cargo to spoil and ships to stand idle.

A more inventive solution had been long ago envisaged by the medieval philosopher Nicole Oresme (Lutz 1975: 70). Although writing a century before the first circumnavigation, Oresme was intrigued by the effect circumnavigation would have on time keeping. To ensure that *everybody* progresses through the week uniformly, Oresme proposed an imaginary line along some remote region of the Earth from the North Pole to the South Pole. Nations should then declare that the day begins when an observer along that line would first encounter the Sun. And indeed, in 1878 the International Date Line was declared at 180 degrees from Greenwich, England.

The International Date Line slices through our intuitive sense of day and night, swallowing up birthdays and holidays as we travel west over the Pacific Ocean. The new rule immediately became a favorite loophole. Edgar Allen Poe's "Three Sundays in a Week" features a curmudgeonly father who will only permit his daughter Kate to marry his nephew Bobby "when there are three Sundays in a week". Clever Kate invites, to a Sunday family gathering, two captains who have just circumnavigated in opposite directions. Upon union with the seafarers, Kate leaps to the conclusion her father's impossible demand has been met:

> Captain Smitherton says that yesterday was Sunday: so it was; he is right. Cousin Bobby, and uncle and I say that to-day is Sunday: so it is; we are right. Captain Pratt maintains that to-morrow will be Sunday: so it will; he is right, too. The fact is, we are all right, and thus *three Sundays have come together in a week.*

Her father is persuaded. I am not. Under no relativization of 'Sunday' *to a single individual* is there three Sundays in a week. From every perspective, there is only one Sunday. Fortunately, Jules Verne did not notice Kate's equivocation.[1] Verne was inspired by Kate's syllogism to write *Around the World in Eighty Days*. The climatic conclusion is modeled on Kate's fallacious deduction—but manages to be valid.

7. Thales and the Demystification of the Night

I have argued that the day and night riddle sprang from an early insight into the stability of the day and night. Thales was well positioned to infer that the night is the Earth's shadow. Hieronymus of Rhodes reports that Thales made a careful study of shadows, thereby devising the simplest way of measuring the height of an object. To measure a pyramid, Thales found a spot next to the massive monument. He then marked his length on the ground (perhaps by simply lying in the sand). Once Thales' shadow had equaled his height, Thales knew that at this moment of the day, the length of each object's shadow equaled its height. Accordingly, Thales strode over to the tip of the pyramid's shadow. Marking that point, he paced out the length and deduced the height.

Hieronymus' anecdote is geometrically suspicious (Casati 2003: 80–2). A pole will cast a shadow that matches its length when the sun is at a 45-degree angle. But a squat object will not; the potential shadow is, as it were, stuck inside the object. The Egyptian pyramids are not much steeper than 45 degrees so they cast little shadow when the sun is at 45 degrees. Possibly, Thales performed the calculation with an obelisk and the feat was retold to include a grander monument.

According to Patricia O'Grady, Thales understood that eclipses consist of the Moon covering the Sun (2002: 142–5). In addition to seeing that the Moon is round, says O'Grady, Thales also believed that the Earth is spherical (O'Grady

[1] Cherry (1930: 233) notes that reasoning occurs in an unsigned article Poe probably read: "Three Thursdays in One Week", *Philadelphia Public Ledger* (October 29, 1841). Poe's story first appeared in the *Philadelphia Saturday Evening Post* in November 1841. The perspectival fallacy is earlier committed by a medieval riddler who asked "When is a man both right side up and upside down?—*When at the center of the Earth!*"

2002: 95–100). Ancient observers of lunar eclipses noted that the shadow cast on the Moon's surface was neatly curved like part of a circle. This suggests that the object casting the shadow, the Earth, is a ball (Aristotle, *De Caelo* 297 b25-298 a8). If the Earth were a flat disk, its shape would be elongated and elliptical (except in the special case in which the Sun is directly under the center of the disk). If the Earth were drum-shaped 🥁, the shadow would be distorted in other ways. So, it would have been natural of Thales, also a gifted mathematician, to infer that the Earth casts a shadow in the shape of a cone. (For reservations about O'Grady's spherical attribution, consult Dirk Couprie (2011: 64–8).)

Given O'Grady's interpretation, Thales would have welcomed the identification of the Earth's shadow with the night; he pursued a program of secular explanation. Whereas Homer had explained earthquakes as the shaking of the ground caused by Poseidon angrily stamping his feet, Thales explained them as collisions between bodies of earth drifting on water. Thales substituted natural explanations for superstition and fables. Equating the spooky night with the Earth's shadow would have been a great stroke of demystification. Nyx was nixed!

The stability of the night is not affected by the question of whether the Earth goes around the Sun or vice versa; Thales had all he needed to grasp that night and day have a hidden permanence and that they are each of superlative antiquity. He may have concluded that the night is older because he believed that the Sun had not always existed. Thales counted the dark beginning as a long night and added a second sense of 'day' to signify an alternation of the night–day cycle. This analysis makes sense of the answer Diogenes attributes to him: 'The night is older by one day'.

The underlying equation of night with darkness would ensure that night will always be older than the day. For Thales appears to have pictured the universe as eventually lapsing back into darkness. In that case, the final stanza of "The Garden of Proserpine" somberly expresses the end:

Then star nor Sun shall waken,
Nor any change of light:
Nor sound of waters shaken,
Nor any sound or sight:
Nor wintry leaves nor vernal,
Nor days nor things diurnal,
Only the sleep eternal
In an eternal night. (Swinburne 1909–14)

References

Buddensiek, Friedemann (2014). "Thales Down the Well: Perspectives at Work in the Digression in Plato's Theaetetus", *Rhizomata* 2 (1): 1–32.

Carroll, Lewis (1850). *The Rectory Umbrella and Mischmasch* (New York: Dover, repr. 1971).

Carroll, Lewis (1897). *Through the Looking-Glass and What Alice Found There* (Aubier-Flammarion).

Casati, Roberto (2003). *The Shadow Club* (New York: Knopf).

Cherry, Fannye N. (1930). "The Source of Poe's 'Three Sundays in a Week'", *American Literature* 2 (3): 232–5.

Couprie, Dirk L. (2011). *Heaven and Earth in Ancient Greek Cosmology: From Thales to Heraclides Ponticus* (New York: Springer).

Diogenes Laertius (1925). *Lives of Eminent Philosophers*, trans. R. D. Hicks, 2 vols (Cambridge, MA: Harvard University Press).

Harrison, Edward R. (1987). *Darkness at Night: A Riddle of the Universe* (Cambridge, MA: Harvard University Press).

Herodotus (1920). *The Histories*, trans. A. D. Godley (Cambridge, MA: Harvard University Press).

Heyd, Thomas (2014). "And Yet She Moves!—The Earth Rests on Water: Thales on the Role of Water in Earth's Mobility and in Nature's Transformations," *Apeiron* 47 (4): 485–512.

Jaki, Stanley L. (1969). *The Paradox of Olbers' Paradox: A Case History of Scientific Thought* (New York: Herder and Herder).

Krauss, Lawrence M. (2012). *A Universe from Nothing* (New York: Free Press).

Lutz, Cora E. (1975). 'A Fourteenth-Century Argument for an International Date Line', *Essays on Manuscripts and Rare Books* (Hamden, CT: Archon Books), pp. 63–70.

Lynch, David, K. and William Livingston (2001). *Color and Light in Nature*, 2nd edn (Cambridge: Cambridge University Press).

O'Grady, Patricia F. (2002). *Thales of Miletus: The Beginnings of Western Science and Philosophy* (Burlington, VT: Ashgate).

Pigafetta, Antonio (1969). *Magellan's Voyage: A Narrative Account of the First Circumnavigation*, ed. and trans. R. A. Skelton, 2 vols (New Haven, CT: Yale University Press).

Priou, Alex (2016) "'...Going Further On Down the Road...': The Origin and Foundations of Milesian Thought", *Review of Metaphysics* 70: 3–31.

Sorensen, Roy (2008). *Seeing Dark Things* (New York: Oxford University Press).

Stephenson, F. Richard and Louay J. Fatoohi (1997). "Thales' Prediction of a Solar Eclipse," *Journal for the History of Astronomy* 28: 279–82.

Swinburne, Algernon Charles (1909–14). "The Garden of Proserpine", in *English Poetry III: Tennyson to Whitman*, Harvard Classics XLII (New York: P. F. Collier & Son), pp. 1251–3.

Ustinova, Yulia (2009). *Caves and the Ancient Greek Mind: Descending Underground in the Search for Ultimate Truth* (Oxford: Oxford University Press).

4

Something from Nothing

Why Some Negative Existentials are Fundamental

Fatema Amijee

When we inquire into the nature of the fundamental, it seems obvious to many that the fundamental facts—those facts in virtue of which all other facts obtain—are all positive. There are good questions about how we ought to characterize the distinction between positive and negative facts, and whether a precise distinction between the two kinds of facts is even possible. But let us say, as a working characterization, that negative facts are about absences or lacks, whereas positive facts are not.[1] Thus, for example, the facts that there are Komodo dragons and that Justin Trudeau is the prime minister of Canada are positive facts, whereas the facts that there are no unicorns and that Justin Trudeau is not the prime minister of Australia are negative facts. The fact that there are no unicorns is about the absence of unicorns, and the fact that Justin Trudeau is not the prime minister of Australia is about a property that Justin Trudeau lacks. The dogma that the world is fundamentally positive can be traced as far back as Parmenides, and to the thought that there is nothing in the world that could possibly correspond to a negation. I argue in this paper that the dogma is mistaken: at least some negative facts *are* fundamental.

To say that some negative facts are fundamental is to say that they are ultimate explainers. This follows from the view on which the fundamental facts are just those in virtue of which the non-fundamental facts obtain, or which explain the non-fundamental facts. The relevant notion of explanation at work here is metaphysical explanation. The fact A metaphysically explains another fact B just in case A *makes it the case* that B. Thus, for example, the fact that my sweater is maroon makes it the case that it is red.

Metaphysical explanation has become closely associated with the notion of ground. Some insist that a single metaphysical dependence relation ('Grounding', with a big 'G'), with a unified set of formal features, backs metaphysical explanation.[2] Others argue that Grounding just is, rather than backs, metaphysical

[1] See Barker and Jago (2011) for a similar characterization.
[2] Cf. Schaffer (2012; 2016a) and Audi (2012).

Fatema Amijee, *Something from Nothing: Why Some Negative Existentials are Fundamental* In: *Non-Being: New Essays on the Metaphysics of Non-Existence.* Edited by: Sara Bernstein and Tyron Goldschmidt, Oxford University Press (2021).
 DOI: 10.1093/oso/9780198846222.003.0004

explanation.[3] Opponents of Grounding argue that no single metaphysical dependence relation can play this role, and instead deploy a (formally and substantively) diverse set of metaphysical dependence relations ('grounding relations', with a small 'g').[4] I remain neutral with respect to each of these views about grounding and explanation. In what follows, for convenience I use 'metaphysically explains' (or henceforth just 'explains') interchangeably with 'grounds', with the caveat that these terms should not be used interchangeably in all contexts.

Metaphysical explanation is widely taken to be irreflexive, asymmetric, transitive, and necessitating, though I will not presuppose those formal features here. If a fact is unexplained, it does not obtain in virtue of any other fact, and it will thus qualify as fundamental. It may turn out however that some fundamental facts are explained: perhaps they are explained reflexively, or 'zero-grounded', or symmetrically explained by other fundamental facts, or members of an infinitely descending explanatory sequence of fundamental facts.[5] In what follows, I argue that at least some negative facts are fundamental by showing that they are unexplained.

Negative facts come in many guises. Some negative facts—such as that Socrates does not exist—concern non-existent individuals. Others—such as that Justin Trudeau is not the prime minister of Australia—concern individuals lacking specific properties. Yet other negative facts—such as that there are no unicorns or that there is no greatest prime number—are negative existentials. Of these negative existentials, some are necessary, whereas others are contingent. For example, that there is no greatest prime number is arguably logically necessary, whereas that there are no 10ft tall humans is contingent: metaphysically (or perhaps even just logically speaking), there *could* have been 10ft tall humans. I will focus on contingent negative existentials. This is because, arguably, necessary negative existentials can be explained without requiring a further negative existential in their *explanans*: such existentials may be explained by essences or strong laws.[6] For example, what makes it the case that no triangle has

[3] Cf. Fine (2001), Litland (2013), and Dasgupta (2014). Wilson (2016a) argues that proponents of such views are guilty of conflating metaphysical explanation—a partly epistemic notion—with metaphysical dependence.

[4] Cf. Wilson (2014) and Koslicki (2015). Wilson argues against both the posited formal features of Grounding and the explanatory utility of positing a single relation to underwrite metaphysical explanation.

[5] See Wilson (2016b: 197). Fine (2001) also sketches a view on which each of a sequence of infinitely many explained facts is fundamental (which on Fine's view just is to be part of 'reality'). Fine argues: "Suppose, to take one kind of case, that Aristotle is right about the nature of water and that it is both indefinitely divisible and water through-and-through. Then it is plausible that any proposition about the location of a given body of water is grounded in some propositions about the location of smaller bodies of water (and in nothing else). The proposition that this body of water is here, in front of me, for example, will be grounded in the proposition that the one half is here, to the left, and the other half is there, to the right. But which of all these various propositions describing the location of water is real? We cannot say some are real and some not, since there is no basis upon which such a distinction might be made." (2001: 27) Fine concludes that all the various propositions describing the location of water are real, or part of reality, where reality is a primitive notion.

[6] See Rosen (2010).

four angles is that it is part of the essence of a triangle to have three angles. And for proponents of strong laws, such as many anti-Humeans, what makes it the case that a given (negative) regularity obtains is the fact that a specific law of nature obtains.

What about negative facts that are expressed by sentences that involve empty names or negated predicates? In such cases, the relevant facts either can be straightforwardly explained by positive facts, or are identical to negative existentials, or are such that an explanation for them involves negative existentials in its *explanans*. There is thus no special problem that arises when explaining negative facts that are expressed by sentences that involve empty names, or negative facts that are expressed by sentences that involve negated predicates. Any puzzle that arises in explaining such facts reduces to the more general puzzle of explaining negative existentials. For example, a fact expressed by a sentence involving an empty name—such as that Santa Claus does not exist—can be taken to be a negative existential if 'Santa Claus' is just a definite description, or equivalent to a predicate like *being called 'Santa Claus'*. Likewise, that Justin Trudeau is not the Australian prime minister—a negative fact expressed by a sentence involving a negated predicate—may be explained by the positive fact that Justin Trudeau is the Canadian prime minister and that one cannot be both the Australian prime minister and the Canadian prime minister at the same time, where the latter fact can either be taken to be a modal fact or a negative existential. Moreover, on an Armstrongian view, any subject-predicate sentence implicitly quantifies over properties, such that the sentence 'Justin Trudeau is not the Australian prime minister' says that there is no property which is the property of being the Australian prime minister and which is instantiated by Justin Trudeau. Such an analysis turns every negative fact expressed by a sentence involving a negated predicate into a negative existential.[7]

I proceed as follows. In section 1, I discuss motivations and arguments for the view—a view that I ultimately reject—according to which there can be no negative existentials at the fundamental level. In section 2, I show that there is good reason to include a totality fact in the *explanans* for any contingent negative existential. But totality facts are themselves contingent negative existentials, which makes it difficult to see how we might be able to avoid positing at least some negative existentials at the fundamental level. As part of my argument for the claim that some negative existentials are fundamental, in section 3 I argue against candidate alternative accounts for eliminating the tension between the claim that no negative existential is fundamental and the claim that every negative existential is partially explained by a negative existential. Finally, in section 4, I show that the arguments for not positing negative facts—and specifically totality facts—at the fundamental

[7] See also Parsons (2006) for detailed discussion of the Armstrongian view.

level are inadequate. This completes my case for the view that totality facts are fundamental.

1. Are Negative Existentials Fundamental?

In *The Philosophy of Logical Atomism*, Russell writes:

> When I was lecturing on this subject at Harvard I argued that there were negative facts, and it nearly produced a riot: the class would not hear of there being negative facts at all. (1940: 42)

And in his essay 'On Propositions', he writes:

> There is implanted in the human breast an almost unquenchable desire to find some way of avoiding the admission that negative facts are as ultimate as those that are positive. (1919: 4)

For Russell, a fact is a worldly entity, a complex made up of constituents. In the context of the first passage quoted above, Russell does not draw a distinction between fundamental and less-fundamental facts. Yet if there are no negative facts at all, then *a fortiori*, there can be no fundamental negative facts. By contrast, the second passage is more clearly about the question of whether negative facts can be fundamental. But what motivates the general consensus that negative facts—and in particular negative existentials—cannot figure at the fundamental level, a consensus so strong that opposition to it (as Russell reports) nearly produced a riot?

First, one might argue that positing fundamental negative facts violates a version of Ockham's Razor, namely the claim that facts at the fundamental level should not be posited without necessity. Ockham's Razor implies that when given the choice between two ontologies that explain all the same facts at the non-fundamental level, we should prefer the ontology that posits fewer facts at the fundamental level.[8] This version of Ockham's Razor implies that there should be no redundancy at the fundamental level. But negative existentials seem clearly redundant: after God brings about the existence of humans, penguins, sharks, and all the other creatures that populate the earth, did he also have to bring about the non-existence of unicorns and centaurs? Intuitively, 'no': God didn't have to do anything extra to make it the case that unicorns and centaurs don't exist. That

[8] Schaffer calls a principle in the neighborhood the "bang for the buck" principle. According to this principle, "[w]hat one ought to have is the strongest theory (generating the most derivative entities) on the simplest basis (from the fewest substances)" (Schaffer 2009: 361). Della Rocca (2014), however, argues that Ockham's Razor cannot apply to fundamental entities without also applying to non-fundamental entities.

unicorns and centaurs don't exist 'comes along for free'. This consideration extends to other types of negative facts. Suppose that God brought about the fact that zebras are mammals. Did God then have to bring about the fact that zebras are not fish? Intuitively, 'no'. The fact that zebras are not fish comes along for free.

Secondly, one might worry that positing negative existentials at the fundamental level risks violating a version of Hume's Dictum, the widely endorsed principle according to which there are no necessary connections between distinct entities. Hume's Dictum underlies recombination, a principle for generating the space of possible worlds. According to recombination, there is a possible world corresponding to any combination of the fundamental entities. Suppose we include facts in the class of fundamental entities. Now suppose that a negative existential—such as the fact that there are no humans over 10ft tall—was a fundamental fact. Then by recombination, there is a possible world *w* where all the same positive facts obtain, yet the fact that there are no humans over 10ft tall does not obtain. But if it is not the case that there are no humans over 10ft tall, then there *are* humans over 10ft tall. So, it turns out that the same positive facts cannot obtain after all. This sort of argument has been taken to support the view that negative existentials cannot figure at the fundamental level.[9]

I argue in section 4 that both these arguments against positing negative existentials at the fundamental level fail. I show that at least some negative existentials are not redundant, and that the argument from Hume's Dictum rests on a misapplication of that principle. There may be an argument that seeks to show that negative existentials cannot be fundamental that I have not canvassed here. But if there is no good argument available, then the intuition that there can be no fundamental negative facts is just that—an intuition. And we should not put much stock in an intuition that cannot be substantiated by argument.

However, let us grant for the sake of argument that negative existentials are not fundamental. How might they be explained? I turn to this question in the next section.

2. Explaining Negative Existentials

Negative existentials may be either necessary or contingent. If they are necessary, they may be explained by essences or laws. However, contingent negative existentials cannot be explained in the same way.[10] The worry with respect to contingent negative existentials in particular is that, at least on the face of it, they cannot be explained without appealing to yet another contingent negative

[9] See Muñoz (2019) for a version of this argument.

[10] At least not if we assume necessitation, for laws and essences are plausibly necessary. Necessitation is the thesis that explanation carries modal entailment, such that if some facts explain a fact *p*, then necessarily, if those facts obtain, then so does *p*.

existential. And this is problematic because it suggests that contingent negative existentials can never be eliminated from any explanatory sequence of facts that grounds a contingent negative existential. At least on the face of it, this result is in tension with the claim that negative facts—including negative existentials—cannot be fundamental.

If we assume that conjunctions are explained by their conjuncts, then perhaps *some* contingent negative existentials are explained by other contingent negative existentials. For example, the fact that there are no humans over 10ft tall partially explains the fact that there are no unicorns and no humans over 10ft tall. But there is an argument for the stronger result that *every* contingent negative existential is at least partially explained by a contingent negative existential. Suppose for example that F is the predicate 'is a unicorn'. Then we can formalize the claim that there are no unicorns as follows: $\sim\exists x\, Fx$. A negative existential is logically equivalent to a universal generalization. But if we also take a negative existential to be *the same fact as* the equivalent universal generalization (as is standardly supposed), then the fact that $\sim\exists x\, Fx$ is identical to the fact that $\forall x{\sim}Fx$.[11] The instances of a universal generalization, however, do not entail it (and thus do not fully explain it) unless we fix the domain in advance. For entailment, a further *totality fact* is required. In our example, let us say that the instances consist in the negative facts expressed by 'Sam is not a unicorn', 'Dawn is not a unicorn' and 'Evelyn is not a unicorn'. But the full explanation also seems to require the following totality fact: $\sim\exists x$ (x is not identical to Sam and x is not identical to Dawn and x is not identical to Evelyn). This is just the fact that Sam, Dawn, and Evelyn exhaust the domain of the quantifier. But a totality fact is itself a contingent negative existential!

We thus have a tension between two claims. On the one hand, we have the strong intuition that negative existentials, as negative facts, cannot be fundamental. On the other hand, on the standard way of explaining contingent negative existentials, the *explanans* for every contingent negative existential contains a totality fact, which is itself a contingent negative existential.

In the next section, I discuss some alternative ways to eliminate the tension between these two claims, before arguing for the view that totality facts are simply unexplained, and thus fundamental.

3. Why Totality Facts are Fundamental

My overall aim in this paper is to show that we should take at least some contingent negative existentials—namely, totality facts—to be fundamental. Part

[11] While taking negative existentials to be identical to the equivalent universal generalizations is the standard view, Fine (2012) rejects this view for the case of totality facts. I discuss Fine's view in more detail in section 3.

of my case for this claim rests on showing that it is the only adequate way to eliminate the tension between the claim that no contingent negative existential is fundamental and the claim that every contingent negative existential is at least partially explained by a contingent negative existential.

Opponents must argue that there are other viable ways to eliminate this tension. If contingent negative existentials are not fundamental, they are explained. If all negative existentials are explained, then explaining a contingent negative existential either requires a totality fact, or it does not. If explaining a contingent negative existential requires a totality fact, then either the totality fact is part of the *explanans* (or ground), or the explanation requires a totality fact without needing it to be part of the *explanans*. Finally, if explaining a contingent negative existential does not require a totality fact, then it is either zero-grounded—i.e. grounded by zero-many facts—or it has an alternative explanation in terms of non-zero-many facts.

The above options exhaust the possible alternatives for someone committed to the claim that no contingent negative existential is unexplained, and they generate the following alternative possibilities for eliminating the tension between the claim that no negative existential is fundamental and the claim that every contingent negative existential is partially explained by a contingent negative existential:

(a) Admit a regress of negative existential facts, where a totality fact figures as a partial ground for every negative existential.
(b) Accept that contingent negative existentials are grounded in their instances but deny that a totality fact also figures as a partial ground when explaining any contingent negative existential.
(c) Claim that contingent negative existentials are grounded in something other than their instances, such as the universe.
(d) Claim that contingent negative existentials are grounded, but in nothing—i.e. they are zero-grounded.

I show that the above alternatives for eliminating the tension are inadequate, and thereby defend the view that totality facts are fundamental.

3.1 The Regress Account

The regress account gives us a way to accept both the claim that no negative existential is fundamental and the claim that every contingent negative existential is partially grounded in a contingent negative existential. On this account, we simply have an infinitely descending regress of contingent negative existential facts.

Suppose for the sake of argument that there is no incoherence in admitting such a regress. The obvious cost of the regress account would then be that it does away with a fundamental level altogether. Insofar as many philosophers would like to admit a fundamental level, this seems to be a significant cost. There are, of course, accounts on which an infinitely descending explanatory regress is compatible with fundamental facts.[12] But since the goal of the regress account is to preserve the non-fundamentality of negative existentials, while allowing that every contingent negative existential is at least partly explained by a contingent negative existential, I put such views aside.

However, even if we accept the cost of doing away with a fundamental level, I argue that the regress account fails. This is because the totality facts that are part of the *explanans* for every contingent negative existential are in fact *one and same* totality fact. The regress account thus violates the irreflexivity of explanation. To see why, let us return to our earlier example and suppose that $\sim\exists x\, Fx$, where F is the predicate 'is a unicorn'. This negative existential is equivalent to $\forall x{\sim}Fx$. On the assumption that universal generalizations are grounded in their instances and a totality fact, let us say that $\forall x{\sim}Fx$ is grounded in $\sim Fa$, $\sim Fb$, and $\sim Fc$, and the claim that a, b, and c are all the things. That is, $\sim\exists x\, (x{\neq}a$ and $x{\neq}b$ and $x{\neq}c)$. But what grounds this further negative existential? This negative existential is equivalent to $\forall x{\sim}(x \neq a$ and $x \neq b$ and $x \neq c)$, which is equivalent to $\forall x\, (x = a$ or $x = b$ or $x = c)$. The instances that ground the preceding universal generalization are $(a = a$ or $a = b$ or $a = c)$, $(b = a$ or $b = b$ or $b = c)$ and $(c = a$ or $c = b$ or $c = c)$. Now the grounds for any negative existential (or universal generalization) must also include a totality fact. The totality fact that grounds our universal generalization is just this: $\sim\exists x\, (x \neq a$ and $x \neq b$ and $x \neq c)$. It is the very same totality fact as the totality fact that partially grounds $\sim\exists x\, Fx$, our original negative existential! The regress account thus violates the irreflexivity of explanation.[13]

At the outset, I claimed neutrality with respect to whether explanation is irreflexive. However, even if explanation is not irreflexive across the board and there are some instances of reflexive explanation, at least on the face of it, it is implausible that totality facts can partially explain themselves. The burden of proof lies with the proponent of such a view: they would need to show not only that explanation is not irreflexive as a rule, but also that totality facts in particular can be partially explained by themselves. Absent further argument, the regress account thus fails to adequately accommodate both the intuition that no negative existential can be fundamental, and the intuition that every contingent negative existential is grounded in a contingent negative existential.

In order to block a violation of the irreflexivity of explanation in explaining totality facts, Fine (2012) argues that while a totality fact is equivalent to a universal generalization, it is not the same fact as a universal generalization, and so need not be grounded in the same way that a universal generalization is

[12] See, for example, Wilson (2016b) and Amijee (ms), "Relativism about Fundamentality".
[13] Fine (2012) also acknowledges this worry.

grounded. However, while Fine can then deny that a totality fact is grounded in its instances and a totality fact, Fine does not provide a positive proposal for what, if anything, grounds the totality fact.[14]

3.2 Conditional Grounding Account

The second option for opponents is to allow that contingent negative existentials are grounded in their instances, but deny that a totality fact also figures as a partial ground when explaining any contingent negative existential. How might such an account work? The best candidate for such a view—and the only one that I am aware of—treats the totality fact as a background condition for explanation rather than a partial ground. According to this conditional grounding account, a contingent negative existential can be grounded in just its instances, so long as a condition—namely, the totality fact—obtains. This account relies on there being a principled distinction between a ground and a condition. If a totality fact does not figure as a partial ground of a contingent negative existential, then a contingent negative existential can be straightforwardly grounded in just its instances, and there isn't a worry that we will end up with a contingent negative existential at the fundamental level.

However, the conditional grounding account succeeds only if good sense can be made of the distinction between a partial ground and a condition without merely appealing to intuition, or salience in a given context. To be sure, there is an analogous distinction in the causal case between a cause and a background condition.[15] But it is unclear whether a similar distinction can be plausibly drawn in the case of metaphysical explanation. While the distinction in the case of causation seems intuitive and even familiar (in most contexts we would be inclined to say that the presence of oxygen did not *cause* the fire but was a mere background or enabling condition), it does not in the case of metaphysical explanation.

Bader (ms) proposes the following sufficient condition for something's counting as a condition for metaphysical explanation, as opposed to a partial ground: if what is required for a grounding relation to obtain is an absence, then that absence is a mere condition, rather than a ground. This criterion relies on there being a

[14] Fine (2012: 62) writes: "The issue of the ground for universal truths has caused a great deal of puzzlement in the philosophical literature, going back to Russell (1918) and continuing to this day (Armstrong 2004). But if I am right, there is a purely logical aspect to the problem which is readily solved once one draws a distinction between the totality claim and the corresponding universal claim. Of course, this still opens the question of the grounds, if any, for the totality claim. But this is a question that lies on the side of metaphysics, so to speak, rather than of logic; and it should not be supposed that there is anything in our general understanding of the quantifiers or of the concept of ground that might indicate how it should be answered."

[15] See Schaffer (2016b) for discussion.

substantive metaphysical difference between presences and absences. It also relies on the idea that absences cannot figure as grounds, because absences do not exist as such—they are nothing. Let us grant for the sake of argument that there is indeed a robust metaphysical distinction between presences and absences, and that absences do not exist and thus cannot figure as grounds. It is still not clear that the conditional grounding proposal can extend to totality facts, for totality facts are not themselves absences or non-existent, even if they are *about* absences.[16] When a totality fact figures as a partial ground for a negative existential, the work it does is not the work of nothing (which is nothing!), but of the fact that *a*, *b*, and *c* exhaust what there is.

Moreover, if a totality fact does not figure as a partial ground for a negative existential, then we get a failure of necessitation (the thesis that if some facts explain a fact *p*, then necessarily, if those facts obtain, then so does *p*).[17] To see why, let us return to our toy example, the fact that there are no unicorns. The instances that partially ground this fact consist in the negative facts expressed by 'Sam is not a unicorn', 'Dawn is not a unicorn' and 'Evelyn is not a unicorn'. But these instances do not, on their own, make it the case that there are no unicorns, for it is possible that Sam, Dawn, and Evelyn exist (as non-unicorns), and yet Ed, who *is* a unicorn, also exists. The instances thus fail to entail, and so fail to metaphysically explain, the fact that there are no unicorns. For entailment, a totality fact is required—the fact that Sam, Dawn, and Evelyn are all the beings.

The above objection to the conditional grounding proposal does not presuppose that metaphysical explanation is governed by necessitation. It does presuppose that if metaphysical explanation is *not* governed by necessitation, then, given that the general consensus is in favor of necessitation, the burden of proof for showing that it does not lies with those who deny necessitation.[18] In particular, unless there is an independent argument for conditional grounding, it would not do to reject necessitation as a principle that governs explanation.

But let us suppose for the sake of argument that a good case can be made for a robust distinction between conditions and grounds. Then, at least on the face of it, it seems that conditional grounding allows us to say that there are no totality facts at the fundamental level, for totality facts figure as mere conditions, rather than as

[16] I return to this point in the next section.

[17] Necessitation is widely taken to govern metaphysical explanation. See for example Rosen (2010), Audi (2012), Bliss and Trogdon (2014), and Dasgupta (2014).

[18] In Amijee (forthcoming), I argue that we should reject the claim that no necessary facts can, on their own, explain a contingent fact. Rejecting this claim entails a rejection of necessitation (though not vice versa). However, one might reject necessitation and allow that in some cases necessary facts can fully explain a contingent fact, while still subscribing to a restricted necessitation principle, according to which there is entailment whenever the *explanans* is contingent.

grounds of a contingent negative existential. It also allows us to capture the intuition that totality facts are somehow involved in the grounding of contingent negative existentials.

I argue, however, that the conditional grounding account still fails, for it involves a violation of a principle closely related to the irreflexivity of explanation. Let us call this principle irreflexivity*. According to irreflexivity*, a fact cannot figure as a condition for the grounding of itself. To see why irreflexivity* is plausible, let us consider the case of causation, where the distinction between conditions and grounds is more intuitive. Suppose that the striking of a match causes it to be lit, and that the presence of oxygen in the atmosphere is a background condition for the striking causing the match to be lit. The presence of oxygen in the atmosphere can itself be taken to be a causal event, about which we can ask: what causes *it*? Now there would surely be circularity of a problematic variety if we then cited the presence of oxygen in the atmosphere as a background condition for the presence of oxygen in the atmosphere. Irreflexivity* thus seems fairly plausible for causal explanation. What about metaphysical explanation and grounding? Given the intuitive appeal of irreflexivity* for causal explanation, the burden of showing that irreflexivity* does not hold for metaphysical explanation lies with those inclined to reject the principle.

Conditional grounding falls afoul of irreflexivity*. Suppose that totality facts figure as mere conditions in explaining a contingent negative existential. What might explain the totality fact, which, as discussed above, is itself a contingent negative existential? Just as it is implausible to cite the presence of oxygen in the atmosphere as a background condition in explaining the presence of oxygen in the atmosphere, it would be implausible, and a violation of irreflexivity*, to cite a totality fact as a condition in explaining the very same totality fact. Thus, if irreflexivity* holds for metaphysical explanation, then conditional grounding cannot help us resolve the tension between the claim that no negative existential can be fundamental and the claim that every contingent negative existential is partially grounded in a contingent negative existential.

The type of worry I have raised here is analogous to my argument against the regress account in section 3.1. There I argued that the regress account involves an illegitimate violation of irreflexivity. This suggests that the mere fact that a totality fact plays a different kind of role—in this case, the role of a background condition rather than a partial ground—is not enough to resolve the tension between the claim that no negative existential can be fundamental and that every contingent negative existential is partially grounded in a contingent negative existential. The tension arguable arises because totality facts must play some *role* in grounding contingent negative existentials, even if that role is not a straightforward grounding role. Having the totality fact play a non-grounding role does not allow us to avoid a vicious explanatory circle.

3.3 Actuality Account

Yet another option for opponents is to claim that contingent negative existentials are grounded in something other than their instances. A seemingly suitable candidate explanation for negative existentials appeals to the way the universe actually is. On this view, there are no humans over 10ft tall because the universe—the totality of all that exists—is such that there are no 10ft tall humans in it. Let us call the universe 'The One'. On this view then, there are no humans over 10ft tall because The One exists, where The One does not contain humans over 10ft tall. This view is endorsed by Cheyne and Pigden, who write:

> Our answer is that *the (first-order) way the universe actually is* (a very large and complex fact, but a positive fact nonetheless) *makes it true* that there are no unicorns. For (on the assumption that there are no unicorns) the universe would have to be a different way for unicorns to exist. Thus the *way the universe actually is* would *not* exist and some other way the universe might have been *would* exist (namely a way which involved existing unicorns).
>
> (Cheyne and Pidgen 2006: 257).

However, this account does not succeed in doing away with a totality fact. Unless a totality fact to the effect that The One is the totality of all that exists is also part of the grounds, the existence of The One does not on its own explain why there are no humans over 10ft tall. This is because 'The One' picks out the world as it actually is, and absent a totality fact that stipulates that The One is all that exists, it is possible that in addition to The One, there also exist humans over 10ft tall. Thus, appealing to the existence of the universe as it actually is in explaining a contingent negative existential does not get rid of the need for a totality fact.[19] At best, it smuggles that totality fact into the grounds. The worry remains if we replace talk of "the universe" with talk of "the actual world", for if "the actual world" is taken referentially, then it simply picks out what actually exists, and it is consistent with what actually exists—say, *a*, *b*, and *c*—that there also exists a further thing, *d*.[20]

On a slight variant of the view endorsed by Cheyne and Pidgen, there are no humans over 10ft tall because there are no humans over 10ft tall in *w*, and *w* is actual. One might then add that it is essential to *w* that it lacks humans over 10ft tall. This essentialist fact explains why there are no humans over 10ft tall in *w*, which in turn explains the fact that there are no humans over 10ft tall. The worry

[19] Josh Parsons elegantly makes this point in Parsons (2006).

[20] By contrast, "the actual world" may be treated attributively, in which case it picks out whatever happens to the be the totality of what exists. However, if treated attributively, the grand fact that explains a negative existential would itself involve a negative existential. Cf. Parsons (2006).

with such an account is that the most plausible examples of essential properties (if any such properties exist) are positive properties: presences, rather than absences. Thus, for example, it might be essential to me that I have the parents I do, but intuitively, it is not essential to me that I do not have frog DNA, even if what is in fact part of my essence entails that I do not have frog DNA.

I close this subsection with a brief discussion of accounts that seek to explain why our world is the actual world. Such accounts would seem to explain both why our world exists—where "our world" referentially picks out whatever in fact exists—*and* why nothing else exists, i.e. why what exists is *all* that exists. Consider, for example, Leibniz's 'optimist' account, according to which our world is the actual world because it is best of all possible worlds. I am here putting aside the question of whether Leibniz's account of why our world is actual is correct. The question I am interested in is whether such an account could serve to explain contingent negative existentials without smuggling a contingent negative existential into the *explanans*. While the optimist account of why our world is the actual world may be a bit too optimistic for contemporary tastes, Leibniz also has a different, less popular account. On this alternative account, our world—and no other possible world—is actual because nothing prevented our world from coming into existence. This is Leibniz's 'striving possibles' account, on which all possibles strive for existence, and unless there is something that prevents *x* from coming into existence, *x* will come into being.[21] Again, I am not interested in the question of whether Leibniz's explanations succeed. I am instead interested in whether they can help us avoid positing negative existentials at the fundamental level.

I argue that they cannot. Both Leibnizian explanations for the actuality of our world also involve a totality fact. To say that our world is the best of all possible worlds is just to say that it is better than any other world in a given domain. But notice now that we also require a domain-specifying totality fact.[22] Likewise, that our world is actual because nothing prevented it from coming into existence also clearly involves a negative existential, and one that would need to be contingent if it is to help explain a contingent negative existential.[23]

[21] Cf. Look (2011) and Leibniz's 1697 essay "On the Ultimate Origination of Things" in Leibniz (1989).

[22] A proponent of this Leibnizian explanation might argue that the relevant totality fact is necessary, rather than contingent, and (as discussed above) a necessary negative existential need not be partially grounded in a negative existential. If right, this might allow the Leibnizian to avoid positing contingent negative existentials at the fundamental level. Of course, few contemporary metaphysicians would accept the resulting escape route, since it relies upon highly controversial Leibnizian assumptions.

[23] Spinoza, too, has an account of makes our world the actual world. On Spinoza's view, our world is the only possible world, and consists in only one substance—God. Since God and everything that follows from God exists necessarily, the world could not have been any other way. God is also the most powerful substance because God has infinitely many attributes. This explains why the only possible world (our world) is the world that God inhabits rather than some other necessarily existing substance. Unlike Leibniz's accounts, it is not obvious that Spinoza's account of why our world is actual appeals to a contingent negative existential. After all, on Spinoza's view, there isn't a possible world or possible entity that could have existed but didn't. However, since on a standard interpretation of Spinoza's view

It is of course possible that there are viable alternative explanations of contingent negative existentials that avoid explaining them in terms of their instances (and which provide them with non-zero-many grounds). I here canvassed only variants on explanations which appeal to the way the universe actually is. But absent any good alternative candidates, it is safe to assume that this general style of grounding contingent negative existentials is a non-starter.

3.4 Zero-Grounding Account

A final proposal for explaining contingent negative existentials is inspired by Kit Fine's notion of 'zero grounding'. According to Fine, a fact may lack a ground either because it is ungrounded, or because it is zero-grounded, where to be zero-grounded is to be grounded, but in nothing. But what exactly does it mean to say that something is zero-grounded? Fine (2012) draws an analogy with sets:

> Any non-empty set {a, b,...} is generated (via the "set-builder") from its members a, b,.... The empty set {} is also generated from its members, though in this case there is a zero number of members from which it is generated.
>
> An urelement such as Socrates, on the other hand, is ungenerated; there is no number of objects—not even a zero number—from which it may be generated. Thus "generated from nothing" is ambiguous between being generated from a zero number of objects and there being nothing—not even a zero plurality of objects—from which it is generated; and the empty set will be generated from nothing in the one sense and an urelement from nothing in the other sense.
>
> (Fine 2012: 47)[24]

A zero-grounded fact is then a fact that is grounded in zero facts, rather than one that is ungrounded. According to a recent proposal defended by Muñoz (2019), contingent negative existentials are zero-grounded.

Muñoz highlights a worry with the zero-grounding proposal for contingent negative existentials, namely that contingent negative existentials are *contingent*, whereas their zero-many grounds obtain at all possible worlds. The zero-grounded proposal thus entails a failure of necessitation when applied to contingent negative existentials. Like in all cases where necessitation fails, a question

there are also no contingent facts, the problem of explaining contingent negative existentials does not arise at all. (See, for example, Della Rocca (2008); it is worth noting, however, that there is some disagreement in Spinoza scholarship over whether Spinoza is really committed to a full-blown necessitarianism: see especially Curley and Walski (1999).)

[24] Litland (2017) further and more rigorously develops the notion of zero-grounding.

arises: what explains why a fact *p* fails to obtain in world w_1 but obtains in world w_2, when its zero-many grounds obtain at both w_1 and w_2? Muñoz's solution to this worry relies on a distinction between background conditions and grounds, and the idea that zero-grounding fails when a disabling condition is present. But Muñoz does not provide much reason to think that this is a metaphysically robust distinction.

However, unlike Muñoz, I do not see the worry as posing a major challenge to the zero-grounding proposal for contingent negative existentials. On my view, the question of why a fact *q* (say) which obtains at both w_1 and w_2 grounds *p* at w_1 but does not at w_2 is just a question about what grounds the grounding facts. Given that *q* grounds *p* at w_1, there are candidate answers available to the question of what grounds the fact that *q* grounds *p*.[25] By contrast, *q* does not ground *p* at w_2, and there is thus no grounding fact about which we can ask 'what grounds it?'. There is of course more to be said in defense of my view that there is no real problem here posed by the failure of necessitation, but it is not necessary for present purposes.

Even if the zero-grounding proposal is not problematic for the reasons just given, it remains implausible when applied across the board to all contingent negative existentials. Intuitively, if a totality fact is grounded, then its grounds must have something to do with which facts there are. This is because a totality fact is a domain-specifying fact. Yet the zero-grounding proposal makes it the case that every possible domain-specifying fact will have the same ground—namely nothing—despite each of these totality facts delineating a different domain.

Moreover, even if we grant the coherence and plausibility of zero-grounding, it is far from clear that zero-grounding can apply to contingent negative existentials. In explaining zero-grounding, Fine appeals to set-membership and the construction of sets. Litland (2017) further develops the notion of zero-grounding. Litland writes:

> The seemingly mysterious distinction between being *ungrounded* and being *zero-grounded* is a special case of the more familiar distinction between not being derivable and being derivable from the empty collection of premisses.
> (Litland 2017: 280)

Neither the analogy with sets nor the distinction between being derivable and being derivable from the empty collection of premisses seems particularly applicable in the case of contingent negative existentials: a contingent negative existential cannot (or at least not obviously) be treated like an empty set. It also does not obviously make sense to say that a contingent negative existential is 'derived' from, and so grounded in, zero-many facts.

[25] The options include those defended by Bennett (2011), deRosset (2013), Dasgupta (2014) and Litland (2017).

While my remarks in this section do not provide a definitive case against the zero-grounding proposal for explaining contingent negative existentials, I hope to have raised distinctive and significant concerns with the proposal as it stands. Given these concerns, we should not opt for zero-grounding when an alternative remains available for resolving the tension discussed at the outset: we should take some contingent negative existentials—the totality facts—to be fundamental.

4. Negative Fundamental Facts: Revisited

I have argued that we should take totality facts to be fundamental. In this section, I show that the reasons canvassed in section 2 for taking negative existentials to be non-fundamental—namely, that negative existentials are redundant and lead to a violation of Hume's Dictum—do not extend to totality facts.

First, totality facts do not seem redundant in the way that facts about things that don't exist might seem redundant. Totality facts simply say "that's it, and no more!", and thus specify a negative limit. They are boundary facts that carve out domains.

Second, it is far from obvious that fundamental totality facts—or indeed any kind of fundamental negative facts are in tension with Hume's Dictum. Recall that according to the objection from Hume's Dictum, including negative existential facts at the fundamental level is in tension with free modal recombination: it precludes a scenario—one that apparently corresponds to a possible world—on which a fundamental contingent negative existential is removed while all the positive facts stay the same. However, this objection neither succeeds on its own terms nor involves a correct application of Hume's Dictum.

It does not succeed on its own terms, for an analogous line of argument can be taken to show that we should do away with fundamental positive facts at the fundamental level, since it is not possible to remove a fundamental positive fact from a given world while keeping all its positive facts the same.[26] To see why, suppose that a positive fact—such as the fact that there are butterflies—is a fundamental fact. Then by recombination, there is a possible world *w* where all the same negative facts obtain, but the fact that there are butterflies does not obtain. But if it is not the case that there are butterflies, then an additional negative fact obtains at *w*, namely, there are no butterflies. So, it turns out that—contra our hypothesis—the same negative facts cannot obtain after all.

The argument from Hume's Dictum also rests on a misapplication of that principle. Free modal recombination only makes sense when applied to entities—particulars

[26] Alternately, we might then conclude that this shows that there are no fundamental facts at all, whether negative or positive, especially in the absence of any other argument that might tip the balance in favour of only positive facts (or only negative facts) at the fundamental level. Thanks to Michael Della Rocca for this point.

and the properties that instantiate them—and not facts. It is the fundamental *entities* that are recombined in order to generate the space of possible worlds.[27] Indeed, if we conceive of worlds as entities that are either identical to or correspond to maximally consistent sets of propositions, then the propositions that are true at the actual world cannot be 'recombined', i.e. cannot have a proposition added to or subtracted from the set while maintaining consistency.

5. Concluding Remarks

My goal has been to show that, contrary to popular dogma, at least some negative existentials are fundamental. My case had two parts. First, I argued against the extant candidate solutions for eliminating the tension between two claims: the claim that no negative existential is fundamental and the claim that every contingent negative existential is partially explained by (or grounded in) a contingent negative existential. I argued that the alternatives available to us if we do not take totality facts to be fundamental are, at least at present, inadequate. Second, I showed that the standard arguments against positing any negative facts at the fundamental level—including negative existentials—fail.

My survey of potential attempts to eliminate the tension was perhaps not exhaustive. For instance, there may be yet another way of grounding a contingent negative existential in something other than its instances. I also did not rule out Fine's suggestion that we seek an explanation for totality facts that does not depend upon those facts being universal generalizations. Moreover, I did not provide a definitive case against every option I discussed. More can be said, for example, in favour of the zero-grounding proposal as it might apply to negative existentials, and perhaps my criticisms of that approach could be rebutted by its proponents. And the two Leibnizian proposals for explaining the actual world—or variants on them—might be pursued in more depth within a contemporary framework.

These are areas where there is much room for future work. My case for the claim that some negative existentials are fundamental does not depend on a definitive refutation of every other option for explaining contingent negative existentials, but on a rejection of these options as they currently stand. Thus, while I have presented myself as defending the radical view that some contingent negative existentials—namely, totality facts—are fundamental, my ultimate stance is more nuanced. Nevertheless, those who still feel Russell's 'almost unquenchable desire to find some way of avoiding the admission that negative facts are as ultimate as those that are positive' must now look harder and further afield to satisfy that desire.

[27] Cf. Wilson (2010).

Acknowledgments

For discussion and written comments, thanks to Michael Della Rocca, Dominic Alford-Duguid, and Jon Litland. Thanks also to the editors of this volume, Sara Bernstein and Tyron Goldschmidt.

References

Amijee, Fatema (forthcoming). "Explaining Contingent Facts", *Philosophical Studies.*

Amijee, Fatema (ms). "*Relativism about Fundamentality*", manuscript.

Audi, Paul (2012). "A Clarification and Defense of the Notion of Grounding", in *Metaphysical Grounding: Understanding the Structure of Reality*, ed. Fabrice Correia and Benjamin Schnieder (Cambridge: Cambridge University Press), pp. 101–21.

Bader, Ralf M. (ms). "*Conditional Grounding*", manuscript.

Barker, Stephen and Mark Jago (2011). "Being Positive about Negative Facts", *Philosophy and Phenomenological Research* **85** (1): 117–38.

Bennett, Karen (2011). "By Our Bootstraps", *Philosophical Perspectives* 25: 27–41.

Bliss, Ricki and Kelly Trogdon (2014). "Metaphysical Grounding", *Stanford Encyclopedia of Philosophy*, ed. Edward N. Zalta (Winter 2016 edn).

Cheyne, Colin and Charles Pigden (2006). "Negative Truths from Positive Facts", *Australasian Journal of Philosophy* **84**: 249–65.

Curley, Edwin and Gregory Walski (1999). "Spinoza's Necessitarianism Reconsidered", in *New Essays on the Rationalists*, ed. Rocco J. Gennaro and Charles Huenemann (New York: Oxford University Press), pp. 241–62.

Dasgupta, Shamik (2014). "The Possibility of Physicalism", *Journal of Philosophy* **111** (9/10): 557–92.

Della Rocca, Michael (2008). *Spinoza* (New York: Routledge).

Della Rocca, Michael (2014). "Razing Structures to the Ground", *Analytic Philosophy* 55 (3): 276–94.

deRosset, Louis (2013). "Grounding Explanations", *Philosopher's Imprint* **13** (7): 1–26.

Fine, Kit (2001). "The Question of Realism", *Philosopher's Imprint* **1** (2): 1–30.

Fine, Kit (2012). "Guide to Ground", in *Metaphysical Grounding: Understanding the Structure of Reality*, ed. Fabrice Correia and Benjamin Schnieder (Cambridge: Cambridge University Press), pp. 37–80.

Koslicki, Kathrin (2015). "The Coarse-Grainedness of Grounding", in *Oxford Studies in Metaphysics*, vol. **9**, ed. Karen Bennett and Dean W. Zimmerman (Oxford: Oxford University Press), pp. 306–44.

Leibniz, G. W. (1989). *Philosophical Essays.*, trans. Roger Ariew and Daniel Garber (Indianapolis: Hackett).

Litland, John Erling (2013). "On Some Counterexamples to the Transitivity of Grounding", *Essays in Philosophy* **14** (1): 19–32.

Litland, John Erling (2017). "Grounding Ground", in *Oxford Studies in Metaphysics*, vol. **10**, ed. Karen Bennett and Dean W. Zimmerman (Oxford; Oxford University Press), pp. 279–316.

Look, Brandon C. (2011). "Grounding the Principle of Sufficient Reason: Leibnizian Rationalism versus the Humean Challenge", in *The Rationalists: Between Tradition and Revolution*, ed. Carlos Fraenkel, Dario Perinetti, and Justin E. H. Smith (Dordrecht: Springer, **2011**), pp. 201–19.

Muñoz, Daniel (2019). "Grounding Nonexistence", *Inquiry* **63** (2): 209–29.

Parsons, Josh (2006). "Negative Truths from Positive Facts?", *Australasian Journal of Philosophy* **84** (4): 591–602.

Rosen, Gideon (2010). "Metaphysical Dependence: Grounding and Reduction", in *Modality: Metaphysics, Logic, and Epistemology*, ed. Bob Hale and Aviv Hoffmann (Oxford: Oxford University Press), pp. 109–36.

Russell, Bertrand (1919). "On Propositions: What They Are and How They Mean", *Proceedings of the Aristotelian Society, Supplementary Volumes* **2**: 1–43.

Russell, Bertrand (1940). *The Philosophy of Logical Atomism* (London: Routledge).

Schaffer, Jonathan (2009). "On What Grounds What", in *Metametaphysics: New Essays on the Foundations of Ontology*, ed. David Chalmers, David Manley, and Ryan Wasserman (Oxford: Oxford University Press), pp. 347–83.

Schaffer, Jonathan (2012). "Grounding, Transitivity, and Contrastivity", in *Metaphysical Grounding: Understanding the Structure of Reality*, ed. Fabrice Correia and Benjamin Schnieder (Cambridge: Cambridge University Press), pp. 122–38.

Schaffer, Jonathan (2016a). "The Metaphysics of Causation", *Stanford Encyclopedia of Philosophy*, ed. Edward N. Zalta (Fall 2016 edn).

Schaffer, Jonathan (2016b). "Grounding in the Image of Causation", *Philosophical Studies* **173** (1): 49–100.

Wilson, Jessica (2010). "What is Hume's Dictum, and why believe it?", *Philosophy and Phenomenological Research* **80** (3): 595–637.

Wilson, Jessica M. (2014). "No Work for a Theory of Grounding", *Inquiry* 57 (5-6): 535–79.

Wilson, Jessica M. (2016a). "Grounding-Based Formulations of Physicalism", *Topoi* **37** (3): 495–512.

Wilson, Jessica M. (2016b). "The Unity and Priority Arguments for Grounding", in *Scientific Composition and Metaphysical Ground*, ed. *Kenneth Aizawa & Carl Gillett* (London: Palgrave Macmillan), pp. 171–204.

5
Against Gabriel
On the Non-Existence of the World

Filippo Casati and Naoya Fujikawa

1. Introduction

It is undeniable that, during the last five or six years, the philosophical world met a new, controversial thinker: Markus Gabriel.[1] Among the many controversial ideas he endorses, he is famous for believing that *the world does not exist*. In this paper, we critically examine his arguments against the existence of the world. In section 2, we will introduce Gabriel's philosophy by focusing our attention on some key notions in his philosophy like existence, fields of sense, and the world. Toying with these notions, we will start to explain why, according to Gabriel, the world does not exist, that is, the field of all fields of sense does not appear in any field of sense. In section 3.1, we will summarize Gabriel's arguments; in section 3.2, we will introduce both their formalization based on mereology and the relative criticisms presented by Priest (ms); finally, in section 3.3, after briefly examining Priest's reformalization, we will present an alternative formalization of Gabriel's arguments, which is also based on mereology. We will also show how they fail nonetheless. To conclude, in section 4, we will argue that, even though both Priest's and our criticisms are somehow focused on the formal aspects of Gabriel's arguments, from a more substantial metaphysical point of view Gabriel's arguments remain insufficient to establish that the world does not exist.

2. Fields of Fields

Fields. There are all sorts of fields: fields of corn near my house, battlefields with thousands of soldiers fighting during the Second World War, rugby fields in

[1] Whoever is familiar with Continental philosophy will immediately associate this name with the recent ongoing fight against postmodernism. It is well known that European philosophy had been dominated by the idea that (an important part of) reality is somehow constructed. Foucault, Derrida, Baudrillard, and Lyotard advocated similar ideas in different terms and manners. However, by carrying the flag of the so-called *New Realism*, Gabriel and many other philosophers have tried to resurrect the idea that we should not abandon the notions of 'reality', 'objectivity', and 'truth' (Gabriel 2013; 2015).

Filippo Casati and Naoya Fujikawa, *Against Gabriel: On the Non-Existence of the World* In: *Non-Being: New Essays on the Metaphysics of Non-Existence*. Edited by: Sara Bernstein and Tyron Goldschmidt, Oxford University Press (2021).
 DOI: 10.1093/oso/9780198846222.003.0005

which the All Blacks scored 140 tries, and mathematical fields represented by algebraic structures used in number theory. There are magnetic fields, electric fields, visual fields, fields of research and many others. Now, according to Marcus Gabriel, there are also *fields of sense*.

Gabriel (2013) introduces the notion of sense and the notion of field of sense by echoing Frege's notion of sense. Frege famously claims that the sense of an expression is a *mode of presentation* of its referent—a way its referent is presented. By replacing 'presentation' with 'appearance', Gabriel defines a sense as a "way in which an object appears" (2013: 69). Then, he defines a field of sense as a domain "in which something—a determinate object—appears in a certain way" (Gabriel 2013: 69).

Fields of sense are *ontologically* relevant: that is, an object exists if and only if it appears in, at least, one of them. Gabriel writes: "[Fields of sense] are an essential part of how things are in that without fields, nothing could exist" (2015: 158). Therefore, Gabriel "understand[s] existence to be the fact that some object or objects appear in a field of sense" (2015: 158).

Given this initial characterization, someone might be tempted to understand Gabriel's idea as a version of Takashi Yagisawa's deflationary account of existence (Yagisawa 2014): Someone might think that, as the latter takes existence to be a relation between a thing and a set, the former takes existence to be a relation between a thing and a field of sense. Unfortunately, such an analogy is misleading in so far as it presupposes the equation between fields of sense and sets. In fact, Gabriel claims that "the concept of a field is more neutral than the concepts of domains or sets" (2015: 160) because "to belong to a set is only one way of appearing in a field of sense" (2015: 158). Even though he would be certainly happy to claim that all sorts of sets are fields of sense, he believes that not all fields of sense are sets. According to Gabriel, physical objects appear in a field of sense, that is, the universe; unicorns appear in a field of sense, that is, colouring books; democratic elections appear in a field of sense, that is, constitutions. Even though the universe, colouring books, and constitutions are taken to be fields of sense, they are not sets (cf. Gabriel 2015: 160). And, of course, this leads us to the crucial question: What is a field of sense, if not a set?

According to Gabriel, a field of sense is anything that can help us to understand, imagine, engage, or think about something. It is, so to speak, the necessary condition for something to appear in any way whatsoever. The universe is the field of sense in which physical objects appear because it is the frame which any physical object *must* be part of: Both the Eiffel Tower and Sirius occupy a part of the universe. Some colouring books are the field of sense in which unicorns appear because they are the place in which a girl can colour those fictional animals. Finally, constitutions are the field of sense in which democratic elections appear because, if we are interested in how elections are conducted in Italy, better for us to read the Italian constitution.

Now, fields of sense do not need to be physical objects such as the universe, books, and constitutions. They can be abstract as well. They can be theories, stories, thoughts, and intuitions. They can be fields of vision as well. Consider Wittgenstein's famous duck-rabbit (cf. Gabriel 2015: 160). In one field of sense, that is, in the field of vision of Person A, an animal with a beak appears; in another field of sense, that is, in the field of vision of Person B, an animal with long fluffy ears appears. Strictly speaking, fields of vision are not physical objects; nonetheless, Gabriel is happy to understand them as fields of sense. Broadly construed, fields of sense are what allows an object to "appear", "come forward", "stand apart form a certain background" (Gabriel 2015: 166). They are the frames used to make sense of all the objects that inhabit them. For this reason, Gabriel writes: "'Appearance in a field of sense' is just a technical version of 'being in a context"' (2015: 158).

As we have already suggested, these fields of sense "are relevant ontological terms" (Gabriel 2013: 50) because they determine what exists. If something appears in a field of sense, then it exists. Gabriel writes: "Existence is appearing in a field of sense" (2015: 166). "To exist is to be found in a field of sense" (2013: 65–6). If so, both the Eiffel Tower and Sirius exist because they appear in a first field of sense (i.e. the universe); unicorns exist because they appear in a second field of sense (i.e. colouring books); democratic elections exist because they appear in a third field of sense (i.e. constitutions). Of course, this does not mean that there exists something that is both a duck and a rabbit: It simply means that, in a field of sense (i.e. the field of vision of Person A), a duck exists and, in another field of sense (i.e. the field of vision of Person B), a rabbit exists.

What about fields of sense, though? Do they exist? And, if so, how do they exist? According to Gabriel, fields of sense must exist because they always appear in other fields of sense. At the end of the day, in thinking about the universe (i.e. the field of sense in which physical objects exist), we create another field of sense (i.e. our thoughts about the universe). In the same way, our thoughts about the universe exist because, in articulating them, we create another field of sense (i.e. our thoughts about our thoughts about the universe). Generalizing this example, Gabriel concludes that "every field of sense is an object" because "for every field of sense, there is a field of sense in which it appears" (2013: 79).

At this point, given this ontological proliferation of objects and fields of sense, it might come as a surprise that, according to Gabriel, something does not exist. Such an exception is represented by the world. Of course, Gabriel does not want to argue that the planet we live on does not exist. That would be silly, indeed! He is well aware that the Earth must exist because it appears in the field of sense in which all physical objects appear, that is, the universe. Contrary to the term 'Earth', the expression 'world' refers to *that unique field of sense in which all fields of sense appear*. Echoing Heidegger, Gabriel takes the world to be the domain of all domains. He writes: "The world is [...] that field of sense in which all fields of sense appear and, for this reason, it is the domain in which everything belongs" (Gabriel 2013: 73). Now, according to Gabriel, *this* world, namely the world understood as the field of all fields, does not exist. But why? He argues as follows.

To being with, Gabriel tries to stimulate our philosophical appetite with an analogy. Consider our field of vision. In our field of vision, we can see all sorts of objects: houses, clouds, dresses, and paintings. However, in our field of vision, we never see our field of vision. The field of vision does not appear in itself. It remains hidden. In the same way, Gabriel's world is the fields in which all sorts of different things are, so to speak, visible: houses, clouds, dresses, and paintings. However, in the world, no trace of the world can be found. As the field of vision, the world remains hidden as well. Gabriel writes: "The world does not appear on the stage of the world" (2013: 76). And, at this point, someone might ask: Why shall we postulate its existence, then?

If you do not find these analogies convincing, you are in good company. Luckily enough, Gabriel has two argument which he calls 'formal'. In the next section, we examine these arguments. First we review his two 'formal' arguments, and give a reconstruction of them partly based on Priest's (ms) interpretation.

3. Gabriel's Arguments and their Formal Reconstruction

3.1 Gabriel's Formal Arguments

Gabriel's first argument goes as follows (2013: 74). Let's assume that the world exists. If so, the world needs to appear in a field of sense. Call it S1. Then S1 is a field of sense among many other fields of sense. As such, S1 appears *next to* another field of sense.[2] Call it S2. Moreover, since the world is the fields of all fields, everything (literally everything!) appears in it. If so, S2 does not simply appear next to S1, it also appears in S1 because, first of all, the world appears in S1 and, secondly, everything appears in the world.[3] However, S2 can neither exist nor appear *next to* the world because, since the world is the field of all fields, nothing can exist or appear outside it. Gabriel concludes: "It is impossible for the world to appear in a field of sense that appears next to other fields of sense" (2013: 74).

Let's continue with the second argument (Gabriel 2013: 74–5). Once again, we assume that, if the world exists, the world appears in a field of sense. Call it S1. Moreover, since everything appears in the world, S1 needs to appear in the world

[2] One may hold that this misreads Gabriel's argument since he says "(S2), (S3), and so on do not appear *next to* (S1), but they also appear *in* (S1)" (Gabriel 2013: 74). We think that Gabriel intends to say that (S2), (S3), and so on do not [*only*] appear *next to* (S1), but they also appear *in* (S1), since if not, it is hard to see why the conclusion is "[i]t is impossible for the world to appear in a field of sense *that appears next to other fields of sense*" (Gabriel 2013: 74; our emphasis).

[3] Here Gabriel seems to assume the transitivity of the appearing-in relation.

as well. Therefore, the world appears in S1 and S1 appears in the world. And Gabriel takes this outcome to be impossible for two reasons. First, if the world appears in S1 and S1 appears in the world, then the world appearing in S1 is not identical with the world in which S1 appears.[4] Second, if the world appears in S1 and S1 appears in the world, some other fields of sense S2, S3, and so on, which, by definition, appear in the world, also appear outside the world (the world that appears in S1). But, as we have seen in the first argument, this is impossible (this second argument is stated in the original German edition, but is missing in the English edition). So, he concludes: "The world is not found in the world" (Gabriel 2013: 74).

Contrary to its appearance, it is not clear what exactly these "formal line of argumentation[s]" (Gabriel 2013: 75) are. What does the phrase 'appear/exist next to' exactly mean? Why can nothing appear/exist next to the world? Why is the world appearing in S1 not identical with the world in which S1 appears? To answer these questions, it would be helpful to reformulate Gabriel's formal arguments in an even more formal way. Priest (ms) provides us such a formal reformulation, and, based on his reformulation, criticizes Gabriel's argument. In section 3.2, we review Priest's reformulation and criticism. Then, in section 3.3, partly based on Priest's reformulation, we will give another formulation of Gabriel's arguments and see how they fail, nonetheless.

3.2 Priest's Mereological Interpretation

Priest's reconstruction of Gabriel's argument is based on the following three assumptions: (i) The world is the totality, that is, the mereological sum of all objects; (ii) x is a proper part of a field of sense of x; and (iii) Gabriel's notion of existence is Priest's notion of objecthood G (G for *Gegenstand*), which is defined as follows:

$$Gx := \mathfrak{S}y\, y = x \tag{1}$$

where $\mathfrak{S}$ is the existential-*un*loaded particular quantifier.

Based on these assumptions, Priest reconstructs Gabriel's argument against the existence (objecthood) of the world (the totality of everything) as follows. The world is the mereological sum of all objects. Let us call it e. Suppose e exists (that is, e is an object). Then, there is a field of sense of e, that is, $f(e)$, and e is a proper part of $f(e)$ (that is, $e < f(e)$). However, since e has everything as its parts, and e is

[4] It is not clear how this shows the impossibility in question. One speculation is that because it contradicts the assumption that the world appears in S1 is identical with the world in which S1 appears.

different from $f(e)$, $f(e)$ is a proper part of e (that is, $f(e) < e$). So we have the following infinite chain of proper parts.

$$\ldots < f(e) < e < f(e) < e < \ldots \tag{2}$$

Gabriel holds that such an infinite chain is impossible, so e is impossible. Therefore, e does not exist (that is, e is not an object). The world doesn't exist.

Construing Gabriel's argument in this way, Priest specifies a reason why the argument fails: The infinite chain of proper parts in question is, *pace* Gabriel, *possible*. First, the infinite chain of parts represented in (2) is generated by the loop of parthood: $e < f(e)$ and $f(e) < e$. And such a loop is possible for two reasons. There are consistent mereological theories that do not have anti-symmetricity of the parthood relation and thus in which such a loop consistently holds. Moreover, some examples show that it is conceivable and thus possible that the parthood relation constitutes such mereological loops. One of these examples is a pair of propositions such that one of the pair has the other as its part and *vice versa*.

a. (3b) or snow is white (3)
b. (3a) or snow is white

It is clear that (3a) has (3b) as its part, and *vice versa*. And there is nothing incoherent here.[5]

3.3 Another Mereological Interpretation

So far we have seen that Priest presents Gabriel's arguments by appealing to a mereological framework. Furthermore, given this framework, Priest criticizes Gabriel because he takes a non-wellfounded sequence of proper parts to be impossible while it is, otherwise, well known to be mereologically possible. *Prima facie*, Priest seems to be right. However, if we sit on Priest's argument for long enough, the situation will appear to be more complicated than at first it looks. In particular, it will be clear that Priest's criticism is both exegetically and philosophically problematic.

Let's begin by discussing an exegetical issue. Priest's argument succeeds only if the switch from fields of sense to a mereological framework is somehow compatible with the philosophical ideas that Gabriel wants to defend. However, not only does Gabriel reject the analogy between his ontological theory and the set-theoretical framework, he explicitly rejects the analogy between his ontological theory and the mereological framework too. He writes: "When I say 'appearance

[5] Priest further offers an argument *for* the objecthood of the world to the effect that given that intentionality forces us to take what is thought as an object and the fact that we (including Gabriel himself!) can think about the world, it follows that the world is an object.

[in a field of sense]', I use this term technically in order to avoid 'belonging' or 'being part of', as this might invite the set-theoretical or mereological conceptions I am avoiding" (Gabriel 2015: 44). From this point of view, Priest's argument relies on assumptions that Gabriel would not be happy to grant him. As such, it might be taken to be an unfair criticism.

Having said that, the matter is complicated by the fact that Gabriel gives several (at least conceptually) different characterizations of the world. As we have seen, he defines the world as the field of sense of all fields of sense. However, he also characterizes the world as

> any kind of unrestricted or overall totality, be it the totality of existence, the totality of what there is, the totality of objects, the whole of beings, or the totality of facts or states of affairs. (Gabriel 2015: 187)

This characterization tempts us to take the world as the mereological sum of all objects, that is, Priest's *e*. Moreover, as we have seen, according to Gabriel, any field of sense (except the world) is an object. Then, why shouldn't we use mereology to formally describe the appearing-in relation between the world and the fields of sense, *pace* what Gabriel explicitly says? At the end of the day, whichever is the framework we might prefer to employ, the point made by Priest is straightforward.[6]

Another exegetical issue of Priest's mereological reformulation of Gabriel's argument arises from the fact that Gabriel's argument against the existence of the world does not depend on the impossibility of loop of the appearing-in relation. Indeed, he explicitly says "I do not rule out that some field can appear within itself" (Gabriel 2015: 188).[7] But, if the appearing-in relation can loop, what feature of the world as the totality does prevent itself from appearing in itself? He specifies that the problem is "in the combination of totality and self-containment" and this is "loosely connected to issues from set theory, albeit different in that not all fields are set" (Gabriel 2015: 188). As far as we can see, what is crucial in his arguments, which we have reviewed in section 3.1, is the following premise.

[6] The third characterization of the world by Gabriel is as follows: "the world would be the object which has all properties" (2013: 53). The world in this sense is called the *super-object* (Gabriel 2013: 55). According to Gabriel, the super-object cannot exist. To show this, he appeals to mereology. He says: "If there were a super-object, it would be the mereological sum of all properties" (2013: 59). Unfortunately, this equation is really confusing. Having a property is not the same as having the property as a mereological part. The Empire State Building has the property of a material object, but the property is not a mereological part of it (at least in the sense where contemporary mereology understands the term 'part'). Nor is an object the mereological sum of all properties it has. Gabriel is not a mere mereological sum of the properties he has.

[7] Priest (ms) has already realized that Gabriel doesn't rule out loop of the appearing-in relation, and he claims that this weakens Gabriel's argument and leads to another problem.

$$\text{Nothing appears/exists outside the world.} \tag{4}$$

Now, by taking 'x appears outside/next to y' as 'x is not a proper part of y',[8] we can formulate this premise within the framework of mereology as follows:

$$\mathfrak{A}x \neg\neg x < e, \text{ equivalently, } \mathfrak{A}x(x < e) \tag{5}$$

where $\mathfrak{A}$ is the unrestricted universal quantifier. In what follows, we reformulate Gabriel's two arguments against the existence of the world to explicate how his arguments depend on (5).

Henceforth we use the non-wellfounded mereology defined in Cotnoir and Bacon (2012) as a formal framework of our reconstruction (Priest (ms) appeals to it to show how the loop of the parthood relation is possible). We reproduce an axiomatization of it in Appendix. It should be mentioned here that the non-wellfounded mereology has a specific feature which makes it appropriate for a formal framework to interpret Gabriel's argument. In classical extensional mereology, any loop of the proper part relation is ruled out: that $x < y \rightarrow y \nless x$ is a theorem of it. However, it is neither an axiom nor a theorem in the non-wellfounded mereology in question. This feature of $<$ properly traces Gabriel's explicit statement that loop of the appearing-in relation is not ruled out.

The first argument goes as follows. In the first place, it assumes that the world appears in S1. In addition to this, it also assumes that S2, S3, and so on, appear outside of S1. For simplicity, let us consider only S2 as a field of sense that appears outside S1. Then, these assumptions are stated as follows:

$$e < S1 \text{[9]} \tag{6}$$

$$\neg S2 < S1 \tag{7}$$

It is easy to see that these assumptions are indeed inconsistent: From (5) it follows that $S2 < e$. Then, since $<$ is transitive, $S2 < S1$. This contradicts (7). However, the incompatibility of these assumptions doesn't show that (6) is impossible (and indeed Gabriel doesn't claim that the first argument shows this).

The second argument assumes that the world appears in S1, that is, (6). Gabriel tries to give two reasons in order to show that this assumption is untenable (he seems to try to run two *reductio* arguments). The first reason goes as follows: From (5), it follows that $S1 < e$. This and (6) entail that e is not identical with e (and this

[8] An alternative way of understanding the former relation is to take it as 'x is disjoint from y', which is stronger than 'x is not a proper part of y'.

[9] Here we use the proper parthood to represent the appearing-in relation, as Priest does. One may think that Gabriel's endorsement of the possibility that the appearing-in relation can loop allows us to use the parthood relation $\leq$, instead of the proper parthood, to represent the relation. However, $\leq$ is too strong: $\leq$ is reflexive (for any x, $x \leq x$), and, if we use $\leq$ to formally represent the appearing-in relation, it follows that everything appears in itself. This is obviously stronger than the claim that something can appear in itself.

is absurd). However, in the non-wellfounded mereology in question, they don't. Indeed, $e < S1$, $S1 < e$ (and therefore $e < e$), and $e = e$ are compatible in this mereology.[10] So the first reason fails to show that (6) is impossible. The second reason goes as follows: Again, from (5), it follows that $S1 < e$. From this and (6), it follows that some field of sense is next to S1 (let's call it S2), that is, $\neg S2 < S1$. However, from (5), $S2 < e$. And, as we have seen, this leads to a contradiction. Now, the problem of this argument is that the non-wellfounded mereology in question does not validate the inference from $e < S1$ and $S1 < e$ to $\neg S2 < S1$.[11]

The possibility which Gabriel fails to properly assess is one that both $e < S1$ and $e = S1$ hold. This is the possibility that the world appears in itself, which is possible in our mereological reformulation. This point is helpful to reply to Gabriel's worry concerning *fractal ontology* (2013: 83). An object appears in a field of sense, and the latter appears in a further field of sense, and so on. According to him, in searching for a field of sense of a field of sense "[w]e never come to an end; in this way we never achieve the last field of sense in which everything appears—the world" (Gabriel 2013: 82). This worry is dismissed once we accept that the world appears in itself: The world is the terminal point of the search for a field of sense of a field of sense.

Gabriel's arguments seem to appeal to the idea that if the world appears in S1, there is a field of sense that appears in the world (and S1) and next to the world. It is worthwhile mentioning that in classical mereology there is a principle, *Weak Company*, which says that if something has a proper part, it has another proper part ($x < y \rightarrow \exists z(z < y \wedge z \neq x)$), and this principle will ensure the idea to which Gabriel seems to appeal. However, Weak Company rules out the reflexive proper part (which is a case of loop) and in the non-wellfounded mereology it doesn't hold (cf. Cotnoir and Bacon 2012: section 3.1)

In this way, (5), which is supposed to be a special property of the totality, doesn't establish that (6) is impossible, in the sense that there is a consistent mereology in which (5) and (6) are compatible.

A final remark: On the uniqueness of the fusion. The non-wellfounded mereology does not ensure the uniqueness of fusion. In this case, the mereology doesn't ensure the uniqueness of the totality: There may be two or more things that has everything as its (proper) parts. One may take this as an argument against the existence of *the* world. However, first of all, even though the non-wellfounded mereology doesn't ensure the unique existence of the totality, it is compatible with the unique existence of the totality.[12] Secondly, it is hard to see this as an

[10] It is worthwhile mentioning that in Cotnoir and Bacon's system, $<$ is taken as primitive and $\leq$ is defined as: *xy iff* $x < y \vee x = y$.

[11] Another way of formulating the second reason is given by taking 'x is next to y' as 'x is disjoint with y'. Under this understanding, the claim that S2 is next to S1 is rephrased as $\neg \mathfrak{S} z(z \leq S2 \wedge z \leq S1)$. But, again, this doesn't follow from the assumptions.

[12] Any model of classical mereology validates all axioms of the non-wellfounded mereology, and a model of classical mereology has the top element, which has everything as its part.

interpretation of Gabriel's arguments against the existence of the world: He doesn't have a quarrel with the *uniqueness* of the world. This should be taken as an another argument against the existence of the world, if it is indeed an argument against it.

4. Beyond Formality

Until now, we have shown how Gabriel's formal arguments seem to be problematic. In order to see if Priest's and our objections are successful, it is important to properly understand what kind of impossibility Gabriel refers to. We focused on formal impossibility, in the sense that, in our reply, we argue that Gabriel is wrong by showing that there is a non-wellfounded mereology in which the totality can be a proper part of something. Having said that, someone sympathetic with Gabriel's philosophy might try to save Gabriel's project/arguments from our criticism in the following way: Non-wellfounded mereology only shows that it is *conceivable* that the world appears in a field of sense. Now, though it is contentious, some philosophers hold that conceivability is broader than metaphysical possibility: Something conceivably possible may be metaphysically impossible. Given this distinction, one may argue that Gabriel is concerned with not pure conceivability, but a different, more robust, kind of impossibility, that is, the metaphysical one. This seems to be correct if we consider that Gabriel's philosophy is supposed to be an organic part of what has been currently labelled New Realism (see Ferraris 2014; 2015; Sparrow 2014). In opposition to postmodernism and any form of constructivism, New Realists are concerned with what is actually there. They aim at uncovering reality as such. Following New Realism, Gabriel writes that his theory is concerned with "facticity"—with "what exists" (2013: 133). Then, if Gabriel's arguments aim to show that it is *metaphysically* impossible that the world appears in a field of sense, then showing that it is conceivable that the world appears in a field of sense is not enough to show that Gabriel's arguments fail.

In order to see this point in a clearer way, let us discuss a brief analogy. Consider the difference between paraconsistent logic and dialetheism. The former is a logical system in which some contradictions are acceptable; the latter is the metaphysical view according to which some contradictions are true. Now, let's consider whether dialetheism is metaphysically possible or not. One may claim that the fact that a paraconsistent logic tolerates some true contradictions is not enough to show that dialetheism is metaphysically possible, since this fact at most shows that it is logically possible (conceivable), and something logically possible (conceivable) may be metaphysically impossible. If so, in order to show that dieatheism is metaphysically possible, one needs to do more than simply showing that some formal systems might tolerate true contradictions. In the same way, one may hold that, in order to criticize Gabriel, one needs to do more than simply showing that some formal systems might work as counterexamples to Gabriel's theory.

Gabriel may have in mind this line of argument. If so, he needs to do more than what he actually does. In particular, he has to argue for the distinction between conceivability and metaphysical possibility. More importantly, he needs to show that the non-wellfounded mereology captures only conceivability and fails to track what is metaphysically possible. Since there is no trace of these arguments about these topics, Gabriel seems to be in trouble nonetheless. If Gabriel runs this line of argument, the burden is on his shoulder.

5. Conclusion

To conclude, let's summarize what we have done until now. To begin with, we have analyzed all the notions that Gabriel employs in arguing that the world does not exist. These notions are existence, fields of sense, the appearing-in relation, and the world (section 2). After that, we have summarized Gabriel's reasons to believe that the world does not exists, that is, the field of all fields of sense does not appear in any field of sense (section 3.1). Moreover, we have presented how Priest formalizes Gabriel's arguments by appealing to mereology (section 3.2). Then, we introduced an alternative formalization, and based on it, we have shown that Gabriel's arguments fail (section 3.3). To conclude, we have argued that, even though Priest's and our arguments employ more technicalities than Gabriel's ones, the latter remain unsuccessful from a more substantial metaphysical point of view as well (section 4).

This paper exhaustively focuses on examining Gabriel's arguments against the existence of the world. However, we are well aware that there are many interesting features of Gabriel's theory of fields of sense which are worthwhile mentioning here. Let's briefly list some of them up. They should be interesting topics for further research in either metaphysics or the history of philosophy.

From a metaphysical point of view, it might be interesting to explore the relation between Gabriel's fields of sense and contemporary theories of grounding/metaphysical dependence. In particular, it seems appropriate to say that the existence of an object is grounded/metaphysically depends on the field of sense in which the object appears. If so, the appearing-in relation would be a good example of a grounding/metaphysical dependence relation.[13]

From a historical point of view, it could be worthwhile mentioning that there is striking similarity between Gabriel's theory of fields of sense and Nishida Kitaro's logic of place. Gabriel holds that to exist (to be an object) is to appear in a field of sense; Nishida holds that to be an object is to be within a place. For Gabriel, a field of sense exists (is an object) if it appears in a field of sense; for Nishida, a place is an

[13] For a useful overview of grounding/metaphysical dependence, see Bliss and Trogdon (2016) and Tahko and Lowe (2016). Bliss (2014) discusses the circularity of grounding.

object if it is within a place. Gabriel claims that the world, the field of sense in which everything appears, does not appear in any field of sense, and thus, does not exist (is not an object); Nishida claims that absolute nothingness, the place within which everything is, is not within any place, and thus, is not an object.[14]

Finally, in order to see how the first and the second point are related, it might be helpful to see some recent researches on Nishida's logic of place from the view point of contemporary theory of grounding and metaphysical dependence (Priest, this volume Chapter 2; Casati and Fujikawa ms).

6. Appendix: Non-Wellfounded Mereology Defined in Cotnoir and Bacon (2012)

Cotnoir and Bacon (2012) formulates several different axiomatizations of the same non-wellfounded mereology. For convenience for the readers, we reproduce one of their axiomatizations. An axiomatization of the non-wellfounded mereology in Cotnoir and Bacon (2012) consists of the following three axioms (base logic is classical logic):

$$\textbf{NWA1} \quad (x < y \wedge y < z) \rightarrow x < z$$

$$\textbf{NWS4} \quad y \nleq x \rightarrow \exists z (z \leq y \wedge \neg z \circ x)$$

$$\textbf{NWE1} \quad \exists z \phi \rightarrow \exists x \forall y (x \circ y \leftrightarrow \exists z (\phi \wedge y \circ z))$$

Here $<$ is taken as a primitive symbol. $\leq$ and $\circ$ are defined as follows:

$$\text{P} \quad x \leq y := x < y \vee x = y$$

$$\text{O} \quad x \circ y := \exists z (z \leq x \wedge z \leq y)$$

It is worthwhile mentioning that adding **NWA2**, which rules out any loop of the proper-parthood relation, to these axioms results in classical mereology (Cotnoir and Bacon 2012: 197–8).

$$\textbf{NWA2} \quad x < y \rightarrow y \nless x$$

[14] Nishida's reason for the non-objecthood (nonexistence) of absolute nothingness, the place within which everything is, is different from Gabriel's reason for the nonexistence of the world. See Nishida (1926). Moreover, Nishida also seems to suggest that absolute nothingness is within itself. This immediately leads to a contradiction: Absolute nothingness is an object and not an object. See Casati and Fujikawa (ms) for a dialetheic interpretation of this aspect of Nishida.

References

Bliss, Ricki (2014). "Viciousness and Circles of Ground", *Metaphilosophy* 45 (2): 245–56.

Bliss, Ricki and Kelly Trogdon (2016). "Metaphysical Grounding", *Stanford Encyclopedia of Philosophy*, ed. Edward N. Zalta (Winter 2016 edn).

Casati, Filippo and Naoya Fujikawa (ms). "Inconsistent Metaphysical Dependence: Cases from the Kyoto School", manuscript.

Cotnoir, Aaron and Andrew Bacon (2012). "Non-Wellfounded Mereology", *Review of Symbolic Logic* 5 (2): 187–204.

Ferraris, Maurizio (2014). *Manifesto of New Realism*, trans. Sarah De Sanctis (Albany, NY: State University of New York Press).

Ferraris, Maurizio (2015). *Introduction to New Realism*, trans. Sarah De Sanctis (London: Bloomsbury Academic).

Gabriel, Markus (2013). *Why the World Does Not Exist* (Cambridge: Polity Press).

Gabriel, Markus (2015). *Fields of Sense: A New Realist Ontology* (Edinburgh: Edinburgh University Press).

Nishida, Kitarō (1926). "Place", in *Nishida Kitaro Zenshu* [The Complete Works of Nishida Kitaro], new edn, vol. III, ed. Atushi Takeda, Klaus Riesenhuber, Kunitsugu Kosaka, and Masakatsu Fujita (Tokyo: Iwanami Shoten, 2002–9).

Priest, Graham. this volume. "Nothingness and the Ground of Reality."

Priest, Graham (ms). "Everything and Nothing", manuscript.

Sparrow, Tom (2014). *The End of Phenomenology: Metaphysics and New Realism* (Edinburgh: Edinburgh University Press).

Tahko, Tuomas E. and E. Jonathan Lowe (2016). "Ontological Dependence", *Stanford Encyclopedia of Philosophy*, ed. Edward N. Zalta (Winter 2016 edn).

Yagisawa, Takashi (2014). "Deflationary Existence", *Annals of the Japanese Association for Philosophy of Science* 22: 1–16.

6

How Can Buddhists Prove That Non-Existent Things Do Not Exist?

Koji Tanaka

1. Buddhist Philosophers and Non-Existence

Buddhist philosophers are critical of the generous ontology advocated by non-Buddhist Indian philosophers.[1] They reject the existence of all sorts of things which are important for the non-Buddhist. For instance, the Buddhist holds that the self does not exist. That there is no self is an important Buddhist doctrine. Whatever this doctrine is supposed to mean, the Buddhist has to argue against those who are convinced that the self does exist—and there are plenty of those who implicitly or explicitly believe in the self. The problem with Buddhist philosophers in arguing against their opponent is that they agree with non-Buddhists that one cannot prove a thesis whose subject is non-existent. This agreement has put Buddhist philosophers in a difficult situation. How can they argue against their opponents and show that the self and all other things that Buddhists take to be non-existent do not exist, while agreeing that one cannot prove a thesis whose subject is non-existent?

The problem for Buddhist philosophers is not confined to their central doctrines. They are actually much worse, as they tend to be *global error theorists*. An error theorist typically holds that there are no facts of the matter in a specific context. An error theorist about morality holds that there are no moral facts. Some error theorists further hold that there are no truths about morality. For error theorists about morality, any moral discourse is, strictly speaking, not true (Joyce 2016). Similarly, an error theorist about fictions hold that there are no truths about fictional characters. For them, it is not true that Sherlock Holmes lived in 221B Baker Street, London. Some error theorists also hold that there are no facts or truths about fictional characters even within fictions, as the properties attributed to them are not factual matters. According to those error theorists about fictional characters, it is, strictly speaking, not true that Harry Potter was

[1] This seems to change as Buddhism went East to China (and Korea and Japan). See Tanaka (forthcoming). In this paper, I am mostly concerned with Buddhist philosophers who flourished in India and Tibet.

Koji Tanaka, *How Can Buddhists Prove That Non-Existent Things Do Not Exist?* In: *Non-Being: New Essays on the Metaphysics of Non-Existence.* Edited by: Sara Bernstein and Tyron Goldschmidt, Oxford University Press (2021).
 DOI: 10.1093/oso/9780198846222.003.0006

pretending to be dead when Voldemort cast the killing curse (*Avada Kedavra*) with the Elder Wand, even within the Harry Potter books. For them, any claim about fictional characters is not truth-apt (Brock 2002).

If we generalize an error theory to any context, we get *global error theory*. A global error theorist holds that there are no facts or truths about anything. Many Buddhist philosophers, at least those in India and Tibet, can be characterized as global error theorists who held that there are no facts or truths about anything, as there is not anything that really exists. This does not make them nihilists, as they do think that things do have quasi-existence (conventional existence (*saṃvṛtisat*)).[2] But non-Buddhist philosophers are typically not global error theorists and hold that a lot of things do really exist. Buddhist philosophers, thus, have to argue against them about many things which they do not think really exist and about which there are not really any truths.

Demonstrating that Buddhist philosophers can be described as global error theorists is beyond the scope of this paper.[3] For the sake of this paper, I will assume that Buddhist philosophers are largely global error theorists. The question then is: How can they claim that it is true that something which does not exist does not exist when the opponent holds that it exists?

In this paper, I will first present a difficulty that Buddhist philosophers have faced in proving that what they take to be non-existent does not exist. I will then survey two main solutions that they have provided. Those 'solutions' may not solve the problem, or may solve the problem but create other problems. I will not survey the Buddhist treatment of the problem of proving about non-existence in order to present a new solution that we have yet to see. Instead, I will articulate a problem about non-existence that is unique to Buddhist philosophers. I will do so in order to present an interesting puzzle about non-existence that has largely escaped attention in the 'Western' literature.

2. *Āśrayāsiddha*

Buddhist discussion of non-existence centers around *āśrayāsiddha* (unestablished basis). It is (or was) commonly accepted in India that when the 'basis' (*āśraya*) or subject (*dharmin*) of a thesis (*pakṣa*) is unestablished, that is to say, non-existent, the thesis cannot be established or proved. The fallacious nature of *āśrayāsiddha* is commonly accepted by both Buddhist philosophers and their philosophical

[2] What exactly this means is a complicated issue. See, for instance, Cowherds (2011) for a discussion.

[3] This is a difficult and controversial task and I do not pretend that describing Buddhist philosophers as largely global error theorists is widely accepted. However, see Tillemans (2016) who describes some Buddhist philosophers as global error theorists even though he does not use the phrase 'global error theory' or 'global error theorist'.

opponents (i.e. non-Buddhist Indian philosophers). In everyday situations, we may not be puzzled when someone claims that the pretension of Harry Potter when he was struck by Voldemort cannot be proved to be real because of his non-existence. But, about things that matter philosophically such as the reality or unreality of the self, the threat of the fallacy of *āśrayāsiddha* is a real issue.

For Buddhist philosophers, *āśrayāsiddha* becomes particularly problematic as they are typically global error theorists. The problem is historically acute in two contexts which are concerned with existence generally: Buddhist 'proofs' of momentariness and Buddhist—particularly (later) Madhyamaka—proofs of the absence of intrinsic nature (*niḥsvabhāvatā*).[4] In both contexts, Buddhist philosophers want to prove a thesis whose basis or subject is non-existent. In the context of momentariness, Buddhists want to prove that whatever exists is momentary (i.e. if something exists, it exists only momentarily). If we contrapose it within the universal quantifier (most, if not all, Buddhist philosophers accept contraposition as valid), what they want to prove is that whatever is non-momentary is non-existent. But, for the Buddhists, there cannot be anything which is non-momentary. So the thesis of momentariness has a subject (or subjects) that is (or are) non-existent. Hence Buddhist philosophers have to come up with a way around *āśrayāsiddha* (Matilal 1970). In the context of the absence of intrinsic nature (*niḥsvabhāvatā*), if something lacks intrinsic nature, it does not exist for the opponent. Buddhists, in particular Mādhyamikas, would have to agree to this to a certain degree since, for them, there are not really any facts about anything; and so they cannot claim, for instance, that the self does not really exist. So the thesis that everything lacks intrinsic nature has a subject (or subjects) that does not (or do not) exist.[5]

Given the importance of momentariness and the absence of intrinsic nature (or emptiness (*śūnyatā*)), Buddhist philosophers have to take the fallacy of *āśrayāsiddha* seriously. Historically, Buddhist philosophers have suggested mainly two solutions to *āśrayāsiddha*. In the rest of the paper, I will explain what I take those solutions to be and how they developed in the Buddhist philosophical tradition.

3. Avoiding the Fallacy of *Āśrayāsiddha*, Part I

Buddhist discussions on *āśrayāsiddha* tend to start with the definition of the (valid) thesis (*pakṣalakṣaṇa*) that Dignāga (480–540 CE) provided in his *Pramāṇasamuccaya*:

[4] The Madhyamaka school consists of Buddhist philosophers who are the followers of Nāgārjuna (2nd cent. CE) known for his doctrine of emptiness. Following modern convention, I will use 'Madhyamaka' to refer to the school or thought and 'Mādhyamika' to refer to people who belong to the school or hold the thought.

[5] See Kamalaśīla's (740–95 CE) *Madhyamakāloka*. A translation can be found in Keira (2004).

> [A valid] thesis is one which is intended by [the proponent] himself as something to be stated in its proper form alone; [and] with regard to [the opponent's] own subject, it is not opposed by perceptible objects, by inference, by authority or by what is commonly recognised. (*Pramāṇasamuccaya* III, k. 2)[6]

Dignāga was the main figure, alongside Dharmakīrti (7th cent. CE), who laid down the foundation for the Buddhist study of epistemology and logic, in particular the study of the methods for proof and acquisition of knowledge (*pramāṇa*). The particular concern for Dignāga was to undermine and disprove the scriptural authority of the *Vedas*, the authoritative texts which are said to prove a number of things for non-Buddhist philosophers in India. Dignāga argued against the authoritative words (*āptavāda*) as proving anything on their own and argued for only two means of acquiring knowledge: perception (*pratyakṣa*) and inference (*anumāna*). Dignāga, Dharmakīrti and those Buddhist philosophers who have followed their lead have generally assumed that what we can know depends on how we can know it. That is, to use modern epistemological terminology, they are generally *reliabilists*.[7] They do not think that the cognitive state you happen to arrive at counts as knowledge. Cognition has to go through particular transformations for the resulting cognitive state to count as a knowledge state.[8]

Buddhist philosophers who came after Dignāga have generally taken his definition to mean that the property to be proved "with regard to [the proponent's] own subject", i.e. the thesis that the proponent is arguing for should not be opposed by any (valid) means of acquiring knowledge. More importantly for our purpose, they have also taken the definition to mean that the proponent's subject as well as the property to be proved (*sādhyadharma*) must be existent (Tillemans 1999: 172). Dignāga himself did not seem to have implied that the proponent's subject must be existent by his definition. However, he did seem to have thought that *āśrayāsiddha* is a fallacy to be avoided, as he was aware of the difficulty involved in proving that what he took to be non-existent does not indeed exist.

The issue of *āśrayāsiddha* comes out distinctly for Dignāga in the context of discussing Primordial Matter (*pradhāna*) whose existence the non-Buddhist Sāṃkhya philosophers accept but Buddhist philosophers reject. He is concerned with discussing two different arguments in connection with Primordial Matter of the Sāṃkhya school. First, he discusses the Sāṃkhya arguments that allegedly show the existence of Primordial Matter. In those arguments, Sāṃkhya philosophers argue for the existence of Primordial Matter from the general characteristics

[6] This translation is from Tillemans (1999: 172).

[7] 'Reliabilists' and 'reliabilism' are terms that I am attributing to the Buddhists rather than the translations of any of the terms they use.

[8] See, for instance, Patil (2009), Tillemans (1999), and Tanaka (2013).

that all individual things share in common. Second, he discusses the Buddhist arguments showing the non-existence of Primordial Matter. For the Buddhist, even talking about these two kinds of arguments invites the charge of *āśrayāsiddha* as the subject matter of both kinds of arguments is non-existent. Dignāga seems aware of this and tries to avoid the charge. He offers different solutions in relation to the two kinds of arguments. Let's look at how he tries to avoid the fallacy of *āśrayāsiddha* in relation to the two kinds of arguments involving Sāṃkhya's Primordial Matter.

In response to the Sāṃkhya arguments that purportedly show the existence of Primordial Matter, Dignāga says:

> [Sāṃkhya philosophers] should formulate the thesis as "The various individuals certainly possess one and the same cause [i.e., *pradhāna*], in which case they do not prove [directly the existence of] the Primordial Matter."[9]

He is suggesting here that the Sāṃkhya thesis that Primordial Matter exists can be *paraphrased* in a way that the Buddhist can accept (Tillemans 1999: 174–5). Buddhist philosophers can accept the existence of cause and so they do not have any trouble talking about 'one and the same cause' of various individuals. They do, of course, reject the existence of such a cause; nevertheless, they can understand what the Sāṃkhya thesis states when it is paraphrased in this way. Once they establish the possibility of engaging with the Sāṃkhya thesis, the Buddhist can go on to reject it. In relation to the Sāṃkhya arguments for the existence of Primordial Matter, thus, Dignāga offers the strategy of paraphrase in order to avoid the charge of *āśrayāsiddha*. This is also the strategy that Dharmakīrti employs in his discussion of the proponent's own intended subject (*svadharmin*) in his *Pramāṇavārttika* IV.

In relation to the Buddhist argument for the non-existence of Primordial Matter, Dignāga offers a different solution, however:

> When they [i.e., the Buddhists] argue that [Primordial Matter] does not exist [because of nonperception], 'nonperception' is a property of the imagined object.[10]

The Buddhist metaphysical framework that is assumed here is that only particulars exist.[11] Even though inference is a valid means of acquiring knowledge, it can give us only conceptual knowledge as it operates on the general characterizations

[9] *Nyāyamukha*: the translation is from Tillemans (1999: 175).

[10] *Nyāyamukha*: the translation is from Tillemans (1999: 175).

[11] I should note that not all Buddhist philosophers accept this metaphysical view. Mādhyamikas, for instance, reject the existence of not only universals but also particulars when the existence is understood as real, as opposed to quasi, existence. The metaphysical view that Dignāga is operating with, however, accepts the existence of particulars.

of objects. Having conceptual knowledge is, however, not a proof of existence as only particulars exist, according to the metaphysical view that Dignāga assumes, and conceptual knowledge relies on universals. The proof of the existence of objects ultimately rests on perception, which is the means by which one can acquire knowledge about particulars. So if Primordial Matter can be said to exist, we have to be able to perceive it. However, we cannot perceive Primordial Matter, so Dignāga claims. Hence, so the argument goes, Primordial Matter does not exist. This argument is valid assuming that *modus tollens* is valid. But, in order for it to be effective, Dignāga has to be able to talk about Primordial Matter. He does this by identifying it as an imagined object, meaning that Primordial Matter is a conceptual object. As inference operates on the conceptual objects that are characterized by general features, Primordial Matter can then be talked about meaningfully and, thus, it can be the subject of a proof showing that it does not exist.

While Dignāga does offer the introduction of conceptual objects as a solution to the charge of *āśrayāsiddha* in relation to the positive proof that Primordial Matter is non-existent, he seems to abandon it as a way to avoid the charge. Dignāga might have raised it as a possible solution, but he did not seem to have thought that invoking conceptual objects is a good way to prove anything, and he avoided proving anything via conceptual objects in his later writings (Katsura 1992).

Dharmakīrti the main philosopher who developed on the work of Dignāga on epistemology and logic, also tended to use the strategy of paraphrase rather than conceptual object in his discussion on the topic about non-existent things. In his *Pramāṇavārttika* IV, k. 144–5, Dharmakīrti considers the anti-Sāṃkhya argument that 'pleasure, pain, and bewilderment' are not the permanent nature of the 'transformations' (*vikṛti*) taking place in the world. Sāṃkhya school takes them to be the (essential) qualities (*guṇas*) of Primordial Matter. But, because Dharmakīrti rejects the existence of Primordial Matter, he cannot be talking about pleasure, pain, and bewilderment that are the qualities of what he takes to be non-existent, so the Sāṃkhya retorts. In response, Dharmakīrti paraphrases them and claims that the subject matter of his anti-Sāṃkhya argument are not the qualities of Primordial Matter but the pleasure, pain, and bewilderment that ordinary folks feel. Because everyone, Buddhist or otherwise, accepts these ordinary feelings are real entities (*vastubhūta*) (they are in our experiences), Dharmakīrti goes on to refute the Sāṃkhya thesis that pleasure, pain, and bewilderment are permanent (Tillemans 1999, 178–9). In *Pramāṇavārttika* IV, k. 141–2, Dharmakīrti offers a parallel discussion about space (Tillemans 1999: 179–80).

The solution to the fallacy of *āśrayāsiddha* for early Buddhist epistemologists/logicians such as Dignāga and Dharmakīrti was, thus, the method of paraphrase. Buddhist philosophers who came after Dignāga and Dharmakīrti, however, largely took the introduction of conceptual objects as the main route to avoid *āśrayāsiddha*. Before considering how the method of conceptual objects became the main solution to the fallacy of *āśrayāsiddha*, I will say a few words about the two approaches. That will help us see the distinction between them as well as what is at stake for Buddhist philosophers in choosing which solution to adopt.

4. Paraphrase vs Conceptual Objects

The method of paraphrase Dignāga and Dharmakīrti appeal to in responding to the Sāṃkhya charge of *āśrayāsiddha* might not be thought to be effective. Dharmakīrti claims that he can legitimately entertain the thesis that pleasure, pain, and bewilderment are permanent in order to refute it by paraphrasing 'pleasure, pain, and bewilderment' and rendering it as 'pleasure, pain, and bewilderment of ordinary folks'. Of course, if that's the strategy, the Sāṃkhya philosopher can respond by saying that Dharmakīrti simply changed the subject and has not refuted the Sāṃkhya thesis. The thesis that Dharmakīrti rejects is that the pleasure, pain, and bewilderment that ordinary folks experience are permanent. But the thesis that the Sāṃkhya is putting forward is that the pleasure, pain, and bewilderment which are the qualities of Primordial Matter are permanent. Paraphrasing alone does not allow Dharmakīrti to refute the Sāṃkhya thesis.

The situation is analogous to the following Cookie Monster scenario. Suppose that your child says that there is Cookie Monster in the closet. You might try to convince her that that is not the case by saying that Cookie Monster is in the room by holding a soft toy version of Cookie Monster. She then says: "No, daddy, that's a toy Cookie Monster and the real Cookie Monster is in the closet!" Sure enough, the fact that a toy Cookie Monster is not in the closet does not prove that the real Cookie Monster is not in the closet. "Silly daddy!"

An analogy like this might be thought to show that the method of paraphrase is not really a solution to the fallacy of *āśrayāsiddha*. But what would you say in response to the child who essentially accuses you of changing the subject? I think the answer should be: nothing. You know that there is no such thing as the real Cookie Monster. It is a character enacted by a puppet on a TV show. Even if you got hold of the puppet actually used on the show and held it in your hand saying: "Look! Cookie Monster is here and not in the closet!", you would, most likely, still get the same response. So there is not much point in arguing at that point.

This may just be a silly story about Cookie Monster. I am sure that every parent simply shrugs off by this point. But, for a global error theorist like a typical Buddhist philosopher, a situation like this is ubiquitous. For a global error theorist, nothing can be said to really exist. They can talk about 'replicas' like a toy Cookie Monster. But there is nothing other than those 'replicas' that they can claim to exist. So when someone claims that something really exists, all a global error theorist can do is to put their hands up and surrender. To be a consistent global error theorist is to be a quietist.[12]

The device that Buddhist philosophers use to theorize about this aspect of global error theory is the *two truths* (or two realities) theory: ultimate truth/reality (*paramārthasatya*) and conventional truth/reality (*saṃvṛtisatya*).[13] What exactly they are is a matter of debate both for traditional Buddhist philosophers and contemporary scholars of Buddhist philosophy.[14] Depending on which Buddhist philosopher or Buddhist scholar we have in mind, what counts as ultimately true and what can be said to ultimately exist differ. One thing is clear, however. For Buddhist philosophers, the number of things that can be said to ultimately exist and have ultimate truth about is very limited, if any. Some Madhyamaka philosophers such as Candrakīrti (570–650 CE) seem to think that the realm of ultimately existing entities is empty. In contrast, this realm is rather vast and a very important one for non-Buddhist philosophers. Primordial Matter of the Sāṃkhya, for instance, is part of the realm of ultimately existing entities. Buddhist philosophers disagree among themselves how small this realm should be but they all agree that it is a lot smaller than non-Buddhist philosophers hold it to be. *In nuce*, Buddhist philosophers advocate a very sparse ontology and truth in the context of ultimate truth/reality. If someone insists that something ultimately exists and there are ultimate truths about them, all Buddhist philosophers can do is . . . (being silent), since, for them, there is nothing that can be said about it.

The above discussion may show that the Buddhist philosophers who advocate the method of paragraph as a solution to *āśrayāsiddha* can be quietists and be consistent with their other commitments. However, it is not clear that it fares well as a way to avoid the fallacy of *āśrayāsiddha*. This is because there is no guarantee that the opponent's thesis can *always* be paraphrased. If someone makes up an entity and claims that it really exists, there may not be anything in their conventional reality that they can appeal to in order to paraphrase it. Paraphrasing is, thus, not a guaranteed method of avoiding the fallacy of *āśrayāsiddha*.[15]

The method of introducing conceptual objects, on the other hand, can guarantee that there is an entity that the Buddhist can talk about and, thus, that the fallacy of *āśrayāsiddha* can be avoided. For whatever the subject matter the proponent's thesis is about, one can posit a conceptual object that corresponds to it.

[12] See a quietist treatment of Madhyamaka in Tillemans (2016).

[13] '*Satya*' can be rendered as 'truth' or 'reality' depending on the context.

[14] For an introduction to the two truths, see, for instance, the Cowherds (2011).

[15] Dignāga and Dharmakīrti seem to assume that they can always paraphrase their opponents' theses. To be fair, they and other Buddhist philosophers have developed enough conventional resources to paraphrase all kinds of theses about the things that are crucial to Buddhist philosophers and non-Buddhist philosophers. I do not see, however, that there is a way of showing that paraphrasing as a general method is always available.

So, if the issue is only about *āśrayāsiddha*, the method of positing conceptual objects may be a better way to deal with the issue. However, it essentially reintroduces the entities that the Buddhist wants to avoid. Subscribing this method is, thus, like Occam (or Ockham) multiplying the number of entities in order to demonstrate the law of parsimony: do not multiply entities beyond necessity. In order to show that the Buddhist can entertain the proponent's thesis and that they are, thus, entitled to reject it, they are reintroducing more entities albeit of a different kind. How did this come about? As we will see in the next section, it was Dharmakīrti who planted the seeds for this development.

5. Avoiding the Fallacy of *Āśrayāsiddha*, Part II

As we saw above, in the context of talking about the proponent's own subject matter (*svadharmin*), Dharmakīrti appeals to the method of paraphrase.[16] However, he provided two strands of thought that, when they are put together, might be taken to imply the method of conceptual object as the solution to the fallacy of *āśrayāsiddha*. This seems to be what some Buddhist philosophers who came after Dharmakīrti did. In this section, I will present how the method of conceptual object became the main solution to *āśrayāsiddha* after Dharmakīrti. I will then present the problems with the method as identified in the tradition.

(1) In commenting on Dignāga's discussion of *svadharmin* (the proponent's intended subject), Dharmakīrti distinguishes two kinds of a thesis' subject: the subject actually intended by the proponent (*svadharmin*) and the subject which is 'unrelated, isolated' (*kevala*) (*Pramāṇavārttika* IV k. 136–48). The subject which is 'unrelated, isolated' is a 'nominal' subject (using the gloss Tibetans often give to *kevala*) in the sense that it is a subject that can be talked about even though it is not the actual subject (Tillemans 1999: 172–3).

(2) Dharmakīrti came up with the principle that a word in the subject place of a sentence always signifies a conceptual representation (*kalpanā*) (e.g. PV I k. 205–12). He then applied this principle to the case of Primordial Matter and claimed that while Primordial Matter did not exist, the object of the word did exist as a conceptual object (Tillemans 1999: 175–6).

Buddhist philosophers who came after Dharmakīrti combined these two strands of thought in the following way. First, in the context of (1), two of the prominent commentators of Dharmakīrti's *Pramāṇavārttika*, Devendrabuddhi (630–90 CE) and Śākyabuddhi (660–720), explain that the subject of a thesis which the proponent takes as existent but the Buddhist takes as non-existent is

[16] See also Prajñākaragupta's (750–810 CE) commentary on Dharmakīrti's *Pramāṇavārttika* IV k.141–2 in his *Pramāṇavārttikabhāṣya*. A translation of the relevant passage can be found in Tillemans (1999: 177–8).

a nominal subject (or *kevala* (unrelated or isolated)). In this way, they help themselves to talk about the subject even though it is non-existent.

Śākyabuddhi and Devendrabuddhi recognize that they cannot simply stipulate such a subject to be nominal. The problem for them (and for most Buddhist philosophers) is that they are global error theorists. Being global error theorists means that they do not think that properties can be attributed to anything, whether existent or non-existent, as they are not factual matters as discussed before. How can they show that it is false that Primordial Matter exists, for instance?

They do this by characterizing properties (and reasons given for a thesis) to be mere exclusions (*vyavacchedamātra*). Śākyabuddhi (though not Devendrabuddhi) explains mere exclusions to be non-implicative negations (*prasajyapratiṣedha*). A negation is non-implicative if it does not entail anything positive. For instance, the negation involved in the statement "There are not any flowers that grow in the sky" is non-implicative as it does not imply anything that grows in the sky. These negations are invoked commonly in Buddhist philosophy as a way of legitimizing the lack of any positive commitment. In contrast, a negation is implicative if it does entail something positive. For instance, the negation involved in 'The rose is not red' is implicative as it implies that the rose has some other colour.[17] So mere exclusions mean that there is nothing that is implied, stated or presupposed (Śākyabuddhi's *Pramāṇavārttikaṭīkā* D.269a4–5.) Dharmakīrti's commentators, in particular Śākyabuddhi, can then claim that the proponent's thesis can be denied without implying anything positive. And, because nothing positive is asserted by denying or negating the opponent's thesis, they do not face the fallacy of *āśrayāsiddha* (Tillemans 1999: 173).

Second, the later Buddhist philosophers have often wheeled in Dharmakīrti's principle (2) about conceptual objects as the signifiers of words generously and applied it to an understanding of the proponent's own intended subject matter (*svadharmin*). In particular, they took the introduction of conceptual objects as the main route to avoid *āśrayāsiddha*. Prajñākaragupta (750–810 CE) and Kamalaśīla (740–95) as well as Tsong kha pa (1357–1419), Śākya mchog ldan (1428–1507) and other Tibetans further developed on the invocation of non-implicative negations (*prasajyapratiṣedha*) and claimed that the proponent's intended subject (identified as *kevaladharmin*) is what the proponent takes to be real and that the Buddhist's intended subject (*svadharmin*) is not just a nominal subject but the conceptual object representing that non-existent entity. Since the subject is a conceptual object which may not have any corresponding existent entity, the Buddhist can then say that the proponent's intended subject is non-existent. And because the negation involved in the rejection of the proponent's

[17] For the contrast between non-implicative negations (*prasajyapratiṣedha*) and implicative negations (*paryudāsapratiṣedha*), see Kajiyama (1973).

thesis is non-implicative, no existence of any entity or property is committed. By identifying the subject matter as conceptual object, the danger of the fallacy of *āśrayāsiddha* can be avoided (Tillemans 1999: 173–4). This is how the method of conceptual object became the main solution to *āśrayāsiddha* after Dharmakīrti.

6. Difficulties of Positing Conceptual Objects

Is the method of conceptual objects a better solution than the method of paraphrase? By positing conceptual objects in place of the proponent's subject, the Buddhist can 'guarantee' (though it is basically a guarantee by stipulation) that the fallacy of *āśrayāsiddha* can always be avoided. The method of paraphrase favoured by Dignāga and Dharmakīrti lacks such guarantee as there is no guarantee that paraphrase is always available. However, introducing conceptual entities and inflating one's ontology straight after deflating it cannot be free of any difficulties. We can see some of the difficulties and complicated attempted solutions in the writing of the Mongolian scholar writing in Tibetan, A lag sha ngag dbang bstan dar (or Ngag dbang bstan dar) (1759–1840 CE), in particular in his *bCig du bral gyi rnam bzhag*.[18]

Ngag dbang bstan dar shows that the use of two types of negations, implicative and non-implicative negations, does not clearly distinguish between the cases of subject failure which are harmless because they are in the context of the non-implicative negations and genuinely fallacious *āśrayāsiddha* which is problematic.[19] First, he shows that the negation involved in a thesis does not have to be non-implicative in order to avoid *āśrayāsiddha*. He uses the following example to show this: "Take as the subject, a rabbit's horn; it is fitting to be designated by the word 'moon,' because it exists as an object of conceptual thought" (Tillemans and Lopez 1998: 102).[20] The property of being fit to be designated by the word 'moon' is a positive 'entity'. So, the property attributed to the subject, a rabbit's horn, is a positive 'entity' even though the subject is non-existent. Thus, the property that predicates a non-existent subject does not have to be a mere exclusion (*vyavacchedamātra*); it can be a positive 'entity' or an implicative negation. Second, he shows that the property attributed to the subject and the reason given for the attribution being non-implicative negations does not show that the fallacy of *āśrayāsiddha* is avoided. For instance, consider

[18] The section on *āśrayāsiddha* of this text has been translated in Tillemans and Lopez (1998). The discussion below will follow their translation and their extensive explanatory notes.

[19] For the development of the Buddhist discussions about *āśrayāsiddha*, see also Klein (1991), Kobayashi (1989), and Lopez (1987).

[20] Ngag dbang bstan dar attributes this example to 'Jam dbhyangs bzhad pa'i rdo rje (1648–1721/2 CE). However, the example cannot be found in the works of 'Jam dbhyangs bzhad pa'i rdo rje. See Tillemans and Lopez (1998: 117 n. 9).

proving 'that [something nonexistent like a rabbit's horn] is the subtle selflessness of the elements by means of the reason, 'being the consummate [nature]'. The subject, a rabbit's horn, is non-existent even though the reason (i.e., being the consummate nature) and the property to be proved (i.e., being the subtle selflessness of the elements) are non-implicative negations. Using examples like this, Ngag dbang bstan dar concludes that the fallacy of *āśrayāsiddha* is avoided *not* when the reason and the property to be proved are non-implicative negations but when the reason and the property do not imply existence (Tillemans and Lopez (1998: 102). That is, subject failure is problematic not necessarily when the subject is non-existent but when the properties attributed to the subject imply existence. When those properties do not imply existence, subject failure is not a problem (Tillemans and Lopez 1998: n. 11).

Finally, Ngag dbang bstan dar points out that the conceptual object approach to *āśrayāsiddha* is no better than the paraphrasing approach when it comes to the issue of changing the subject. As we saw before, the adversary can legitimately complain that the Buddhist is changing the subject when the Buddhist paraphrases their thesis. Ngag dbang bstan dar points out that the same problem arises with the method of introducing conceptual objects. For instance, for the non-Buddhist Vaiśeṣika, sound is permanent. The Buddhist rejects this and wants to argue that sound is impermanent. The thesis the Buddhist wants to assert has a subject that is non-existent. The Buddhist method of conceptual objects stipulates that sound is a conceptual object;[21] it is not sound itself but "what appears as sound to conceptual thought... a real entity (*dngos po*) that is independent (*rang dbang ba*) and is a positive phenomenon (*sgrub pa*)" (Tillemans and Lopez 1998: 104). Ngag dbang bstan dar implies that the Buddhist and the Vaiśeṣika are talking past each other. So the method of conceptual object faces the same difficulty as the method of paraphrasing.

This difficulty is something to which Ngag dbang bstan dar himself responds. He claims that "a mere object grasped by the auditive consciousness" concordantly appears to both parties upon hearing the word 'sound' (Tillemans and Lopez 1998: 105). When the Buddhist and the Vaiśeṣika argue about sound (or space which attracts a parallel argumentation) or when the Buddhist and the Sāṃkhya argue about Primordial Matter, they are arguing about a mere verbal designation or a verbal object (*sgra don*). The Vaiśeṣika takes sound as more than just a verbal object: it is permanent and fully real and the Sāṃkhya takes Primordial Matter as

[21] There is a complicated story as to what exactly it means to say that sound is a conceptual object in terms of *apoha* (exclusion). Since the introduction of *apoha* does not add anything substantive in the present context, I refrain from spelling it all out. For discussions on *apoha*, see Siderits, Tillemans, and Chakrabarti (2011).

more than just a verbal object, i.e. fully real. Nevertheless, there are 'concordantly appearing subjects' (*chos can mthun snang ba*) for both parties.[22]

What is crucial here is the distinction between 'the exclusion *qua* thing itself (*rang ldog*) and the exclusion *qua* basis [for the thing] (*gzhi ldog*)' (Tillemans and Lopez 1998: 106). The distinction is between the thing itself and an instance of it under a description. Ngag dbang bstan dar uses this distinction to argue that the Buddhist and their opponent are both arguing about a verbal object, an object under some description, even though neither party recognizes it as such (Tillemans and Lopez 1998: 125 n. 30). He then concludes that there is no problem in changing the subject. Both parties are talking about the same thing. It is just that the opponent wrongly thinks that the object in question is something more than this and thinks that it is fully real.

Is this a plausible solution? In order to deal with a problem that arose by the introduction of conceptual objects, Ngag dbang bstan dar appeals to yet another conceptual apparatus: concordantly appearing subjects. Perhaps, constantly introducing new conceptual apparatus and continuously keeping the conceptual realm fine-grained, the Buddhist can come to prove that non-existent things do not exist. But how rich our conceptual life would have to be to prove that the self, a non-existent entity, does not exist? I will leave this question unanswered.

7. Conclusion

How can Buddhists prove that non-existent things do not exist? With great difficulty. For the Buddhist, this is not a laughing matter as they are largely global error theorists and, thus, many things are non-existent. The difficulty gets compounded as the Buddhist and their opponent, the non-Buddhist of various kinds, both agree that one cannot prove a thesis whose subject is non-existent. Buddhist philosophers have developed mainly two strategies to avoid this difficulty: the method of paraphrase and the method of conceptual objects. Early Buddhist philosophers proposed the method of paraphrase as the solution to the difficulty of talking about non-existent things. This proposal, however, was not taken up by the later Buddhist philosophers. What became the main approach to argue about non-existent things is the method of conceptual objects. But this strategy imports all of the problems associated with a separate issue into what is already a minefield. Rather than addressing the question of how to argue about non-existence, the later Buddhist philosophers have, in addition, taken themselves to

[22] The notion of 'concordantly appearing subject' seems to be a Tibetan development, though there is an Indian precedence of problematizing the lack of commonly acknowledged (*ubhayaprasiddha*) subjects in debates. For the development of 'concordantly appearing subject', see Lopez (1987), Hopkins (1989), and Tillemans (1990).

be tasked with explaining why it is not puzzling to bring non-existence into existence of some sort. As we saw, that multiplied the problems requiring them to come up with ingenious attempts to solve the difficulties. What would the history of Buddhist philosophy look like if the method of paraphrase was further developed and adopted widely? We would never know. That history is non-existent, and that is where we have to be silent.

References

Brock, Stuart (2002). "Fictionalism about Fictional Characters", *Noûs* **36** (1): 1–21.

Cowherds, The (2011). *Moonshadows: Conventional Truth in Buddhist Philosophy* (New York: Oxford University Press).

Hopkins, Jeffery (1989). "A Tibetan Delineation of Different Views of Emptiness in the Indian Middle Way School", *Tibet Journal* **14** (1): 10–43.

Joyce, Richard (2016). "Moral Anti-Realism", *Stanford Encyclopedia of Philosophy*, ed. Edward N. Zalta (Winter 2016 edn).

Kajiyama, Yuichi (1973). "Three Kinds of Affirmation and Two Kinds of Negation in Buddhist Philosophy", *Wiener Zeitschrift für die Kunde Südasiens* **17**: 161–75.

Katsura, Shoryu (1992). "Dignāga and Dharmakīrti on *adarśanamātra* and *anupalabdhi*", *Asiatische Studien/Etudes Asiatiques* **46** (1): 222–31.

Keira, Ryusei (2004). *Mādhyamika and Epistemology* (Vienna: Arbeitskreis für tibetische und buddhistische Studien Universität Wien).

Klein, Anne Carolyn (1991). *Knowing, Naming and Negation: A Sourcebook on Tibetan Sautrāntika* (Ithaca, NY: Snow Lion Publications).

Kobayashi, M., (1989). "The Mādhyamika Argument for *niḥsvabhāvatā* and the Fallacy of *āśrayāsiddha*: Kamalaśīla's View in the *Madhyamakāloka*", *Bunka* **50**: 218–99.

Lopez, Donald S., Jr. (1987). *A Study of Svātantrika* (Ithaca, NY: Snow Lion Publications).

Matilal, Bimal Krishna (1970). "Reference and Existence in Nyāya and Buddhist Logic", *Journal of Indian Philosophy* **1** (1): 83–110.

Patil, Paramil G. (2009). *Against a Hindu God: Buddhist Philosophy of Religion in India* (New York: Columbia University Press).

Siderits, Mark, Tom J. F. Tillemans, and Arindam Chakrabarti (eds) (2011). *Apoha: Buddhist Nominalism and Human Cognition* (New York: Columbia University Press).

Tillemans, Tom J. F. (1990). *Materials for the Study of Āryadeva, Dharmapāla and Candrakīrti* (Vienna: Arbeitskreis für tibetische und buddhistische Studien Universität Wien).

Tillemans, Tom J. F. (1999). *Scripture, Logic, Language: Essays on Dharmakīrti and His Tibetan Successors* (Somerville, MA: Wisdom Publications).

Tillemans, Tom J. F. (2016). *How Do Mādhyamikas Think?* (Somerville, MA: Wisdom Publications).

Tillemans, Tom J. F. and Donald S. Lopez Jr. (1998). "What Can One Reasonably Say About Nonexistence? A Tibetan Work on the Problem of *Āśrayāsiddha*", *Journal of Indian Philosophy* **26**: 99–129.

Tanaka, Koji (2013). "Buddhist Philosophy of Logic", in *A Companion to Buddhist Philosophy*, ed. Steven M. Emmanuel (Chichester: Wiley-Blackwell), pp. 320–30.

Tanaka, Koji (forthcoming). "Buddhist Shipping Containers", in *Reasons and Empty Persons*, ed. Christian Coseru (Dordrecht: Springer).

7
How Ordinary Objects Fit into Reality

Bryan Frances

An ordinary entity such as a tree *boils down to* a great many tiny things, such as molecules or atoms. But there is a problem. Suppose you have an electron, a proton, and a tree in front of you, and pretend for the moment that electrons and quarks are simples and protons are unified groups of three quarks. The electron is a single simple; 'is an electron' is directly true of the simple. The proton is a unified plurality of three simples; 'is a proton' is directly true of the plurality, or extended temporal part thereof. But the tree isn't either a simple or a plurality of simples; 'is a tree' isn't directly true of either a simple or a plurality of simples. Wherever you have a tree, you have zillions of pluralities of tiny things that are equally good candidates for being a tree—but there is just one tree there. So, it appears that none of those groups of tiny things is literally a tree even if they are arranged in a tree-like fashion. But if the material universe has big things that in some sense "boil down to" little things, then how do trees fit in?

In this paper I sketch a novel theory, Plurality Pointillism (PP), according to which 'is a tree' applies to reality in a way different from how 'is an electron' and 'is a proton' apply. The theory contains a whiff of the idea that trees are "less real" than the tiny entities they boil down to. Roughly put, although they don't have the non-being that is often thought to apply to holes or shadows or cracks, they aren't as substantial as electrons or protons either (given our pretenses about simples).

My theory has several intriguing features. First, if trees and other ordinary things exist, then although each one boils down to pluralities of pluralities of tiny things, it isn't identical, or even "identical-light", to any plurality of tiny dots, any plurality of pluralities of tiny dots, etc. Second, dots are just small things that in some sense big things boil down to: on my theory dots need *not* be simples and there need not be any simples at all. Third, the notions of parthood and thus composition are derivative of a trio of notions, to be elaborated. Fourth, the reduction holds regardless of the extent of composition (always, never, sometimes). The notion 'boil down to' need not involve either parthood or composition. Fifth, parthood and composition are revealed to be much less philosophically important than metaphysicians have thought. Sixth, as hinted at above, one of my theory's intriguing features is the thesis that 'is an F' can apply to reality even though no *entity* is F, in one natural sense of 'entity'. Seventh, our

Bryan Frances, *How Ordinary Objects Fit into Reality* In: *Non-Being: New Essays on the Metaphysics of Non-Existence*.
Edited by: Sara Bernstein and Tyron Goldschmidt, Oxford University Press (2021). © Bryan Frances.
DOI: 10.1093/oso/9780198846222.003.0007

commonsensical claims about reality can be true even if there are no entities at all. We may live in a world of pure smoke even though there are trees, people, etc.

1. Pluralities

In order to see how trees fit into reality, and analyze parthood and composition, we need to go over some basics about pluralities. I'm using 'plurality' in the familiar way so that *what it is* for "the plurality of x1 and x2" to exist is exactly this: x1 exists, x2 exists, and x1 $\neq$ x2. The plurality of x1 and x2 is *two* things, and each thing can be either concrete or non-concrete. An object is an "element" of plurality p $=_{df}$ it's one of p. I don't offer any analysis of 'x is one of y'.

A plurality of wholly physical objects is a wholly physical object—albeit a plural object. Hence, it's not a *set*, if a set is something neither spatial nor temporal.

I'm not saying that the two things x1 and x2 are *parts* of the plurality of x1 and x2. We have yet to say anything about parthood, composition, or fusion. Everything I say about pluralities is consistent with the thesis that there are no proper parts whatsoever. The notion of fusion is never needed for any purpose.

The notion of there being many individual pluralities is familiar. There are the red chairs in room 101; that's one plurality (assuming there are chairs). There are the blue chairs in 101; that's another plurality. Each plurality is *a* plurality; the use of 'a' indicates an individual, single thing—although the thing in question is a plurality. Or consider how physicists talk about entangled pairs in quantum theory: 'Here is an entangled *singlet* pair of electrons', etc.

Some authors use 'plurality' in such a way that even a single object falls into the extension of 'plurality'; they do not adopt my definition that restricts 'plurality' to groups of at least two things. In response, I agree that one thing can be a plurality. In fact, in the previous paragraph I claimed that for every x, if x is a plurality, then x is one thing (and, again, this is not to say that x is a composite; we have yet to say anything about composition). But under the assumption that there are mereological simples, I reject the claim that for every x, if x is one thing, then x is a plurality. For simple s, none of these are true: 's is a plurality', 's is a multiplicity', 's is a group', etc. There might be some theoretical conveniences associated with using 'plurality' in such a way that s falls into the extension, but theoretical convenience has to be handled carefully. For comparison, it might be theoretically convenient to use 'parent' so that someone with zero or more children is a "parent". But 'Everyone is a parent' is false. To get the convenience we use 'Everyone is a parent*', with an appropriate definition. Similarly, although 'Mereological simple s is a plurality' is false (just as 'John is a parent with no children' is false), with the appropriate definition 's is a plurality*' can be theoretically convenient and true.

There are many interesting issues to explore regarding the ontology of pluralities as well as the linguistic characteristics of strings such as 'the plurality of marbles' and 'the marbles in the corner'. In the interests of brevity and focus I aim to avoid as many of these issues as possible.

I will not need to say anything about plural quantification either. We will see in the following section that there may be possible worlds in which there are multiple physical objects but no physical non-pluralities. When we talk about the individual physical aspects of such a world, our quantificatory uses of 'there is' and 'there are' range over pluralities alone, since there is nothing else physical in that world for them to range over. Hence, quantificatory uses of 'there is' can range over pluralities; they merely grab them *as singular objects*, as discussed above.

2. Categories of Pluralities

On PP, the existence of trees "comes from" pluralities of pluralities. But the relation is stranger than expected. In order to see this, we need to make some distinctions amongst pluralities. We will use them to show that trees aren't pluralities (in section 4) and then articulate the sense in which trees fit into reality (section 5).

There are exactly three logical possibilities for a given plurality P: P is a plurality of non-pluralities alone, P is a plurality of pluralities alone, or P is a mixed case, with some elements pluralities and some non-pluralities. If even one of P's elements E is a plurality, we can ask whether E's elements are all pluralities, all non-pluralities, or some of both. The question repeats each time there is a plurality: are its elements all pluralities, all non-pluralities, or some of both?

For any plurality P, then, P may or may not *bottom out*, which happens when and only when either all of P's elements are non-pluralities or the above process of questioning always ends with the answer 'all the elements are non-pluralities'. For instance, consider the three-element plurality P_N of (1) the plurality of positive integers, (2) the number 0, and (3) the plurality of negative integers. Assume for illustration that integers are non-pluralities. Then P_N's bottom consists of all the integers, all of which are non-pluralities, but P_N has three elements, two of which are pluralities (viz. elements (1) and (3)).

Let's say that each element of plurality P is a *1-element of P*, each element of each element of P is a *2-element of P*, each element of each element of each element of P is a *3-element of P*, et cetera; and X is an *or-element of plurality P* $=_{df}$ X is either a 1-element of P, or a 2-element of P, or a 3-element of P, etc. P bottoms out if and only if there is an integer m such that there are no m-elements of P (which entails that there are no (m + n)-elements for positive integer n, since there can't be an (m + 1)-element without an m-element). So, P does *not* bottom out just in case, intuitively, the cascade of elements goes on infinitely, without coming to an end.

If P bottoms out, we say that its *bottom* is nothing but the non-pluralities revealed through the questioning process. This is not to say that all its or-elements are non-pluralities. For instance, the three-element plurality P_N a couple paragraphs back has a bottom (an infinite one) but has two or-elements that are not non-pluralities.

For some purposes, we consider only pluralities that do not have any "repeats" amongst their or-elements: each thing can "show up" in the plurality only once (as a single 1-element, a single 2-element, etc.). For instance, the plurality of Tom, Dick, Harry, and Harry has repeats, as does the plurality of Tom, Dick, and the plurality of Harry and Tom. Pluralities that have a decent chance at being identical with ordinary physical objects have no repeats.

If P does not bottom out, and no object shows up more than once in P (that's the "no repeats" clause), then the plurality of all the or-elements of P has infinitely many elements (since the cascade is infinite and has no repeats). But the plurality of all the or-elements of P may be infinite even if P *does* bottom out, provided that for some positive integer m, there are infinitely many m-elements of P. For instance, the three-element plurality P_N a couple paragraphs back has infinitely many or-elements but it bottoms out as well. Therefore, 'P doesn't bottom out' entails 'P has infinitely many or-elements' (assuming no repeats), but not vice versa. Our definitions:

X is a ***bottom plurality, BP*** $=_{df}$ X is a plurality with a bottom (of non-pluralities, but that's redundant).

X is a ***pure BP*** $=_{df}$ there are some non-pluralities such that X is the plurality of just them.

X is an ***impure BP*** $=_{df}$ X is a BP with at least one plurality as an element. Each impure BP P1 corresponds to a pure BP P2: P2 is the plurality of non-pluralities that are P1's bottom.

X is a ***bottomless plurality, BLP*** $=_{df}$ X is a plurality with no bottom.

X is a ***pure BLP*** $=_{df}$ X is a plurality each of whose or-elements is a plurality. So, each element of X is a plurality, each element of each element of X is a plurality, etc. It is pluralities "all the way down".

X is an ***impure BLP*** $=_{df}$ X is a pure BLP with the sole exception of having at least one or-element that is a non-plurality. Every impure BLP is a pure BLP with at least one non-plurality added and nothing subtracted.

You might think BLPs are metaphysically impossible. I address that issue below.

These definitions are more complicated than they may appear because identity for pluralities is puzzling. Suppose integers exist and ask whether the following pluralities are mutually distinct:

P1: the plurality of 1, 2, and the plurality of 3, 4, and 5
P2: the plurality of 1, 2, 3, and the plurality of 4 and 5
P3: the plurality of 1, 3, 5 and the plurality of 2 and 4

It might be thought that in general, pluralities A and B are identical just in case for any x, x is one of A iff x is one of B. This conception of plurality identity doesn't answer our question about the identity relations amongst P1–P3, however, because for example it isn't clear whether the number 4 is *one of* each of P1–P3. If the plurality of 4 and 5 *just is* the numbers 4 and 5—which it is, right?—then it appears as though P2 has five elements. One might object to that result by saying that P2 has *four* elements: 1, 2, 3, and the *single* object that is the plurality of 4 and 5. That's right! But one should be forgiven for thinking that that "single" object *also* is two objects. In the description 'the plurality of 1, 2, 3, and the plurality of 4 and 5', the term 'the plurality of 4 and 5' picks out two things; so, the description 'the plurality of 1, 2, 3, and the plurality of 4 and 5' is a description of a plurality of five things. The plurality of 4 and 5 is both one thing—one plurality—and two things, 4 and 5.

Or so it might be thought! I'm not taking a stand on that matter, although it will come up again in §4.

If P1–P3 are really the very same plurality, then all impure BPs are pure BPs, so that the two categories aren't mutually exclusive, contrary to the impression one might get upon reading the definitions. To see this explicitly, consider P2, which has as an element the plurality of 4 and 5. Because of that, it is impure. But if P1–P3 are identical, then P2 is identical with the plurality of 1–5. Hence, it is pure as well, assuming for illustration that integers aren't pluralities. Similarly, although P_N has *exactly three* 1-elements, it also has *infinitely* many 1-elements—and since there is no contradiction there, we need to be more careful in understanding the logical form of the relevant sentences.

I won't comment on or make assumptions regarding this issue of the distinctness of P1–P3, since it won't matter for my purposes. No matter where one stands on that issue, we can agree that each of P1–P3 *boils down to* the plurality of 1–5, even if we puzzle over how to precisely understand 'boil down to'. In general, pluralities A and B bottom out into the same non-pluralities means that they bottom out into *the plurality of* those entities.

In order to get a firmer handle on BLPs, consider object D.

> First you see what looks like a dot D, with nothing whatsoever around it. It looks like it's a single, unified, non-plural, thing floating all by itself in empty space. Then you look closer and D starts to look fuzzy around the edges as well as throughout the inside. Then you look even closer, radically shrinking yourself in size, and you realize what appeared to be a dot is really just an enormous number of much smaller dots with lots of space between them (a bit like our solar system). So the big dot D has "dissolved" into a great number of smaller scattered dots.
>
> Next, you zoom in on one of the smaller dots that are included in the D plurality you just discovered and the same thing happens: first it's fuzzy, and then you see that it's really just an enormous number of tiny dots. And then it happens again,

> for *every single dot*: every time you focus in on what seems to be a dot you realize that it's just a bunch of much, much smaller "dots", all of which dissolve into further "dots", all of which dissolve into further "dots", ad infinitum. That is, there is no bottom: it's *infinitely* descending. So we have concluded that D is a plurality of just pluralities of just pluralities of.... But there is no bottom.

Objects like D are *infinitely descending all-pluralities*: pluralities of just pluralities of just pluralities of just pluralities... with no end. D is a pure BLP.

A universe W in which *everything* is like D is one in which there are no non-pluralities.[1] A universe of nothing other than pure BLPs is one without building blocks that fail to have building blocks. That is, that universe *does* have building blocks, but all those building blocks have building blocks, the latter building blocks have their own building blocks, etc. For comparison, philosophers have wondered whether there are possible worlds in which every object is *gunky*: it has proper parts, each of whose parts has proper parts, each of those parts have proper parts, et cetera (Lewis 1991). But a pure BLP world is potentially different, for I have yet to say anything about parthood: I have not said—and it's hardly obvious—that pure BLPs have parts. Suppose that both of the following are true in the world W of my thought experiment involving D: (i) W is metaphysically possible with respect to our world, and (ii) in W none of the dots (which exhaust the physical occupants of W) are unified in any way strong enough for parthood to occur, so in W parthood never happens at all in the physical world (set aside any non-physical objects in W). It follows from (i)–(ii) that it's metaphysically possible for there to be *a physically non-empty world in which there are no physical composites or physical simples*, granting the plausible assumption that it's necessary that simples aren't pluralities (where, again, pluralities are groups of at least two things).[2]

Set aside the secondary point about composition. One might think that BLPs are metaphysically impossible because infinitely descending pluralities are impossible. However, by my lights anyway there is no good evidence against the claim 'It is metaphysically possible that there are material objects but they don't *all* boil down to nothing but non-pluralities'. We should *make room* for the metaphysical possibility of BLPs in our ontology. In what follows I will assume that such a world is possible and needs to be accounted for in the true modally strong story of parthood and composition.

[1] Chen (forthcoming) examines the idea that space is made of infinitesimal gunk. A pure BLP world is different: it's silent on the structure of space or the extent of parthood. But it's similar in suggesting that part of reality "keeps dividing infinitely", so to speak.

[2] Then again, if universalism about composition is false, then a plurality of two simples does seem to lack parts if they have no interesting physical relation to one another (e.g. suppose they are electrons forever separated by a billion light years). So maybe such pluralities are mereological simples, contrary to what was asserted above. Even so, they aren't simples in the sense indicated in the next section and used throughout.

A universe might contain *both* BPs and BLPs; it's not as though a universe can't have both. Object x1 boils down to non-pluralities alone, so it's a BP. Object x2 is exactly like D above: a pure BLP. Object x3 is just like D but with this exception: while looking into its innards one occasionally comes across a non-plurality even though the descent of pluralities goes on infinitely. It's an impure BLP.

3. Non-Pluralities

As far as non-pluralities go, the plurality pointillist holds that some of them are pretty much what you pictured in your mind when you pictured a mereological simple. The *definition* of 'non-plurality' is, of course, parasitic on the definition of 'plurality': x is a non-plurality $=_{df}$ x exists but is not more than one thing. The *intuitive* notion of a simple is this: something that does not have or involve or include, in any reasonable ontological sense, any multiplicity of entities; when you dig anywhere into its innards—if it has any spatial or temporal innards to dig into—you will discover no more entities; it's a bit of homogeneous goo when it comes to entity-hood. As I said, that's not a definition. It's a description of the *picture* we have in our minds when thinking about parthood and multiplicity. I take this intuitive notion as a conceptual primitive and use 'simple' to indicate non-pluralities that fit that picture. I do not use 'is a part of' in the characterization; parthood comes later. In the next section I argue that trees are non-pluralities but are not simples either since they obviously include multiple entities. Hence, not all non-pluralities are simples.

4. What Trees Are Not

Suppose 'There is an F' is true. There are exactly two options for the things that make it true: a non-plurality that's F, or a plurality that is collectively F. Some philosophers have argued against the many–one identity of (say) a tree and a plurality. Those arguments are controversial, to say the least: cf. Wallace (2011a; 2011b), and the papers in Baxter and Cotnoir (2014). In this section I give *original* reasons to think that no plurality satisfies 'is a tree'. The only remaining, and thus correct, option—that a non-plurality satisfies 'is a tree'—will be addressed in the next section.

Suppose that one thinks that plurality P is an excellent candidate for being identical to the tree in my backyard, T. So, one is thinking that P might be *identical* to T. The main reason one thinks P is T, presumably, is that one thinks T and P are materially coincident in this sense: the spatio-temporal volume and material collectively taken up by the or-elements of P is precisely the same as that for T. In what follows I will rule out, in order, the identity of T with a plurality from any of the four exhaustive categories of pluralities: (1) impure BPs, (2) BLPs, (3) pure BPs that bottom out in just simples (call them *pure & simple BPs*), and (4) pure BPs that do not bottom out in just simples (so their bottoms contain at least some non-pluralities that aren't simples). If my arguments are sound, then tree T is not a plurality.

Option (1): T = P and P is an impure BP. Since P is a BP, P has a bottom, which by definition is a plurality P* of just non-pluralities. P* is a pure BP, by definition. Hence, corresponding to the impure BP P there is a pure BP P* that is materially coincident with P in the sense articulated in the previous paragraph. So which is it: $P = P^*$ or $P \neq P^*$, using '=' for strict numerical identity? (Recall that we saw earlier, with P1–P3, that there are reasons for thinking any impure BP is identical with a pure BP.)

If $P = P^*$, then the supposition that T is identical with an impure BP is equivalent to the supposition that T is identical with a pure BP. I will examine that possibility below (since it's covered by the disjunction of options (3) and (4), which treat all pure BPs).

If $P \neq P*$, then P and P* are exactly equally good candidates for being identical with T, at least when it comes to spatio-temporal and material fit. They take up the same material and spatio-temporal region. I don't see how T could be identical to P* but not P, given the similarities of P* and P. For the sake of illustration, suppose P is the plurality of these 1-elements:

$$s_1, s_2, \ldots, s_N, \text{and the plurality } s_{N+1} - s_{N+5}$$

where N is around 10^{30}. Then P* is the plurality of these 1-elements:

$$s_1 - s_{N+5}$$

So P has $10^{30} + 1$ 1-elements whereas P* has $10^{30} + 5$ 1-elements *but they have the same bottom*, which is just P*. I just don't see how a particular tree could be identical with just one of those pluralities, although I have to admit I can't think of a good supporting argument. The advocate of PP concludes that it's not the case that $T = P$ while $P \neq P*$. So, if $T = P$, then $P = P*$. That is, if T is an impure BP, then it's a pure BP; T is a pure BP if it's a BP at all.

Options (2) & (3): T = P and P is either a BLP or a pure & simple BP. Any BLP that has a reasonable chance at being a tree will of course have infinitely many or-elements, since it will have no repeats. In addition, if a pure & simple BP—that is, a plurality of just simples—is going to have a decent chance at being a tree, it will have an enormous number of simples as or-elements. For instance, in a typical full-grown tree that you might find in someone's backyard you can find very roughly 10^{30} molecules (i.e. within a few orders of magnitude). So if you think plurality P—which we are now supposing is a BLP or a pure & simple BP—is T, then you will realize that P has an enormous number of or-elements.

Unfortunately, you will also realize that given the existence of enormously numbered P, there is an extremely similar plurality P* that is distinct from P but *just as good as* P as far as being an *excellent candidate* for "fitting" T perfectly. This is because of vagueness. For instance, P* might contain all of P plus one more electron or microscopic dot that is, intuitively, on the border of T. The fact that trees are extremely large compared to chemical atoms, and trees are "vaguely composed", guarantees the existence of P* given the existence of P. Next, you realize that given the existence of P*, there's no reason to think P but not P* is identical to T. And yet, T cannot be strictly numerically identical with both of them because by the transitivity of strict numerical identity, then P would be strictly numerically identical with P*, which it isn't (by design). In sum:

a) For all x, if x is a pure & simple BP, then there exists a y such that x ≠ y, y is a pure & simple BP, and y is as good a candidate—as far as we can ever tell—as x is for being identical with T.
b) For all x, if x is a BLP, then there exists a y such that x ≠ y, y is a BLP, and y is as good a candidate—as far as we can ever tell—as x is for being identical with T.
c) For all x and y, if x and y are pure & simple BPs, x ≠ y, and x and y are equally good candidates (as far as we can ever tell) for being identical with T, then x is identical with it iff y is identical with it.
d) For all x and y, if x and y are BLPs, x ≠ y, and x and y are equally good candidates (as far as we can ever tell) for being identical with T, then x is identical with it iff y is identical with it.
e) For all x, y, and z, if x ≠ y, then it's not the case that both x = z and y = z.
f) Hence, by (a)–(e), there is no x such that x is a pure & simple BP or BLP and x is identical to T.

In order to better understand the argument, think of a case in which a claim analogous to (a) is false. Take the definite description 'The most massive pure & simple BP in the lab apparatus' and assume that there are a great many pure & simple BPs in the lab apparatus, many of which include an enormous number of simples. Assuming that there are a great many almost entirely overlapping pure & simple BPs in the apparatus, there still is singular reference for 'The most massive pure & simple BP in the lab apparatus' because *even though there are many almost equally good reference candidates, the description has a term that singles one out as the best* (the term 'most massive'). Hence, there exists a pure & simple BP P such that for any other pure & simple BP P*, P* is not as good a candidate as P is for being identical with the most massive pure & simple BP in the lab apparatus. That is, (a) is false when we substitute 'the most massive pure & simple BP in the lab apparatus' in for 'the tree in my backyard' (and we adopt the assumption about the lab apparatus containing a great many pure & simple BPs each of which includes an enormous number of simples).

Matters are otherwise when you have the scenario of having, for a given definite description DD, not merely many *reasonably fitting* candidates for the referent of DD but many candidates that are *equally fitting*—which is what happens when DD doesn't include terms that eliminate all but one of the rival candidates. This is the case with 'The tree in my backyard' and reference candidates that are pure & simple BPs or BLPs: you don't have singular reference, so premise (a) is true. That is, in the vast sea of exceedingly similar treeish pure & simple BPs or BLPs in my backyard, the definite description 'The tree in my backyard' does not latch on to *just one* of them—and it doesn't do so because there is nothing in the description (such as 'most massive' in 'the most massive treeish pure & simple BP or BLP') that could eliminate all but one of the reasonably equally good rivals, which needs to be done for singular reference to occur.

Or so many philosophers will think upon reflection! I'm *not* saying they are right, or that they are wrong. Whether one agrees with them depends in part on one's views on meaning determination: cf. Horgan (1997) and Williamson (1997a; 1997b). One might think that contrary to the considerations of the previous paragraph, ordinary natural language is magically discriminating so that 'The tree in my backyard' refers to just one treeish pure & simple BP or BLP.

Premises (a) and (b) are epistemic and I doubt whether anyone, even epistemicists, will reject them: the use of 'as far as we can ever tell' is crucial. It's (c) and (d) that are controversial.

If the group of philosophers who accept (c) and (d) are correct, then my theory is thereby made more complicated. However, for the same reason I'm allowing for the metaphysical possibility of BLPs, I will also allow for the metaphysical possibility that natural language reference is not magically discriminating, so I will accept that in some metaphysically possible worlds (c) and (d) are true ((a), (b), and (e) are much less problematic). Below I will indicate what happens if (c) and (d) are metaphysically necessarily false. Call the conjunction of (c) and (d) ***Referential Sobriety***, since it is saying that reference for 'the tree in my backyard' isn't magical. Note that this claim is tied specifically to 'the tree in my backyard'; we get principles analogous to Referential Sobriety by substituting some other definite description in for 'the tree in my backyard'.

Thus, given Referential Sobriety the argument (a)–(f) shows that T isn't identical with any BLP or pure & simple BP. So much for options (2) and (3).

Option (4): T = P and P is a pure BP that doesn't bottom out in just simples. This means that P has at least one non-simple, non-plurality amongst its elements. In order to escape the considerations that ruled out BLPs and pure & simple BPs—considerations having to do with vagueness, meaning determination, and the enormous number of or-elements in candidate pluralities for T—P can't have an enormous number of elements. So what might P be, in order to meet the twin requirements of (i) being a pure BP that has at least one non-simple element and (ii) being a reasonable candidate for T?

One might think that P's elements could be the tree's leaves plus the piece/hunk of wood that consists of all the tree's wood. Pretend that there are just four leaves on the tree; so P has five elements total. If one hasn't thought too much about the case, this might seem to be the most obvious way a tree could be a plurality. But the leaves could hardly be non-pluralities when the tree itself is, by hypothesis, a plurality: surely they are on a par when it comes to being or not being a plurality. So this kind of P won't work. I have been unable to think of any plausible way of filling out option (4).

Hence, I have argued that the tree T is not a plurality. Clearly, it's not a simple either. So, if T exists, it's a non-plurality but not a simple.

5. The Foundational Theses of Plurality Pointillism

On the face of it, many possible universes contain trees each of which boils down to a great many tiny things, such as molecules and atoms. So those universes contain pluralities. They may or may not contain mereological simples. In comparison to 'There are trees' and 'There are pluralities', 'There are spatio-temporal mereological simples' is quite uncertain. On the face of it, a tree is intimately related to many pluralities of tiny things. We have seen that it isn't identical to any of them, assuming Referential Sobriety; neither is it a simple, mereological or not. So the question is: how do trees fit into reality?

At this point one could say that they belong to a whole new category: composites, where a composite is not a plurality even though it's intimately related to pluralities. PP says that no new category is needed. In addition, it seems obtuse to think there is a mysterious new category when surely we should be able to account for trees via 'x is a plurality' and 'x is one of plurality y'. When one looks at a tree, it's hard to see how it could be anything over and above pluralities of molecules. PP accepts that judgment and offers an account of trees and similar objects that (i) doesn't have them belong to a new ontological category, (ii) doesn't rely on, but makes room for, the questionable claim that there are simples, and (iii) doesn't rely on, but makes room for, the claim that there is anything other than pluralities (a BLP-only world).

The term 'is/are arranged treeish' is true, in the actual world, of certain pluralities of tiny dots (but we don't assume anything regarding what the dots are). The term 'is/are arranged treeish' applies to a *temporally extended* plurality. So two of the many dots that are collectively arranged treeish might exist years apart from one another, from opposite ends of the tree's life.

When 'There is *one* tree in my backyard' is true of a possible world, there are zillions of treeish pluralities in the appropriate backyard of that world. All those treeish pluralities are related to one another in a certain way—because there is a *single* tree there. For instance, any two of those pluralities are almost entirely "overlapping" in this specific sense: the collective material content and spatio-temporal region of the or-elements of treeish plurality p1 are virtually the same as that for the or-elements of treeish plurality p2. In order to express the idea that for a single tree the relevant treeish pluralities are intimately related to one another (with the overlapping, etc.), let's say that when there is a single tree in the backyard, there are many treeish pluralities in the backyard and they are *tree-unified*. They "belong to" the very same tree. That is, *the plurality of the treeish pluralities* is tree-unified. Three points on 'is/are arranged treeish':

First, the predicate 'is/are arranged treeish' is not the same as 'is/are arranged tree-wise' as van Inwagen uses it (1990: 105, 109). For him, a tree-wise plurality's elements are mutually simultaneous, whereas mine are temporally extended through the tree's existence.

Second, unlike some philosophers who employ terms similar to 'is/are arranged treeish', I am not eliminating trees or rocks or other ordinary things from our ontology. I will design the theory so that it's metaphysically possible that a world "just like ours" microphysically has no trees, but one could jettison that clause.

Third, as for what it takes for a temporally extended plurality of dots to satisfy 'is/are arranged treeish', I would advise against an armchair approach and advocate a fully scientific one. So I recommend we don't start with an armchair and frankly uninformative analysis that runs anything similar to "The sentence 'Plurality P is arranged treeish' is true iff it's true that given the existence of trees, P composes a tree, or iff it's true that if trees existed, P would compose a tree." Instead, have a bunch of tree experts examine a few dozen things they take to be trees and see what they come up with for their treeish pluralities. We would certainly rely on them to figure out an illuminating story for 'is a tree', so they should be the ones to do the bulk of the work on an illuminating story for 'is/are arranged treeish'.

According to PP, 'There is a tree' comes out true in virtue of this fact: *there is at least one tree-unified plurality of treeish pluralities that are either BLPs or pure & simple BPs*. If there are simples in a universe, then perhaps all trees boil down to just them. If so, the treeish pluralities are pure & simple BPs. If not, then the treeish pluralities are BLPs. All we will need to account for trees, at most, are simples, BLPs, and pure & simple BPs.

In sum:

'is a quark' is true of a simple

'is a proton' is true of a pure & simple BP (of three quark-simples)

'is/are arranged treeish' is true of either a pure & simple BP or BLP

'is/are tree-unified' is true of an impure BP or a BLP (since each 1-element of a tree-unified plurality P is a plurality, if P is a BP, then it's an *impure* BP).

The predicate 'is a tree' applies to reality but only in an indirect fashion. 'There are trees' is true because 'are tree-unified' and 'are treeish' are true of certain pluralities but 'is a tree' doesn't apply to reality in the way 'is a quark' or 'is a proton' do.

Generalizing on this idea, we have our first thesis:

> **T1**: There are three possible ways 'There is an F' can come out true: it can derive its truth from (i) a single simple that's F, (ii) a BLP or pure & simple BP that's F, or (iii) many F-ish BLPs or pure & simple BPs that are collectively F-unified.

Five explanatory points on **T1**:[3]

- Roughly put, **T1** says that *the things in the world that make our sentences true either boil down to simples (clauses (i) and (ii)) or are BLPs (clauses (ii) and (iii))*. Nothing else is needed.
- The quantifier for 'F' is limited. **T1** is plausible for a great many Fs; the primary question is whether it's true for *all the ones we care about*. For instance, we shouldn't substitute 'non-existent', 'shadow', or other troublesome predicates (or pseudo-predicates) in for 'F'.
- If Referential Sobriety is necessarily false, then (iii) can be deleted. If simples are impossible, then (i) can be deleted.
- All the 1-elements of an F-unified plurality are pluralities. Hence, the F-unified plurality is either an impure BP or a BLP, depending on the metaphysical details of the possible world in question. In addition, as we saw earlier if it's an impure BP, it could also be a pure BP due to the oddities of plurality identity.
- We have yet to say anything about parthood or composition. It's also worth noticing that we have had no *need* to talk about them yet. This suggests that those notions aren't fundamental.

The thinking behind the next thesis **T2** starts from the observation that although 'is a part of' may or may not latch on to a joint in nature, or metaphysically important universal or species or genera or something similar, it can be precisified in multiple ways—and the precisifications may be philosophically, scientifically, or otherwise useful. Let me elaborate.

We have terms like 'helps compose' and 'is a part of'. The initial question is: do each of these terms "get at" some single property or kind—like how 'electron' does? I'm saying that *whether or not the answer is positive*, there's room for multiple interesting precisifications. I think this should be relatively

[3] We would need another thesis for singular terms, but I ignore it here.

uncontroversial, as one can make the same move for other philosophically key phrases, such as 'free will', 'knowledge', 'miracle', etc. One can, for a specific purpose, just offer a *stipulation* of how one is going to use a philosophically important term in a certain theoretical context, and then go on to *justify the stipulation on utilitarian grounds* by arguing that subsequent use of the stipulation will have such-and-such benefits.

For instance, in ordinary life when we say or imply that x is "a part of" y we virtually always are disposed to think that both x and y are more or less "unified" objects. Object x, the part, won't be some utterly random plurality of things that are widely scattered throughout the universe. If x were like that, then it would hardly be *a* part, as the 'a' in 'a part' indicates, in ordinary use, a single *unified* thing. By 'indicates in ordinary use' I mean, roughly, that virtually all competent users of 'part' are disposed to think that parts are never paradigmatically non-unified things. Similarly, object y has got to be unified. So *if* we want a precisification of 'is a part of' that captures a good portion of everyday thought regarding 'is part of', *then* when designing our precisification we should probably use the fortunately (!) wildly imprecise term 'unified'. Similar points hold for precisifications helpful for the special sciences.

We will be left with serious leeway when coming up with a commonsensical precisification of 'x is part of y'. The same holds for precisifications that are intended to do serious work in a special science. I don't have space to offer any here. However, and this is the key point, any decent precisification we come up with will be built up from the notions we encountered in **T1**. As a result, PP offers a substantive thesis regarding the nature of parthood:

> **T2:** Any reasonable precisification (including the true one, if such exists) for 'x is part of y' uses nothing more than these substantive notions: 'x is one of plurality y', 'arranged F-ish', and 'F-unified'. Those are the notions that parthood reduces to.

With **T2** in hand we have, in some soft sense, "explained" or "reduced" or "analyzed" parthood (and hence composition, although I won't treat that reduction here) in terms of the key notions listed in T2.[4] What is odd, and a virtue, are two facts about T2: it doesn't offer necessary and sufficient conditions for 'x is a part of y', and it doesn't take any position on the *extent* of composition, in any world: always (unrestricted), never (Compositional Nihilism), or sometimes (moderate views). PP is not a theory of parthood in any familiar sense; that's why I said we have analyzed parthood "in a soft sense". But it does tell us what

[4] Kleinschmidt (2019) also appeals to pluralities in treating parthood but not in the way PP does.

notions parthood and hence composition are derivative of. In this paper I don't address the issue further.

If PP is true, then trees are like protons in being no "real addition" to the universe once you've accepted the little bits that make them up—even if there are no simples. But according to PP, if a tree exists, it's not identical to any simple or plurality; so, it sure looks like something *in addition to* simples and pluralities. What is going on?

I think we have the materials to satisfactorily respond to that challenge.

> **T3:** The predicate 'is a tree' has a second-class relation to reality compared to that for 'is a proton'. The term 'is a proton' applies to reality *directly*: it is true of a pure & simple BP (again, with our pop-science assumption just for illustration). But 'is a tree' applies to reality only *indirectly*: although it is not true of any plurality (or simple), 'is tree-unified' directly applies to a plurality of a treeish pure & simple BPs or BLPs—and this is why 'there are trees' comes out true. There are pluralities that literally are protons but none that are trees. *'There are trees' comes out true, but not in virtue of 'is a tree' applying to some thing or some things.*

Trees are not second-class *existents*, as the second-classness is representational, not ontological. **T3** means that trees are no "real addition" to reality once we have listed all the simples (if any), treeish pluralities, and tree-unified pluralities. **T3** is an elaboration of **T1**.

Many philosophers think that there is just one way for 'is an F' to apply to reality: there is a thing x such that x itself is F. For one thing, that's how we have been programmed to read '∃xFx.' Although that is the familiar way to satisfy a predicate, my theory says there is another way: there are some things that are F-unified. Although the second way is theoretically unfamiliar, it is the way predication works in *the vast majority of cases* (e.g. 'tree', 'explosion', 'person', 'baseball', 'earthquake', etc.). I don't think this necessarily means we have to revise logic, but it does mean we need to rethink the predicate–reality relation. We can keep the familiar reading of '∃xFx', with its ontological seriousness (so we don't quantify over flesh-and-blood Harry Potter for instance), but we need to be more relaxed about how to interpret it. I return to that issue in the next section.

6. Five Consequences of Plurality Pointillism

First, when trying to get a grip on this new idea about predication, I find it helpful to imagine looking at a single tree and seeing something akin to a swarm of bees. The predicate 'is a tree' applies, in an indirect fashion, to the stuff in that swarm—so trees exist—but there is no *particular* swarm that satisfies 'is a tree'. It's a fantasy to think that ordinary language is truth-conditionally specific enough to single out exactly one swarm amongst the trillions of trillions of trillions of equally good candidate swarms as 'the tree in my backyard' (this is Referential Sobriety again). But all that's really over there, in the area where the tree is, is swarms of tiny bits. So, 'is a tree' has nothing specific to latch onto. If that offends our conception of predicate-reality relations, well, so be it. If you have done much philosophy at all, especially regarding paradoxes, then you have realized that we have no choice but to be philosophically offended.

Second, there is a crucial ambiguity in the notion 'x is nothing over and above (NOAA) the ys': whereas a proton is NOAA tiny bits because it's identical to a plurality of three such bits, a tree is NOAA tiny bits because certain appropriately characterized pluralities of those bits are unified appropriately.

Third, this means that the Special Composition Question, 'When is it true that there is a y such that the xs compose y?' (van Inwagen 1990: 30) may contain a false presupposition: the claim that for any composite object, there is a plurality of entities that composes it. On the face of it, a tree is composite but there is no plurality (or pluralities) that compose(s) it. But these matters (plus the Problem of the Many) need careful treatment that I can't give here.

Fourth, this distinction in how terms apply to reality—'is a proton' vs 'is a tree'—helps illuminate the thesis of Referential Sobriety. It's natural to initially think the thesis has to do with vagueness, but surprisingly enough this is misleading. The predicate 'is arranged treeish' is vague due to borderline cases but still applies to reality the same way 'is a proton' does: it latches onto a particular plurality. PP says that whereas 'is arranged treeish' is true of pluralities, 'is a tree' is not true of pluralities—even though both predicates have borderline cases, exhibit tolerance, and generate sorites paradoxes. Hence, vagueness is not the primary reason for Referential Sobriety. Just because 'is an F' is vague doesn't mean it applies to reality in a second-class manner.

The reason 'is a tree in my backyard' applies to reality only indirectly is twofold: it is vague and, roughly put, there is *just one* tree in my backyard. There isn't anything terribly counter-intuitive with the idea that in my backyard there are trillions of treeish arrangements. Perhaps it is counter-intuitive, but then it is only mildly so given that no one outside a metaphysics discussion has ever heard of such arrangements. But the idea that in my backyard there are trillions of trees is just plain false.

Fifth, there are of course options for fitting trees into reality other than with PP. To categorize them, consider the following individually plausible yet jointly inconsistent claims.

1. "is a tree in my backyard' applies in a universe in which big things boil down to little things.
2. 'is a tree in my backyard' applies in a universe in which big things boil down to little things É 'is a tree in my backyard' applies to a simple, a pure & simple BP, or BLP in that universe.

3. 'is a tree in my backyard' applies to a simple $\supset$ some tree in my backyard is a simple.
4. No tree in my backyard is a simple.
5. 'is a tree in my backyard' applies to a pure & simple BP or BLP $\supset$ 'is a tree in my backyard' applies to zillions of pure & simple BPs or BLPs.
6. It's not the case that 'is a tree in my backyard' applies to zillions of pure & simple BPs or BLPs.

Due to the inconsistency, one has to reject at least one of (1)–(6). One could reject (1), asserting that there are no trees in universes in which big things boil down to little things. One could say this because one thinks that there are no trees that aren't simples. One could reject (5) by attributing magical powers of discrimination to ordinary language, thereby rejecting Referential Sobriety. One could reject (6) by asserting that there are a *lot* more trees out there than common sense or science says.

Those are three implausible ways of responding to the paradox. (3) and (4) are obviously true. Only (2) is left. By my lights, rejecting (2), as the plurality pointillist does, is more reasonable than rejecting any of (1), (5), or (6). The offensive taste of rejecting (2) is softened by the theory about second-class predication. We reject (2) while keeping our reductive theory of parthood and composition.

7. Plurality Pointillism and Non-Being

It seems that according to PP the only things *really out there in reality* are simples, pure & simple BPs, and BLPs, at most. Shouldn't PP conclude that trees aren't, well, really out there in reality? That they exist only on a linguistic technicality, so to speak?

There are two conceptions of what it is to be "an entity out there in reality":

Conservative Conception: Trees are entities really out there in the world if and only if 'is a tree' applies to reality on the proton model (viz. being true of a plurality). But 'is a tree' fails to fit reality like that, since it fits reality only via 'is tree-unified' and 'is arranged treeish'. So, trees aren't entities out there in our environment despite the truth of 'There are trees' and the fact that there are many tree-unified and treeish entities out there in reality.

Liberal Conception: What we have learned is that there are two ways to be "an entity out there in reality". There is the theoretically familiar and simple way, the way 'is a proton' works; and then there is the other way, the one that applies to 'is a tree' and the vast majority of other predicates from natural language.

Furthermore, *the latter is derivative of the former*: 'is a tree' applies to reality because it's semantically related to both 'is tree-unified' and 'is arranged treeish', both of which apply to reality in the first, familiar way.

Perhaps there are good reasons to choose one of these over the other, but I'm not sure much hangs on it. When it comes to the question 'Are trees out there in reality or not?', I'm inclined to claim ambiguity and refuse to give a yes/no answer.

But others may disagree. One could insist that on PP, simples, pure & simple BPs, and BLPs exist in some fundamental sense, but trees and virtually everything else from ordinary life exist only in a derivative sense. This is tantamount to taking the linguistic point about 'is a tree' versus 'is an electron' and making it into an ontological point. I can only flag the issue here (cf. Tahko 2018).

Acknowledgments

Thanks to Juhani Yli-Vakkuri and the editors for valuable comments.

References

Baxter, Donald and Aaron Cotnoir (eds) (2014). *Composition as Identity* (Oxford: Oxford University Press).

Chen, Lu (forthcoming). "Infinitesimal Gunk", *Journal of Philosophical Logic*.

Horgan, Terrance (1997). "Deep Ignorance, Brute Supervenience, and the Problem of the Many", *Philosophical Issues* 8: 229–36.

Kleinschmidt, Shieva (2019). "Fusion First", *Noûs* 53 (3): 689–707.

Lewis, David (1991). *Parts of Classes* (Oxford: Basil Blackwell).

Tahko, Tuomas (2018). "Fundamentality", *Stanford Encyclopedia of Philosophy*, ed. Edward N. Zalta (Fall 2018 edn).

van Inwagen, Peter (1990). *Material Beings* (Ithaca, NY: Cornell University Press).

Wallace, Megan (2011a). "Composition as Identity: Part 1", *Philosophy Compass* 6 (11): 804–16.

Wallace, Megan (2011b). "Composition as Identity: Part 2", *Philosophy Compass* 6 (11): 817–27.

Williamson, Timothy (1997a). "Imagination, Stipulation, and Vagueness", *Philosophical Issues* 8: 215–28.

Williamson, Timothy (1997b). "Replies to Commentators: [Horgan, Gomez-Torrente, Tye]", *Philosophical Issues* 8: 255–65.

8
The Cosmic Void

Eddy Keming Chen

1. Introduction

One of the hardest questions in fundamental physics and fundamental metaphysics is this: what exists at the fundamental level of reality? There is no consensus in contemporary physics. There is much less consensus in metaphysics.

Nevertheless, on the physics side, we usually assume that the fundamental level of reality will be something physical and material. We assume that the fundamental level will not consist in purely mental ideas. We also assume that the fundamental level will not be completely devoid of matter, since our world is manifestly not empty. Different physical theories postulate different kinds of fundamental matter (and sometimes different versions of the same theory will differ in the fundamental material ontologies). To see some examples, here is a preliminary list of fundamental material ontologies in different physical theories:

- Newtonian mechanics: point particles;
- Maxwellian electrodynamics: charged particles and electromagnetic fields;
- general relativity: matter field on space-time;
- quantum mechanics: a quantum state and/or local ontologies such as matter density field, flashes, or particles;
- quantum field theory: a quantum state and/or local ontologies such as particles, particle number ontology, or fields;
- loop quantum gravity: spin-foams;
- string theory: one-dimensional strings.

(This list is certainly incomplete: future theories may have radically different ontologies; and even theories on this list can be compatible with other ontologies.)

Describing the fundamental material ontology is an important task in theoretical physics, since fundamental matter plays crucial roles in a physical theory. First, the fundamental material ontology in a theory plays an explanatory and metaphysical role. It is the ultimate explanation for non-fundamental ontologies and non-fundamental facts. We use fundamental matter as the ultimate reduction base: we try to reduce the behaviors of macroscopic systems such as tables and

Eddy Keming Chen, *The Cosmic Void* In: *Non-Being: New Essays on the Metaphysics of Non-Existence.* Edited by: Sara Bernstein and Tyron Goldschmidt, Oxford University Press (2021).
DOI: 10.1093/oso/9780198846222.003.0008

chairs to the behaviors of microscopic constituent objects such as particles and fields. How tables and chairs move around can be explained by how the constituent particles and fields behave. How particles and fields behave can be derived from the laws of physics. If the particles and fields are part of the theory's fundamental material ontology, then the reduction is complete. If they are further reducible to something else in the fundamental material ontology, such as the quantum state, then more reduction is in order. In any case, it is hard to see how to carry out the reduction without the fundamental material ontology.

Second, the fundamental material ontology plays a semantic role by being the subject matter of the laws of nature. The terms in the laws of nature refer to properties of the fundamental material ontology. For example, the mass term in $F = ma$ refers to a property of Newtonian point particles. The acceleration term describes the change of their velocities. It is hard to see how fundamental physical laws make sense without the fundamental material ontology.

Third, the fundamental material ontology plays an informational role in the physical theory. The physical laws are supposed to be simple. But how can such simple laws explain such a wide range of complicated phenomena? How does so much information come from such simple laws? It is made possible by appealing to the complicated initial conditions and boundary conditions of physical systems. The complicated data is stored not in the laws but in the matter distribution, such as the locations of particles and configurations of fields. By postulating a fundamental material ontology, complicated phenomena can be derived from simple laws plus certain contingent (and complicated) facts about fundamental matter. There is a tight connection between fundamental material ontology and laws of physics that we will try to make more precise later. But it is worth noting that we usually put the "mess" in the fundamental material ontology so that the laws can be simple.

Those roles are important and seemingly irreplaceable. Hence, the fundamental material ontology seems indispensable to any successful physical theory.[1] Given those roles fundamental matter seems to play in successful physical theories, we also have reasons to postulate them in our fundamental metaphysics.

However, it can be intellectually healthy to reexamine our assumptions. As Turner (2010) puts it, "we can come to better understand a proposition by studying its opposite." We ask the following question: is a fundamental material ontology indispensable to any successful physical theory? Reflecting on this question does not require us to reject scientific realism. It may even help us better appreciate the place of fundamental material ontology in our physical theories and metaphysical frameworks.

[1] Quantum theory presents an apparent counter-example, as the followers of the "Copenhagen interpretation" deny the existence of a determinate microscopic ontology. But we have made sufficient progress about the quantum measurement problem to understand that there are better ways to understand quantum theory. See section 3.1.

In this paper, I discuss a possibility that I call "the cosmic void." In the cosmic void scenario, the universe is devoid of any fundamental matter.[2] I focus on the informational role of the fundamental matter ontology and suggest that, in certain theories, it can be played by the laws of physics. To recover the non-fundamental facts—the manifest image of tables, chairs, and computers—I make use of *strongly deterministic* laws of nature that determine a unique history of the universe. In this scenario, all of the non-fundamental facts are ultimately explained by nomic facts. I discuss a concrete example of the cosmic void scenario that arises when we try to put together a many-worlds theory of quantum mechanics in a time-asymmetric universe. The physical theory turns out to be strongly deterministic. To be sure, there are both philosophical and scientific challenges to such a possibility, especially regarding the metaphysical role and the semantic role of the fundamental matter ontology. I introduce the cosmic void scenario *not to endorse it but to draw our attention to an interesting area of logical space that deserves more scrutiny.*

There are some similarities between this project and the anti-object metaphysics of ontological nihilism (Hawthorne and Cortens 1995; Turner 2010), structural realism (Ladyman and Ross 2007), generalism (Dasgupta 2009), and the bare facts framework (Maxwell ms). However, the cosmic void possibility is at the same time more radical and more modest. It is more radical in that we do not postulate even fundamental properties, general facts, bare facts, or structural facts in the fundamental ontology. It is more modest in that the strategy is not supposed to work in general, but only in some special physical theories where strong determinism holds. Hence, the cosmic void scenario has a more restricted scope of application.

2. The Possibility of the Cosmic Void

What is the cosmic void? It is the scenario in which *nothing material exists at the fundamental level.* Suppose space-time is fundamental. Then, in this scenario, fundamentally speaking the space-time is completely empty. There are no fundamental particles, fundamental fields, fundamental quantum states, any kind of fundamental material distributions, or any non-trivial decorations on space-time.

This cosmic void scenario is allowed by both special and general relativity. However, that is not the intended interpretation here. In the empty solutions allowed by special and general relativity, non-fundamental facts that there are tables and chairs are false. In the cosmic void scenario I would like to describe, those non-fundamental facts remain *true*, even though there is no fundamental material ontology.

[2] Hence, it is different from the notion of the cosmic void in astrophysics that the large-scale structure of the universe can be characterized by mostly empty space between galaxies.

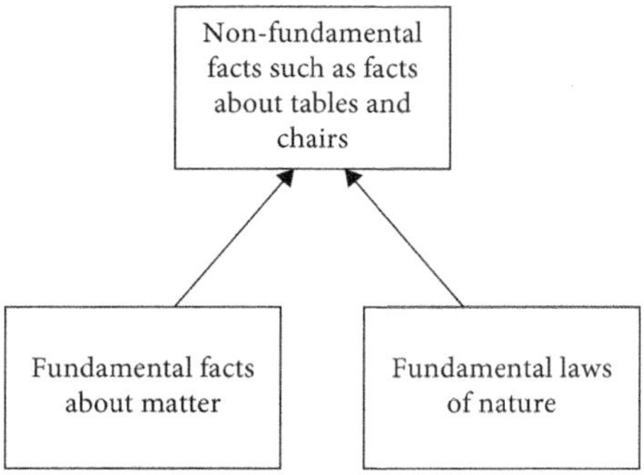

Figure 8.1 The standard picture where non-fundamental facts are explained by fundamental facts about matter and laws of nature.

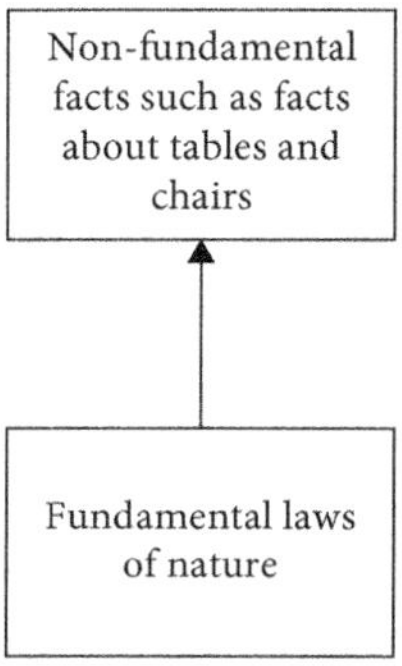

Figure 8.2 The non-standard picture where non-fundamental facts are completely explained by fundamental laws of nature.

On the standard picture (see Figure 8.1), non-fundamental facts about tables and chairs are made true by fundamental laws of nature and fundamental facts about the existence and behavior of fundamental matter, such as point particles. Fundamental particles constitute tables and chairs, and the existence and behavior of tables and chairs can be explained by the existence and behavior of the fundamental material ontology, together with the laws.

On the non-standard picture (see Figure 8.2), since there is nothing material at the fundamental level we have to appeal to something else to explain the non-fundamental facts. What can that explanation be? A candidate is the fundamental laws of nature. For the purpose of this paper, we commit to a non-Humean theory of lawhood. Here we assume that laws are not merely summaries of the mosaic. There can be non-trivial laws even when the mosaic is completely empty and undecorated.

Usually, fundamental laws of nature are *at best* partial explanations for most non-fundamental facts. The existence of a table in front of me at this point is not entailed by the standard laws of physics. At the very least, we need also the complete state of the universe at some time (or during some time interval).

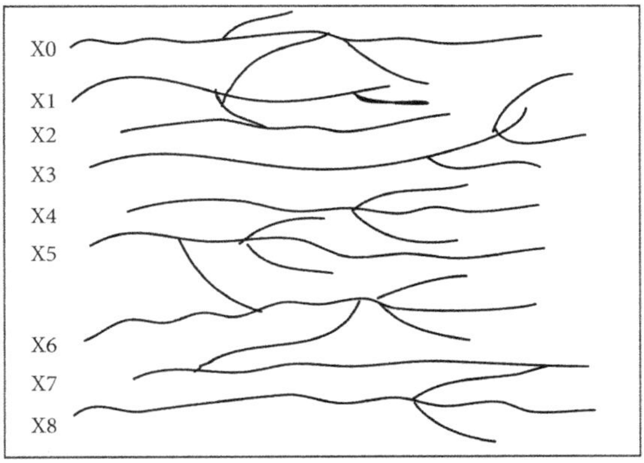

Figure 8.3 Indeterministic laws of physics. *X*0–*X*8 refer to different initial conditions of the universe. (Here the illustration is schematic. Usually there are infinitely many possible initial conditions.) The horizontal curves correspond to different histories of the universe. Different histories of the universe can overlap at some time and diverge later. Fixing an initial condition of the universe does not fix a unique history of the universe.

The complete state of the universe is not encoded in the dynamical laws, for they are compatible with many different states of the universe. In standard physical theories, the complete state of the universe refers to the state of the fundamental material ontology. Hence, the non-standard picture is not accommodated by the usual laws of physics.

It is useful to distinguish between two types of laws of physics: indeterministic laws and deterministic laws. If the laws are indeterministic (see Figure 8.3), then given a complete state of the universe at some time (or some duration of time), the laws can allow multiple different pasts and futures of the universe. In other words, different histories can overlap. If the laws also assign objective probabilities or chances to the histories, then given a complete state of the universe at some time, the laws assign a unique probability distribution over future (or past) histories.

If the laws are deterministic (see Figure 8.4), then given a complete state of the universe at some time (or some duration of time), the laws allow only one past and one future of the universe. In other words, different histories cannot overlap: no two distinct histories of the universe can overlap (be exactly the same in terms of the distribution of microphysical properties and ontologies) at any point in time. Hence, the deterministic theory can be more informative than an indeterministic theory in the sense that the former but not the latter uniquely determines the entire history of the universe given the dynamical laws and a specification of the material ontology at some time (or some duration of time). In a universe governed by deterministic laws of physics, fixing the initial condition of the universe is sufficient to fix the entire history of the universe.

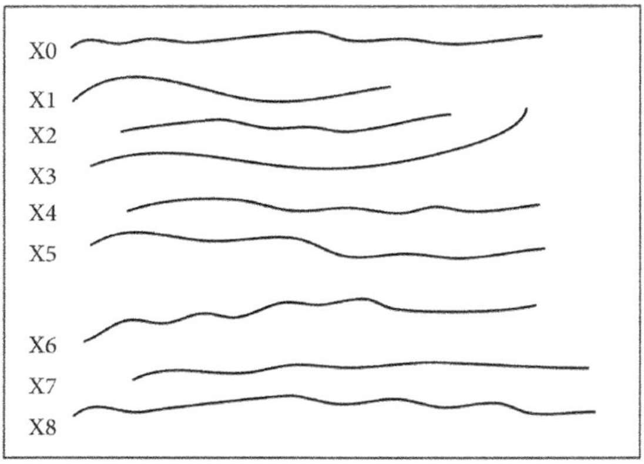

Figure 8.4 Deterministic laws of physics. *X*0–*X*8 refer to different initial conditions of the universe. The horizontal curves correspond to different histories of the universe. Different histories of the universe cannot overlap at any point in time. Fixing an initial condition of the universe fixes a unique history of the universe.

However, information about the exact initial condition itself is not contained in any standard law of physics. Usually it is the distribution of the fundamental material ontology, not the laws, that specify the complete initial condition. After all, usually the exact initial condition is too complicated to express by any simple law. (As we discuss below, this is true even after adding the Past Hypothesis as an additional fundamental law of nature, for it only pins down the macroscopic initial condition of the universe and not the microscopic initial condition.) Hence, it is in this sense that the standard deterministic laws of physics are compatible with many different possible worlds. Usually the laws do not pick out a unique world.

The situation is transformed when the laws are *strongly deterministic* in the following sense:

Strong Determinism The laws of nature pick out a unique history of the universe.

A world is strongly deterministic if its fundamental laws pick out a unique history of the universe (see Figure 8.5).[3] One way the laws can be strongly deterministic is by satisfying two conditions:

1. The laws are deterministic.
2. The laws pick out a unique initial condition.

[3] This notion of strong determinism is introduced in Penrose (1989). This is different from the notion of "superdeterminism" that is sometimes invoked in the context of avoiding Bell non-locality. See Chen (2020a) for an overview.

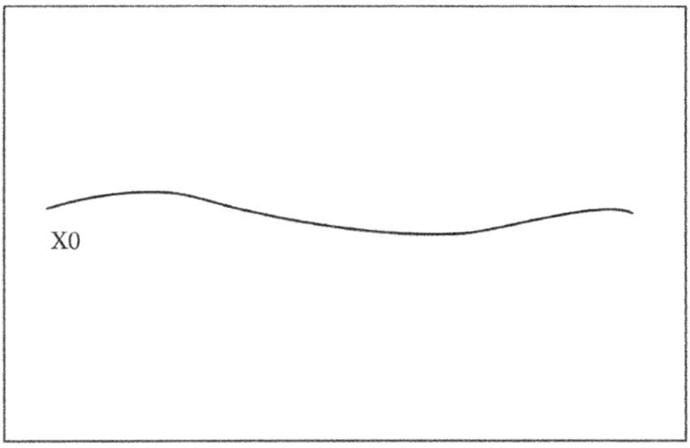

Figure 8.5 Strongly deterministic laws of physics. X_0 refers to the unique initial condition allowed by the laws. The horizontal curve refers to the unique history of the universe allowed by the laws. The laws completely fix the history of the universe.

Here, the unique initial condition refers to an exact specification of the microphysical details of the universe at t_0. Typically, it is difficult to pick out a unique (microscopic) initial condition without making the laws overly complicated (so that they no longer qualify as good candidate fundamental laws of physics). Hence, it is highly non-trivial to write down a *simple* law (or set of laws) that can do that. In section 3, we look at a concrete example of a law—the Initial Projection Hypothesis—that is not only simple but also selects a unique (microscopic) initial condition in the Everettian many-worlds theory of quantum mechanics.

In a strongly deterministic world, at the fundamental level, there is no contingency. There is only one way the universe could be, in the sense of nomic possibility. If the actual world is strongly deterministic, then the actual world is the only one that is nomologically possible. That is, what is actual is also nomologically necessary. Any sense of contingency would have to come out at some non-fundamental level.

Let us contrast that with a more familiar theory—classical mechanics. The only dynamical law $F = $ ma selects a space of possible worlds—the solutions to that equation. It is possible to add to it a macroscopic initial condition law—the Past Hypothesis, which says that the universe started in a low-entropy macrostate.[4] This special macrostate allows the emergence of various asymmetries in time such as the Second Law of Thermodynamics. However, even after adding the Past Hypothesis, there is still an infinity of microscopic initial conditions compatible with the laws. These initial conditions (compatible with both $F = $ ma and the Past Hypothesis) have smaller measure than the original set (whose only constraint is $F = $ ma). In this scenario, even though the laws are deterministic, they do not pick

[4] For discussions of the Past Hypothesis, see Boltzmann (1964), Feynman (1965), Penrose (1979), Albert (2000), Goldstein (2001), Callender (2004), Loewer (2012; forthcoming).

out a unique history of the universe, as there are many possible worlds compatible with the laws. Hence, the laws are not strongly deterministic.

In a strongly deterministic world, the laws specify the entire history of the universe without needing any further input. Hence, it is in principle possible to extract all the information about the world from just the laws alone. It is this feature of strong determinism that makes possible an empirically adequate theory of a cosmic void, in the sense of being informationally complete to recover all the facts. (Is this sense of empirical adequacy strong enough? We return to this question in section 4.) Suppose at the fundamental level we have strongly deterministic laws. There is only one way the fundamental material ontology could be like—they have to be arranged exactly according to the only way allowed by the laws. Such an arrangement of fundamental matter would be nomologically necessary. Facts about fundamental matter would make true non-fundamental facts about the locations and behaviors of tables and chairs.

We can go further. The strongly deterministic laws constrain the fundamental ontology *if it exists*. But now suppose that there is no material ontology at the fundamental level. All we have are the laws of physics, from which we can in principle derive all facts about the non-fundamental. Would it make a difference whether the derivation goes through some postulate about fundamental matter? It depends on one's view about ontological dependence: must non-fundamental facts about tables and chairs bottom out in fundamental facts about matter? (We return to this question in section 4.) Suppose we accept the possibility that non-fundamental facts about tables and chairs can bottom out in fundamental facts about laws of nature. That is, all explanation about the non-fundamental can be purely nomic. We arrive at the cosmic void scenario—in a strongly deterministic universe, the physical theory can be empirically adequate without postulating any fundamental material ontology. All there is at the fundamental level are the laws of nature.

What is this notion of "deriving" non-fundamental facts from laws of nature? I suggest this does not have to differ from the notion of mathematical-physical derivation that links the macroscopic facts such as the positions of tables to the microscopic facts such as the positions of particles. The derivation combines a variety of techniques, such as mathematical proofs, coarse-graining, approximations, and idealizations. The kind of examples I have in mind are those in statistical mechanics concerning the reduction from the macroscopic phenomena to kinetic theory. But they can also be found in high-energy physics and quantum theory. However, in the more familiar examples in physics, since the laws people use are not strongly deterministic, the laws themselves are insufficient, and the derivation must involve further input about the contingent initial conditions.

It is worth emphasizing that, since we are assuming a non-Humean theory of lawhood, strongly deterministic laws are compatible with the existence of fundamental material ontology as well as their absence. If we do not postulate fundamental material ontology in a strongly deterministic universe, if the cosmic void scenario is possible, then there can be non-fundamental facts about the material ontology that can coincide with the content of the fundamental material ontology that could be postulated. However, whether or not the facts about such material ontology are fundamental, they make true the same set of non-fundamental facts—the locations of tables and chairs. So it is best to say that the strongly deterministic universe is compatible with two worlds that coincide in all facts except the fundamental material facts. Hence, we need to revise our earlier definition of strong determinism:

Strong Determinism* The laws of nature pick out a unique way to completely specify the fundamental matter.

However, a strongly deterministic world on this criterion does not need to specify any fundamental matter—it could realize the cosmic void scenario that is devoid of fundamental matter. But if we were to completely specify fundamental matter, then there would be only one way to do so according to the laws. The qualifier "completely" is important here: if we were to only partially specify the fundamental matter, then there could be infinitely many ways to do so. We could, for example, specify only the configuration of matter in space-time region R_1, specify only that in R_2, and so on.

The cosmic void scenario has many interesting features. In such a world, even though the fundamental arena (space-time or something analogous to space-time) is completely empty, we can still recover all the non-fundamental facts from fundamental laws alone, relying on purely nomic explanations. That would be surprising if it turns out to be a successful theory. What are some advantages for maintaining the cosmic void scenario? First, it has a fairly parsimonious ontological base. All else being equal, we have reasons to prefer a more parsimonious theory. Of course, not everyone will be convinced that other things are equal here. Perhaps we lose certain explanatory power. We also should be careful where and how to apply Ockham's Razor for reasons mentioned in Maudlin (2007). We will return to this point in section 4. Second, the cosmic void scenario also shares some of the advantages of radical versions of ontological nihilism, such as the versions developed in Hawthorne and Cortens (1995). For example, since there are no material objects at the fundamental level, the difficult metaphysical disputes about identity of material objects over time and mereological composition of material objects simply do not arise, at least for things that exist at the fundamental level. Nevertheless, neither consideration is decisive. Moreover, the cosmic void scenario may cost us important intuitions or principles that we cherish (see section 4). In the end, it will require a more comprehensive cost/benefit analysis (one that we do not have the space to develop in this paper) to reach a reflective equilibrium about whether the cosmic void scenario should be allowed in logical space. Let us now turn to a case study of the cosmic void scenario.

3. A Case Study

In this section, we provide a concrete example of strong determinism to illustrate the possibility of the cosmic void. The example relies on a non-standard way of combining the Everettian many-worlds interpretation of quantum mechanics with the low-entropy initial condition of the universe that is essential for explaining the arrow of time.

3.1 Many-Worlds Interpretation of Quantum Mechanics

Hugh Everett III's many-worlds interpretation (1957) was an attempt to address a central problem in the foundation of quantum mechanics—the quantum measurement problem.[5] Textbook quantum mechanics suggests that we assign a quantum state, represented by a wave function, to some physical systems of interest, such as a cat in a box. The wave function obeys two dynamics. First, it obeys a linear equation of motion—the Schrödinger equation. That equation has the tendency to spread out the wave function into what is called a superposed state, such as the cat-being-alive state superposed with the cat-being-dead state. In that state, it is hard to understand what is going on physically: is the cat alive, dead, or both? Second, upon "observation" or "measurement," the system's wave function will undergo a random and non-linear collapse to a particular state, such as the state that the cat is alive.

Despite the predictive success, this recipe for making predictions faces foundational problems. How can the wave function obey two different dynamics? If the system is completely described by the wave function, and if both observers and measurement instruments are physical systems interacting quantum mechanically, they would always obey the Schrödinger equation and never collapse into a definite state (such as the cat-being-alive state) from a superposed state (such as the superposed state of the cat-being-alive and the cat-being-dead). We can put together the principles that result in a contradiction:

(P1) The wave function is the complete description of the physical system.
(P2) The wave function always obeys the Schrödinger equation.
(P3) Every experiment has a unique outcome.

There are three types of solutions to the quantum measurement problem. Each of them rejects one of the assumptions. The first type of solutions rejects (P1) and adds additional variables beyond the wave function to represent other aspects of

[5] Below is a rough sketch. For a more thorough discussion about the quantum measurement problem, see Bell (1990), Albert (1992), and Myrvold (2017).

the system. A prominent theory in this type is Bohmian mechanics, originally proposed by De Broglie (1928) and Bohm (1952a; 1952b) independently. See Goldstein (2017) for an overview. On that theory, there are point particles with precise locations in addition to the wave function. The cat was made out of particles guided by the wave function. The particles are always in a determinate macrostate—either the cat is alive or the cat is dead. The wave function plays two roles: it guides the motion of particles and it provides a probability distribution over their precise locations.

The second type of solutions rejects (P2) and replaces the Schrödinger equation by a spontaneous collapse dynamics. A theory in this type is the GRW theory, proposed by Ghirardi, Rimini, and Weber (1986). See Ghirardi and Bassi (2020) for an overview. The GRW theory replaces the measurement-triggered collapse dynamics with something much less anthropocentric—an objective and spontaneous collapse mechanism that collapses the wave function with a fixed probability per unit time per unit particle. Before we open the box to measure the cat, its wave function has already collapsed into one of the two states: either alive or dead.

The third type of solution—Everett's "many-worlds" interpretation—rejects (P3) and embraces the non-uniqueness that comes with it. See Vaidman (2018) for an overview. On this theory, the wave function is the complete description of the physical system. It always obeys the Schrödinger equation. But typically experiments lead to multiple definite outcomes all at once. When we go and measure the cat, we observe that it is in a definite state—say, the cat is alive. But appearances can be misleading. Since nothing has collapsed the wave function and there are no additional independent variables, the part of the wave function in which the cat is dead still exists and is equally real as the part in which the cat is alive. What is going on? On this interpretation, there are at least two "parallel worlds" after the experiment that correspond to the two outcomes. Both are real—they are known as two branches of the wave function. Usually after the experiment, the different branches no longer interact much with each other. Whether the parallel worlds come from a splitting of a single world into two or are merely emergent descriptions is a matter of debate. As with Wallace (2012), I think it is much more plausible to interpret the parallel worlds from the emergence perspective—they are not fundamental properties of the Everettian universe.

Unlike the GRW theory and *like* Bohmian mechanics, the many-worlds interpretation is deterministic: given a complete specification of the state of the universe at one time, all the past and all the future is completely fixed by the law of motion. In this case, the wave function of the universe at one time, by the Schrödinger equation, completely fixes the wave function at all other times. But the many-worlds interpretation differs from Bohmian mechanics in that the latter has a single world while the former entails a multiplicity of (emergent) worlds. The reason is that the wave function, which is ontologically complete in the Everettian picture, introduces a multiplicity that is resolved in the Bohmian picture by the additional postulate of point particles with precise locations.

What is the fundamental material ontology on the many-worlds interpretation? On the most flat-footed way of thinking about it, the fundamental material ontology consists in the quantum state, represented by the wave function. However, what the wave function represents in the physical world is a controversial matter. Realism about the wave function is compatible with many different views about the underlying nature of the physical reality represented by the wave function. See Chen (2019b) for an overview. There are two interpretations that will be relevant below. First, we can use the wave function of the universe to define a matter distribution on physical space-time. This is known as Sm and was proposed first in Allori et al. (2010). The basic idea is that we can use the wave function to define a function $m(x, t)$ whose domain is physical space-time and whose range is the set of real numbers. This function can then represent a physical field called the matter-density field in space-time. The higher the "sum" of values of the $m(x, t)$ function in some region R, the more stuff there is in R. The locations of tables and chairs at different times can be read off from the history of this function $m(x, t)$. Second, we can follow Wallace and Timpson (2010) and postulate local quantum states for sub-regions of the universe. Given a partition of the universe into collections of sub-regions, we can use the wave function of the universe to define sub-region states. (More technically, we can define reduced density matrices for some local region R by tracing out the degrees of freedom of the environment E in the universal wave function.) Because of the holism inherent in quantum entanglement, merely defining the reduced states for each sub-region is not sufficient to describe the history of the universe. In general, even after we postulate the reduced state of R_1 and the state of R_2, we will still need to postulate a state of $R_1 \cup R_2$, as its state is in general not a logical sum of the states of its parts.

We will focus on the many-worlds interpretation of quantum mechanics here because it has the potential to develop into a strongly deterministic theory that realizes the possibility of a cosmic void.

3.2 The Everettian Wentaculus

The many-worlds interpretation is already deterministic. To make it strongly deterministic, it suffices to pick a unique microscopic initial condition of the universe. One idea is to pick a particular wave function and call it the actual and nomologically necessary initial wave function of the universe. But that would not work. Typical wave functions of the universe contain too much information so that they will in general be extremely complicated. It is unlikely that a simple law can pick out any of those complicated wave functions.

In this section, we outline a new version of the many-worlds interpretation whose laws are strongly deterministic *and* simple. The laws will be the von Neuman equation that generalizes the Schrödinger equation, the Initial Project Hypothesis that replaces the Past Hypothesis, and possibly an additional matter-density equation.

To appreciate the new version of the many-worlds interpretation of quantum mechanics, we need to introduce some background about the arrows of time. We see the manifestation of arrows of time everywhere. Many processes in nature typically happen in only one direction: they are time-asymmetric. At room temperature, ice cubes in the past are always larger than ice cubes in the future; bananas are more ripe in the future than in the past; people are more wrinkled in the future than in the past. However, the fundamental dynamical laws of physics are (essentially) symmetric between the past and the future. Take for example the conjunction of Newton's equation of motion $F = \text{ma}$ and law for universal gravitation $F = \text{G}m_1 m_2/r^2$: whatever can happen forward can as easily happen backward. The same is more or less true for the Schrödinger equation and special and general relativity. So the explanation for the arrows of time plausibly comes from elsewhere.

One prominent explanation, first proposed by Boltzmann (1964) (published first in 1898), invokes what is now called the Past Hypothesis—that the universe in the beginning had very low entropy, where entropy is a measure of disorder. Given a low-entropy initial macrostate, the universe will typically develop in such a way to exemplify arrows of time in its typical subsystems. That is, typically, typical subsystems of the universe are entropic. Modern developments of Boltzmann's ideas in the foundations of statistical mechanics further substantiate this idea. The Past Hypothesis is still compatible with infinitely many initial microstates. It is reasonable to assume that some of them will lead to anti-entropic behaviors (e.g. ice cubes will become larger at room temperature). To solve this problem, it is essential to have the Statistical Postulate according to which these anti-entropic initial states have relatively small measure—they are atypical. Albert (2015) and Loewer (forthcoming) call the conjunction of the dynamical law, the Past Hypothesis, and the Statistical Postulate *the Mentaculus*.

Albert and Loewer's Mentaculus theory was developed in the classical domain. In the quantum case, we can postulate something similar to the Mentaculus. In the quantum case, the Past Hypothesis is now a constraint on initial wave functions of the universe—all of the nomologically possible ones started in a low-entropy macrostate $\mathcal{H}_{PH}$, which is a extremely small subset of all available wave functions. The wave functions inside the macrostate are macroscopically similar in entropy, temperature, volume, and pressure. Inside this macrostate, every wave function is equally likely as any other. (To be more rigorous, we say that there is a uniform probability distribution on wave functions compatible with $\mathcal{H}_{PH}$ with respect to the normalized surface area measure on the unit sphere of the Hilbert space subspace picked out by $\mathcal{H}_{PH}$.) Given the Schrödinger equation, the wave function of the universe will rotate continuously inside the full energy shell of the Hilbert space. That will describe (deterministically) the history of the universe for all times.

In the quantum case, however, another framework is possible to combine the low-entropy initial macrostate with the quantum dynamics. The alternative framework, which I call *the Wentaculus*, has surprising consequences, including the realization of strong determinism. I introduce this framework in Chen (2018) and develop it in other work such as Chen (2019a; 2020c; 2020b). Instead of postulating an initial macrostate compatible with many initial microstates, we can use the initial macrostate $\mathcal{H}_{PH}$ to select a unique and natural microstate—the normalized projection onto that subspace. This is represented by a mixed state density matrix W. This statement is captured in what I call the *Initial Projection Hypothesis* (IPH), a candidate new law of nature that replaces the Past Hypothesis in the Mentaculus. (IPH is as simple as the Past Hypothesis.) Given that there is only one initial microstate possible under IPH, we no longer need to impose the Statistical Postulate to neglect abnormal initial states. Moreover, we can replace the wave function dynamics by the corresponding density matrix dynamics, which is still deterministic in the case of Everettian theory. In this version of Everett, then, the dynamics is deterministic and there is only one microscopic initial condition compatible with the laws. Thus, this version of Everett is strongly deterministic: given the laws, there is only one microscopic history of the universe that is possible. That makes the actual history of the Everettian universe nomologically necessary.[6]

We arrived at a strongly deterministic version of *the Wentaculus* ("W" for the initial density matrix). Not all versions of the Wentaculus are strongly deterministic. The Bohmian version is not strongly deterministic because there are additional independent variables representing possible initial locations of point particles. The GRW version is not strongly deterministic because the dynamics is stochastic.

The Everettian Wentaculus theory we outlined above admits two interpretations: one that is compatible with a fundamental material ontology and the other compatible with the cosmic void scenario. For the first one, we can postulate that at the fundamental layer of reality there is matter: the quantum state,

[6] More precisely, we can write down the postulates of this version of Everett. The Initial Projection Hypothesis is as follows:

$$\widehat{W}_{IPH}(t_0) = \frac{I_{PH}}{\dim \mathcal{H}_{PH}} \tag{1}$$

where I_{PH} is the projection operator onto the subspace $\mathcal{H}_{PH}$ and "dim" counts the dimension of that subspace. The Von Neumann Equation is as follows:

$$i\hbar\frac{\partial \widehat{W}}{\partial t} = \left[\widehat{H}, \widehat{W}\right] \tag{2}$$

where the commutator bracket denotes the linear evolution generalized from the Schrödinger equation.

the reduced states, or the matter-density ontology represented by $m(x,t)$.[7] The non-fundamental facts and the manifest image are recovered by using facts about fundamental material ontology as well as the laws.

However, it also supports a second interpretation that is compatible with the cosmic void scenario. Since the laws are sufficient to determine the actual micro-state of the universe at all times, following the arguments in section 2, we no longer need to postulate facts about fundamental material ontology to explain the non-fundamental facts. We can bypass the fundamental ontology and derive facts about the locations of tables and chairs directly from the laws—in this case the von Neumann equation that determines how the density matrix evolves and the Initial Projection Hypothesis that fixes the initial density matrix. However, the language we use here can be misleading. The equation and the hypothesis seem to assume the existence of some material object represented by the density matrix. But it does not have to exist at the fundamental level. All we need is the information stored in the laws of nature. (What are the laws about if they are not about some fundamental objects changing in time? We return to this question in section 4.)

The Everettian cosmic void scenario, if it can be made empirically adequate, has the following features:

1. The fundamental laws are strongly deterministic. There is no contingency at the fundamental level.
2. There is no material ontology at the fundamental level (no quantum state, reduced states, or matter density field).
3. At the non-fundamental level, there is an emergent multiplicity of worlds that can be characterized by non-fundamental material ontologies such as the quantum state, reduced states, or a matter density field.
4. The non-fundamental facts can be derived from facts about the fundamental laws of nature.
5. At the non-fundamental level, contingency can reappear as descriptions of the non-fundamental worlds.

The Everettian Wentaculus provides a concrete example of the possibility of the cosmic void. In such a world, all there is at the fundamental level are the strongly deterministic laws. The non-fundamental facts are derived from them.[8]

[7] For the matter-density ontology, we can postulate the following law—the Matter Density Equation:

$$m(x,t) = tr(M(x)W(t)) \tag{3}$$

where x is a physical space variable, $M(x) = \sum_i m_i \delta(Q_i - x)$ is the mass-density operator, which is defined via the position operator $Q_i\psi(q_1, q_2 \ldots q_n) = q_i\psi(q_1, q_2 \ldots q_n)$.

[8] The status of space-time in this theory is an interesting issue. A defender of the Everettian cosmic void scenario may entertain the possibility that space-time is not fundamental and is also emergent from the strongly deterministic laws—perhaps using ideas from the research program initiated in

4. Challenges

Can the Everettian cosmic void scenario be made empirically adequate? That question depends on a further question: can the Everettian picture be made empirically adequate? Mover, even if the standard Everettian picture can be empirically adequate, the Everettian cosmic void may suffer from additional metaphysical and semantic challenges. We discuss some of them in this section.

4.1 The Semantic Challenge

We start with a semantic challenge: What are the fundamental laws about if there is no material ontology at the fundamental level? What do the theoretical terms refer to?

On the standard picture, the fundamental laws are about the fundamental material ontology. The terms in the fundamental laws refer to properties of fundamental matter (e.g. the mass of point particles). Since there is no material ontology in the cosmic void scenario, this way of thinking about *aboutness* and reference need to be revised.

One possible defense of the cosmic void scenario is to invoke certain non-fundamental objects to make sense of the laws. The Initial Projection Hypothesis, for example, refers to an initial density matrix. But what is the meaning of the initial density matrix if there is nothing material at the fundamental level of which it is about? One possibility is to appeal to the effects it has on *non-fundamental* worlds and objects. To understand the meaning of the Initial Projection Hypothesis, it may suffice to understand it in terms of the behaviors of tables and chairs, which are derived from the hypothesis. After all, we make sense of fundamental laws by their effects on observable objects, such as the behaviors of pointers and measurement instruments. How their behaviors are connected to more fundamental facts requires theoretical postulates. Our epistemic access to the fundamental material ontology is rather limited anyways.[9]

Carroll and Singh (2018). However, this move will be in tension with the arguments discussed in North (2018).

[9] Thinking about the metaphysics of fundamentality, Bernstein (2020) suggests that we should take seriously the possibility that the middle level inhabited by medium-sized dry goods such as tables and chairs is the most fundamental level of reality. Middle-level fundamentalism provides an alternative to the usual debate between microphysical-fundamentalism and cosmos-fundamentalism. This is not the metaphysical picture I favor (and neither does Bernstein endorse middle level fundamentalism). However, if that view is possible, then presumably the fundamental laws in such a world are primarily about the behaviors of tables and chairs. Hence, it should also be possible that we can make sense of the fundamental laws in the cosmic void scenario by looking at their effects on non-fundamental objects.

To be sure, some might object that fundamental laws cannot be about anything other than fundamental things or properties.[10] If one holds this view and thinks this view is justified, one would dismiss the possibility of the cosmic void scenario. However, one might think it is *better* if the laws refer only to fundamental objects and their properties but remain agnostic whether we are justified in imposing such a condition on all physical theories. One can maintain that the principle is a desirable theoretical virtue to be balanced with other considerations. For example, Hicks and Schaffer (2017) suggest that this principle is not always true. However, I think that it is a significant cost if we have to give up this principle to entertain the cosmic void scenario.

4.2 The Metaphysical Challenge

Next, we have a metaphysical challenge: How can non-fundamental facts about tables and chairs depend on facts about non-matter (laws)?

On the standard picture, the non-fundamental facts are ultimately explained (at least in part) by fundamental facts about fundamental matter. For example, a typical reductionist would like to reduce facts about tables and chairs into facts about the arrangements of particles and configurations of fields. In the cosmic void scenario, the fundamental facts do not include such facts. So the old way of thinking about ontological dependence needs to be revised if the non-standard picture is to succeed.

A defender of the cosmic void scenario may point to an ambiguity about "fundamental facts about fundamental matter." At the fundamental level, there may not be any facts about matter. However, at some non-fundamental level, there can be facts about matter that are as fundamental as it gets in the cosmic void picture. For example, in the Everettian cosmic void scenario, the bottom level of reality consists in fundamental laws: the von Neumann equation, the Initial Projection Hypothesis, and the matter-density equation. From those equations, we can derive the values of the matter-density field, which can exist non-fundamentally but at a level that is more fundamental than any other material objects. Tables and chairs would emerge at a much higher level from the matter-density field. Therefore, there can be a vestige remaining about the intuition that non-fundamental facts of tables and chairs should bottom out in terms of fundamental matter: it is just that the most fundamental facts about matter will be explained by even more fundamental facts about laws of nature.

Another metaphysical challenge to the cosmic void scenario is related to the issue of completeness in the foundations of quantum mechanics. Is the wave

[10] For example, see discussions in Lewis (1983), and Sider (2011).

function all there is or is there something else in the material ontology? In the framework of GRW theory, several answers are available (Allori et al. 2008). First, there is the bare GRW theory (GRW0) where the only thing in the material ontology is the wave function. Second, there is the GRW theory with a matter density ontology (GRWm) that is similar to the matter-density version of the many-worlds interpretation (Sm) we discussed in section 3.1. The definition of the matter density is the same in both theories. However, they differ in the dynamics of the wave function. They also have different interpretations of the nature of probability. Third, there is the GRW theory with a flash ontology (GRWf), where the matter-density ontology is replaced by discrete events in space-time.

Both GRWf and GRWm postulate something fundamental and material in physical space-time. (The flash ontology and the matter-density ontology are called the *primitive ontology* in the literature.) The configurations of matter in space-time are derived from the universal wave function. So in principle we just need the information about the wave function to derive the information about everything else. Maudlin (2007) suggests that even though the GRW wave function is *informationally complete* in this sense, it should not be thought of as *ontologically complete* in the sense that all there is in the theory is the wave function. After all, which material ontology is privileged in the GRW theory? Is it the flash ontology or the matter-density ontology? They are two different ways to define the material ontology in space-time. A defender of GRW0 may respond that both ontologies can be taken as derivative physical structures that we can use to make precise the connection between theory and evidence. However, whether the response is acceptable may depend on one's view about primitive ontology.

We can ask a similar question about the cosmic void scenario. Even though we can derive everything from the strongly deterministic laws, it only shows that the laws are informationally complete, and it does not follow that they are ontologically complete. It requires a further assumption and perhaps additional justification to postulate that the laws are all there is.

At this point, a defender of the cosmic void scenario may appeal to Ockham's razor: it is more parsimonious to postulate just the strongly deterministic laws than to postulate both the laws plus some material ontology. This is different from the application of the razor in a merely deterministic universe: since we can derive all facts about the past and the future from the laws and the complete state of the universe at some time, perhaps we can get rid of all the other times (Maudlin 2007: 3153). We can care about minimizing the kinds of things without caring so much about minimizing the number of things in the same category. The application of Ockham's razor in the strongly deterministic universe gets rid of an entire category of things: fundamental material ontology. The application of the razor in a merely deterministic universe only gets rid of the vast number of times saving just one.

There is a strong intuition that no theory can be entirely satisfactory unless it postulates some kind of ordinary material ontology. However, if one is antecedently sympathetic to a *wave function monist* interpretation of GRW and accepts GRW0 as a satisfactory physical theory, perhaps one should be more open to reject the intuition. On GRW0, the quantum state (represented by a wave function) is the only thing that exists. But it is nothing like the ordinary material ontology of particles and fields. It is defined on a vastly high dimensional space, and the recovery of ordinary facts about tables and chairs is through some abstract mathematical derivations that look nothing like the ordinary "causal" explanations in space-time. Someone who accepts GRW0 will be happy to accept that ordinary facts about tables and chairs are just emergent patterns in the high-dimensional wave function. But the cosmic void scenario also locates the non-fundamental facts in some unfamiliar fundamental facts—facts about fundamental laws of nature. If one is happy to accept the emergence of non-fundamental facts from a radical material ontology, then one should be more open to accept the emergence of non-fundamental facts from strongly deterministic laws.

4.3 The Empirical Challenge

Finally, we have an empirical challenge: How can the Everettian Wentaculus theory be empirically adequate when every possible experimental outcome is realized? How can the Born rule of probability make sense in such a world?

This is a familiar challenge to many-worlds interpretations of quantum mechanics. It applies to both the Everettian Mentaculus and the Everettian Wentaculus (as well as the cosmic void scenario). It is also related to the emergence of contingency in a strongly deterministic universe. All versions of the many-worlds interpretation face a general problem of probability: if all outcomes of the experiments actually occur, how do we make sense of non-trivial Born rule probabilities? For example, in the case of Schrödinger's cat, we can set up the experiment such that quantum mechanics says the probability that the cat will be alive is 0.3 and the probability that the cat will be dead is 0.7. But if both outcomes are realized the cat is alive in one branch and the cat is dead in another, what to make of such probabilities? Defenders of Everett, such as Deutsch (1999) and Wallace (2012), appeal to Savage-style decision theory. There is also a line of defense developed by Sebens and Carroll (2016) that appeal to rational norms governing self-locating probabilities.[11]

[11] Everett himself appealed to arguments about long run frequencies and typicality of branches. See (Barrett 2017).

But these defenses of Everett make two kinds of assumptions. First, contrary to the Wentaculus, the quantum state of the universe is assumed to be pure, represented by a wave function. A technical problem for the Everettian Wentaculus is how to adapt the arguments in the case when the universe is in a fundamental mixed state, represented by a density matrix. The technical challenge, in my opinion, is not difficult to overcome but there is work to be done. Second, they also postulate some principles of rationality in order to derive the Born rule. In the case of Wallace (2012), the Savage–Wallace axioms of preferences are far from being rationally obligatory. In the case of Sebens and Carroll (2016), the epistemic separability principle seems to be open to counter-examples. So the hard problem is to make a strong case to justify these assumptions in the proof of the Born rule. I am pessimistic about the prospects of this project, but perhaps the difficulties can be overcome. If the technical problem of extension of the argument to mixed states can be solved, then the Everettian Wentaculus fares no worse than the standard Everettian theories in the literature, at least with respect to the problem of probability.

If this project of justifying Born-rule probability in the Everettian framework can succeed, it can also help answer the question: how can contingency emerge in a strongly deterministic Everettian universe like the Everettian Wentaculus? The answer lies in the emergence of the branching structure with a multiplicity of different macroscopic histories (worlds) and the interpretation of branch weight as probability. We leave this project to future work.

5. Conclusion

I have introduced a non-standard picture of the physical universe called the cosmic void. I have discussed the general possibility as well as a concrete example in the Everettian Wentaculus theory. The possibility of the cosmic void is tightly connected to the possibility of strong determinism. In the cosmic void scenario, at the most fundamental level of reality, there are only fundamental laws of nature and no material ontology. Facts about matter can emerge at a non-fundamental level: they can be derived from the laws if they are strongly deterministic. The possibility is certainly unfamiliar and faces difficult challenges, but it deserves careful and critical scrutiny. I hope the picture outlined above can interest other people and lead to further explorations of this topic.

There is an interesting connection between the cosmic void scenario and the ontologist nihilist's picture of a fundamental ontology without any objects. According to ontological nihilism, there are no objects at the fundamental level. This view is discussed in, for example, Hawthorne and Cortens (1995) and Turner (2010). Dasgupta (2009) develops a related view where facts involving individual objects are paraphrased away. Maxwell (ms) also develops a closely related view according to which objects do not play an indispensable role in metaphysics. In the nihilist/anti-object framework, ordinary sentences involving objects are to be paraphrased in terms of non-objectual language that does not ontologically commit us to the existence of objects.

The cosmic void scenario is compatible with that picture, but the two approaches are quite different. The heavy lifting in the cosmic void scenario is done not by revising logic or semantics, but by using a particular kind of physical theory to derive non-fundamental facts about matter from fundamental laws alone. Hence, the cosmic void possibility is at the same time more radical and more modest. It is more radical in that we do not postulate even fundamental properties, general facts, bare facts, or structural facts (Ladyman and Ross 2007) in the fundamental ontology. It is more modest in that the strategy is not supposed to work in general, but only in some special physical theories where strong determinism holds. Hence, we have two potential strategies for getting rid of fundamental material ontology: the metaphysical strategy and the scientific strategy. Both are difficult positions to defend. On the metaphysical side, we need to show that the relevant kind of semantic and logical maneuver results in a satisfactory paraphrase of the original object-language. On the scientific side, we need to show that (1) the actual laws are strongly deterministic and (2) a strongly deterministic cosmic void scenario can be empirically adequate.

Can they be made to work? I am not sure. Even if neither strategy works, we can learn something from these examples: they may give us additional reasons why material objects are crucial for formulating successful physical theories and writing the book of the world.

Acknowledgements

I would like to thank the editors of this volume—Sara Bernstein and Tyron Goldschmidt—for useful written comments on an earlier draft of this paper. I am also grateful for helpful discussions with David Chalmers, Sam Elgin, Bixin Guo, Mario Hubert, Joseph Martinez, Mark Maxwell, Daniel Rubio, Ayoob Shahmoradi, J. Robert G. Williams, and the audience at the 2020 Central APA Symposium "Metaphysics Without Ontology."

References

Albert, David Z. (1992). *Quantum Mechanics and Experience* (Cambridge, MA: Harvard University Press).

Albert, David Z. (2000). *Time and Chance* (Cambridge, MA: Harvard University Press).

Albert, David Z. (2015). *After Physics* (Cambridge, MA: Harvard University Press).

Allori, Valia, Sheldon Goldstein, Roderich Tumulka, and Nino Zanghì (2008). "On the Common Structure of Bohmian Mechanics and the Ghirardi–Rimini–Weber Theory", *British Journal for the Philosophy of Science*, **59** (3): 353–89.

Allori, Valia, Sheldon Goldstein, Roderich Tumulka, and Nino Zanghì (2010). "Many Worlds and Schrödinger's First Quantum Theory", *British Journal for the Philosophy of Science* **62** (1): 1–27.

Barrett, Jeffrey A. (2017). "Typical Worlds", *Studies in History and Philosophy of Science Part B: Studies in History and Philosophy of Modern Physics* **58**: 31–40.

Bell, John (1990). "Against 'Measurement'", *Physics World* 3 (8): 33.

Bernstein, Sara (2020). "Could a Middle Level Be the Most Fundamental?", *Philosophical Studies*, online first.

Bohm, David (1952a). "A Suggested Interpretation of the Quantum Theory in Terms of 'Hidden' Variables: I", *Physical Review* **85**: 166–79.

Bohm, David (1952b). "A Suggested Interpretation of the Quantum Theory in Terms of 'Hidden' Variables: II", *Physical Review* **85**: 180–93.

Boltzmann, Ludwig (1964). *Lectures on Gas Theory*, trans. Stephen G. Brush (Berkeley: University of California Press; repr. New York: Dover Publications, 1995) [German orig. *Vorlesungen über Gastheorie* (Leipzig: J. A. Barth, 1896–8)].

Callender, Craig (2004). "Measures, Explanations and the Past: Should 'Special' Initial Conditions be Explained?", *British Journal for the Philosophy of Science* 55 (2): 195–217.

Carroll, Sean M. and Ashmeet Singh (2018). "Mad-Dog Everettianism: Quantum Mechanics at Its Most Minimal", *arXiv*: 1801.08132.

Chen, Eddy Keming (2018). "Quantum Mechanics in a Time-Asymmetric Universe: On the Nature of the Initial Quantum State", *British Journal for the Philosophy of Science*, forthcoming.

Chen, Eddy Keming (2019a). "Quantum States of a Time-Asymmetric Universe: Wave Function, Density Matrix, and Empirical Equivalence", *arXiv*: 1901.08053.

Chen, Eddy Keming (2019b). "Realism about the Wave Function", *Philosophy Compass* **14**(7): e12611.

Chen, Eddy Keming (2020a). "Bell's Theorem, Quantum Probabilities, and Superdeterminism", in Knox, Eleanor, & Wilson, Alastair (eds), *The Routledge Companion to the Philosophy of Physics*, ed. Eleanor Knox and Alastair Wilson (London: Routledge) Forthcoming.

Chen, Eddy Keming (2020b). "Nomic Vagueness", *arXiv*: 2006.05298.

Chen, Eddy Keming (2020c), "Time's Arrow in a Quantum Universe: On the Status of Statistical Mechanical Probabilities", in *Statistical Mechanics and Scientific Explanation: Determinism, Indeterminism and Laws of Nature*, ed. Valid Allori (Singapore: World Scientific).

Dasgupta, Shamik (2009). "Individuals: An Essay in Revisionary Metaphysics", *Philosophical Studies* **145**: 35–67.

De Broglie, Louis (1928). "Nouvelle dynamique des quanta", in *Electrons et Photons: Rapports et discussions du cinquième Conseil de Physique tenu à Bruxelles du 24 au 29 octobre 1927 sous les auspices de l'Institut international de physique Solvay* (Paris: Gauthier-Villars), pp. 105–32.

Deutsch, David (1999). "Quantum Theory of Probability and Decisions", *Proceedings of the Royal Society of London Series A: Mathematical, Physical and Engineering Sciences* **455** (1988): 3129–37.

Everett, Hugh, III (1957). "'Relative State' Formulation of Quantum Mechanics", *Reviews of Modern Physics* **29** (3): 454–62.

Feynman, Richard (1965). *The Character of Physical Law* (London: British Broadcasting Corporation; repr. Cambridge, MA: MIT Press, 2017).

Ghirardi, G. C., A. Rimini, and T. Weber (1986). "Unified Dynamics for Microscopic and Macroscopic Systems", *Physical Review D* **34** (2): 470–91.

Ghirardi, Giancarlo and Angelo Bassi (2020). "Collapse Theories", *Stanford Encyclopedia of Philosophy*, ed. Edward N. Zalta (Summer 2020 edn).

Goldstein, Sheldon (2001). "Boltzmann's approach to statistical mechanics", in *Chance in Physics: Foundations and Perspectives*, ed. J. Bricmont, D. Dürr, M. C. Galavotti, G. Ghirardi, F. Petruccione, and N. Zanghi (Berlin: Springer), pp. 39–54.

Goldstein, Sheldon (2017). "Bohmian Mechanics", *Stanford Encyclopedia of Philosophy*, ed. Edward N. Zalta (Summer 2017 edn).

Hawthorne, John O'Leary- and Andrew Cortens (1995). "Towards Ontological Nihilism", *Philosophical Studies* **79**: 143–65.

Hicks, Michael Townsen and Jonathan Schaffer (2017). "Derivative Properties in Fundamental Laws", *British Journal for the Philosophy of Science* **68** (2): 411–50.

Ladyman, James and Don Ross with David Spurrett and John Collier (2007). *Every Thing Must Go: Metaphysics Naturalized* (Oxford: Oxford University Press).

Lewis, David (1983). "New Work for a Theory of Universals", *Australasian Journal of Philosophy* **61**: 343–77.

Loewer, Barry (2012). "Two Accounts of Laws and Time", *Philosophical Studies* **160** (1): 115–37.

Loewer, Barry (forthcoming). "The Mentaculus Vision", in *Time's Arrow and World's Probability Structure*, ed. Barry Loewer, Brad Weslake, and Eric Winsberg (Cambridge, MA: Harvard University Press).

Maudlin, Tim W. E. (2007). "Completeness, Supervenience and Ontology", *Journal of Physics A: Mathematical and Theoretical* **40** (12): 3151–71.

Maxwell, Mark (ms). "Metaphysics Without Ontology: The Bare Facts", 2019.

Myrvold, Wayne (2017). "Philosophical Issues in Quantum Theory", *Stanford Encyclopedia of Philosophy*, ed. Edward N. Zalta (Spring 2017 edn).

North, Jill (2018). "A New Approach to the Relational–Substantival Debate", in *Oxford Studies in Metaphysics*, vol. 11, ed. Karen Bennett and Dean W. Zimmerman (Oxford: Oxford University Press), pp. 3–43.

Penrose, Roger (1979). "Singularities and Time-Asymmetry", in *General Relativity: An Einstein Centenary Survey*, ed. S. W. Hawking and W. Israel (Cambridge: Cambridge University Press), pp. 581–638.

Penrose, Roger (1989). *The Emperor's New Mind: Concerning Computers, Minds, and the Laws of Physics* (Oxford: Oxford University Press).

Sebens, Charles T.and Sean M. Carroll (2016). "Self-Locating Uncertainty and the Origin of Probability in Everettian Quantum Mechanics", *British Journal for the Philosophy of Science* **69** (1): 25–74.

Sider, Theodore (2011). *Writing the Book of the World* (Oxford: Clarendon Press).

Turner, Jason (2010). "Ontological Nihilism", in *Oxford Studies in Metaphysics*, vol. 6, ed. Karen Bennett and Dean W. Zimmerman (Oxford: Oxford University Press), pp. 3–54.

Vaidman, Lev (2018). "Many-Worlds Interpretation of Quantum Mechanics", *Stanford Encyclopedia of Philosophy*, ed. Edward N. Zalta (Fall 2018 edn).

Wallace, David (2012). *The Emergent Multiverse: Quantum Theory according to the Everett Interpretation* (Oxford: Oxford University Press).

Wallace, David and Christopher G. Timpson (2010). "Quantum Mechanics on Spacetime I: Spacetime State Realism", *British Journal for the Philosophy of Science* **61** (4): 697–727.

9
Ballot Ontology

Roberto Casati and Achille C. Varzi

The U.S. presidential election of 2000 was crucially decided in Florida. And, in Florida, the election hinged crucially on a peculiar sort of question: Does this ballot have a hole? "Yes, it does", so the ballot is valid and ought to be counted. "No it doesn't", and the ballot must be discarded. If only one could tell! Where were the hole experts when we needed them? Eventually the matter was thwarted by the Supreme Court and we all gave up. But we did learn something. We learned that even the destiny of a Presidential Election, if not more, might ultimately depend on one's criteria for identifying holes. And we mean the holes themselves, not their material surroundings. The surroundings were all there and everyone could detect them. But what about the rest, the missing bits, those little chunks of immaterial non-being that even the Votomatic was confused about?

1. Too Close to Call

Rewind to the facts. Tuesday, November 7, 2000, election day. The 54th presidential election is coming to an end. By the evening, the incumbent vice president and Democratic nominee, Al Gore, was comfortably ahead in the Northeastern states (except New Hampshire), most of the Upper Midwest, New Mexico, the Pacific Coast, and Hawaii. For his part, the Republican candidate, George W. Bush, was winning the rural Midwest and most Southern states (including Arkansas) along with Indiana, Ohio, the Rocky Mountain states, Arizona, and Alaska. As the final results were tallied on Wednesday morning, it was clear that Gore had 250 electoral votes and Bush 246, with 270 needed to win the election. Three states were still too close to call, but two of them, Oregon and Wisconsin, would only bring 18 votes altogether; the election came down to the third state, Florida, whose 25 electoral votes would be decisive regardless. And then the mess began.

While Oregon and Wisconsin went to Gore, the vote count in Florida turned out to be in favor of Bush, but only by a margin of 1,784.[1] That was well below the margin of error tolerated by the state law, which is set at 0.5 percent of the total

[1] For these and related details we rely on Dionne and Kristol (2001), Gillman (2001), Dover (2003), Whitman (2003), and deHaven-Smith (2005) along with the chronicles by *The New York Times*

Roberto Casati and Achille C. Varzi, *Ballot Ontology* In: *Non-Being: New Essays on the Metaphysics of Non-Existence.* Edited by: Sara Bernstein and Tyron Goldschmidt, Oxford University Press (2021). © Roberto Casati and Achille C. Varzi. DOI: 10.1093/oso/9780198846222.003.0009

votes cast (in this case, 0.5 percent of 5,816,486, hence 29,082), so a statutory machine recount was immediately called in all of Florida's sixty-seven counties.[2] The recount began on Wednesday afternoon and was reportedly completed on Friday, November 10.[3] It confirmed Bush's lead, but the margin dwindled even further: only 327 votes separated the two candidates.[4] Meanwhile the Democrats requested a third, manual recount—again pursuant to Florida state law[5]—in four counties where Gore had prevailed but the ballots were particularly in dispute: Broward, Miami-Dade, Palm Beach, and Volusia. By the statutory deadline of November 14, Volusia County completed its manual recount and certified the results; Bush's lead decreased by another 98 votes. For the other three counties, the deadline had been upheld and, indeed, permission to proceed with the manual recount was still pending.[6] It was granted to Broward and Palm Beach on November 16 and to Miami-Dade the next day. On November 21, the Florida Supreme Court allowed the three ongoing manual recounts to delay certification until 5 p.m. on Sunday, November 26, "in order to allow maximum time for contests".[7] In the meantime, the returns of all overseas absentee ballots had been certified and Bush's lead had grown to 930 votes. On November 22, the canvassing board of Miami-Dade determined that the manual recount was "impossible to complete"[8] and voted unanimously to suspend it, resubmitting the election returns previously compiled by machine (with a slight correction of seven more votes for Gore). Palm Beach County continued its recount and turned in the results on the 26th, but at 7 p.m., two hours after the deadline. Too late. Only the Broward recount (netting Gore 567 votes) was accepted as valid along with Volusia's. Concurrently, Bush had picked up another 181 votes from adjustments

correspondents (2001), the report of the U.S. Commission on Civil Rights (2001), and the essays in Jacobson and Rosenfeld (2002) and Watson (2004).

[2] "If the returns for any office reflect that a candidate was defeated or eliminated by one-half of a percent or less of the votes cast for such office [...] the board responsible for certifying the results of the vote [...] shall order a recount of the votes cast with respect to such office"; *The 2000 Florida Statutes*, Title IX, Chapter 102, Section 141(4).

[3] Reportedly, for it appears that no less than eighteen counties—a quarter of Florida votes—did not rerun their ballots through the machines but simply checked the numbers and the proper functioning of their computer software. See Toobin (2001: 66) and Kaplan (2001: 51).

[4] This result was never released officially and corresponds to the Associated Press's tally. According to the Florida secretary of state's office, Bush's lead was higher (960 votes), but with 66 of 67 counties reporting. See "Cast and Chronology" in Jacobson and Rosenfeld (2002: 26).

[5] "Any candidate whose name appeared on the ballot, any political committee that supports or opposes an issue which appeared on the ballot, or any political party whose candidates' names appeared on the ballot may file a written request with the county canvassing board for a manual recount"; *The 2000 Florida Statutes*, Title IX, Chapter 102, Section 166(4).

[6] On the complex issues concerning the November 14 deadline, see Greene (2001: ch. 3).

[7] *Palm Beach County Canvassing Board v. Harris*, 772 So. 2d 1220, Section X.

[8] *Miami-Dade County Democratic Party v. Canvassing Board*, 773 So. 2d 1179.

to previous calculations.[9] Final verdict: Bush won by 537 votes, taking Florida to become the forty-third president of the United States with just one electoral vote above the 270 threshold.[10]

2. Too Moot to Count

Why a manual recount? Why the extended deadline? And why did the Palm Beach canvassing board *fail* to meet the deadline, or the Miami-Dade board resolve that it *could not* be met? In the latter case, the board's initial decision was only to halt the recount of all the 654,000 Miami-Dade ballots and limit the manual recount to the "undervotes", a total of 10,750 ballots that did not register presidential preferences in tabulating machines. That would have been understandable, especially because the operation was to take place over Thanksgiving; manually examining 654,000 ballots would have been a grueling task.[11] But 10,750? Why did the Miami-Dade board eventually give up on that task, too, halting their recount entirely?[12]

The first question—why a manual recount—is peculiarly interesting. In most countries, votes are *always* counted manually (or so it was in 2000). In the United States, however, the vast majority of electoral districts employ mechanical or electronic voting systems of various sorts, and for this reason the initial count (and, when necessary, the first recount) is machine-based. This is a long-standing tradition, whose origins go all the way back to the late nineteenth century. The first voting machine—a lever device known as the "Automatic Booth"—made its official debut in 1892, at a local election in Lockport, New York. By 1930, twenty-four states had passed laws permitting the use of similar devices and by the 1960s

[9] Specifically, 52 additional votes from Nassau County (which decided to use the election night tabulation instead of the machine recount) and an extra 129 votes out of previously disqualified military overseas ballots.

[10] The verdict itself was not immediate. Florida Secretary of State Katherine Harris announced Bush's victory on the evening of the 26th, but Gore sued to contest it and appealed to the Florida Supreme Court on November 29. On December 8 the Florida justices, on a 4 to 3 vote, resolved to request immediate manual recount of all undervotes. The next day the U.S. Supreme Court, on a 5 to 4 decision, halted the recount on Bush's appeal, pending a hearing from both parties. After the hearing, on December 12, the U.S Court overturned the Florida Court on another 5 to 4 decision (the five conservative justices against the four liberal justices) in the famous *Bush v. Gore* case (531 U.S. 98). It is only at that point and in such terms that Harris's original declaration was confirmed, and the following day Gore appeared on national TV to concede. Needless to say, this administrative coda fueled a huge political and legal debate of its own. For detailed reports and discussion, see Bugliosi (2001), Dershowitz (2001), Gillman (2001), Jarvis, Coleman, and Burris (2001), Posner (2001), and the essays in Sunstein and Epstein (2001) and Ackerman (2002). For a sense of the extensive scholarly literature that followed soon afterwards, see Weinberg (2002) and the legal works surveyed in Cole (2006). For an extended historical perspective, see Foley (2016).

[11] Though not much worse than in Palm Beach (about 463,000 ballots), let alone Broward (588,000). In Volusia, which met the early deadline of November 14, the ballots were 184,000.

[12] The initial decision took place at 6:15 a.m.; the second, four hours later, at 10:14 a.m.

voting machines of various types had replaced the traditional paper ballots in most parts of the country.[13] The general idea has been that a machine-aided voting system carries a number of advantages, both from a practical perspective (fast vote casting, consistent vote tallying, quick returns, etc.) and with regard to a number of concerns about vote fraud that were and continue to be common with paper ballots (rigging, misrecording, spoiling, interpretation biases, etc.). Outdated as they may sound, the opening words of Jacob H. Myers's patent for the Automatic Booth are telling:

> My present invention relates to voting machines, and has for its objects to provide one by the employment of which an honest vote can be had and counted without liability of voters being intimidated, the balloting being secret, or of their voting more than once for the same candidate or different candidates for the same office, and as the votes are counted as fast as the voter indicates his preference the total number cast for each candidate can be ascertained rapidly and accurately at the close of the polls.[14]

On the other hand, there remains the fact that few machines work perfectly all the time. In particular, all voting and counting machines have a margin of error. Usually this margin is so small as to be immaterial when it comes to determining the outcome of an election, since the difference in votes among the top candidates is considerably larger. When the difference in votes is smaller than the margin of error, however, we have a problem, and the standard solution is to revert to traditional means. Machines are faster and less passionate than us, but when the need arises we must step in and take up the job ourselves. As election law experts put it:

> An attempt on election night to manually count and to record votes in multiple races from thousands of ballots could result in substantial human error. A manual recount of machine-counted ballots, however, generally is limited to only one race and takes place in a strictly regulated process designed to ensure that the interests of all candidates are protected and that the most accurate count possible is achieved.[15]

To be sure, U.S. state laws do not *require* a manual recount, and indeed Florida's initial statutory recount was done again by machine. But most state laws, including Florida's, *allow* for a manual recount so long as a written request is submitted containing "a statement of the reason the manual recount is being requested"

[13] For a history, see Saltman (2006) and Jones and Simons (2012). [14] Myers (1889: 1).

[15] We are quoting from Bickerstaff (2001: 443), but similar remarks may be found across the literature, including texts from the heydays of voting machines. See e.g. Zukerman (1925) and Harris (1934: ch. 7) (the same Harris who later patented the Votomatic used in Florida; see below).

(by a candidate or political party, if not by the voters).[16] It is this right that the Gore team exercised in the four counties of Broward, Miami-Dade, Palm Beach, and Volusia, and in each case the motivation was precisely that meticulous human examination is the best solution. In Gore's own words:

> There is a simple reason that Florida law, and the law in many other states, calls for a careful check by real people of the machine results in elections like this one. The reason? Machines can sometimes misread or fail to detect the way ballots are cast. And when there are serious doubts, checking the machine count with a careful hand count is accepted far and wide as the best way to know the true intentions of the voters.[17]

Now, normally humans are of course much slower than machines, so these considerations provide an answer also to the second question raised above—why the extended deadline. One might still wonder how exactly the extension was determined, and on what grounds, but never mind; every deadline is bound to be somewhat arbitrary and manual recounts are no exception. The question that really cries for an answer is the third, concerning Palm Beach's failure and Miami-Dade's refusal to *meet* the deadline. What happened?

Here is where nobody, not even the members of the Supreme Court of Florida, could predict the trouble. To see why, we need to look into the details of the ballots used in those counties. Machine-aided voting systems are widespread in the U.S. but, alas, that doesn't mean they are the same everywhere. Authorized systems may vary from state to state and, indeed, from county to county within a state. During the 2000 election, one of Florida's counties (Union) was still using traditional paper ballots; the other counties used machine-aided systems of four main kinds: optical scan with central tabulation (16 counties), optical scan with precinct tabulation (25 counties), punch cards (24 counties), and machine lever (one county).[18] This variety of systems showed up also in the four counties involved in the manual recount. Volusia used a precinct-based optical-scan device called "AccuVote"; Broward, Miami-Dade, and Palm Beach employed instead a punch-cards system, the "Votomatic". The difference between these systems turned out to be crucial.

[16] See again *The 2000 Florida Statutes*, Title IX, Chapter 102, Section 166(4).

[17] Televised statement aired on the evening of November 15, here quoted as recorded by <IBT>*The New York Times*, November 16, p. A28</IBT>. Cf. Bush's response: "That's why my campaign supported the automatic recount of all the votes in Florida. [...] Manual counting, with individuals making subjective decisions about voter intent, introduces human error and politics into the vote-counting process. Each time these voting cards are handled, the potential for errors multiplies. Additional manual counts of votes that have been counted and recounted will make the process less accurate, not more so." (Ibid.)

[18] See U.S. Commission on Civil Rights (2001: ch. 8 and Appendix).

The AccuVote (based on a 1988 patent)[19] works much like in a standardized test: ballot cards have the candidates' names preprinted next to an empty oval, each voter records their choice by filling in the relevant oval with a pencil, and the ballot is fed into a high-speed optical reader and tabulating device. (The system is precinct-based in that the machine is programmed to immediately "kick out" ballots that appear to have been voted incorrectly and voters have the opportunity to correct any errors before they leave the precinct.) As it happened, on the night of the election the AccuVote didn't work so well,[20] and further problems showed up the next day.[21] When it came to the manual recount, however, the process went smoothly. Counters were instructed to examine each ballot for no more than fifteen seconds and any ballot where the voter's intention was clear—and agreed upon by the Republican and Democratic observers—was counted; questionable ballots were turned over to the canvassing board for a final decision.[22] A tedious job, but an easy one, like checking old-fashioned paper ballots. No wonder Volusia County managed to complete it by the early deadline of November 14.

With the Votomatic (from a 1965 patent),[23] voters express their preferences by punching a hole in a pre-scored ballot on which each candidate's name corresponds to a numbered tab. Candidate names are not printed on the ballot itself, which is a standard-size IBM data processing card, but rather on the pages of the ballot holder, and the hole is punched by pressing a stylus that will remove a tiny rectangular chad. The perforated card is then fed into a computerized tabulating device. This seems like a foolproof system, and in the event of a manual count it might even claim a significant advantage over the AccuVote: in principle, all Votomatic ballots look the same. Whereas filling an oval with a pencil is bound to involve a great deal of variation (partial filling, check signs, X marks, etc.), perforating a card by removing a rectangular chad with a stylus is in principle a uniform and perfectly standardized task and tabulating the results should be utterly straightforward, by machine and by hand alike. Unfortunately, this is only true in principle; in practice it's a gamble.

One problem with this system, absent in the AccuVote, is that the matching between a candidate's name and the corresponding tab depends heavily on the design of the pages of the ballot holder, which is another feature that may vary from county to county. In the 2000 election, this turned out to be especially

[19] See Webb (1988). The AccuVote was originally marketed as the Unisys ES-2000.

[20] In Volusia's 216th precinct, the machine recorded that 412 out of 585 registered residents had voted, but a faulty memory card translated that number into +2,813 votes for Bush and −16,022 for Gore (and an extra +9,888 votes to the Socialist Workers Party candidate, James Harris.) This caused TV networks to momentarily call the election for Bush, but the error was discovered and corrected a few hours later, assigning 22 votes to Bush and 193 to Gore (and eight to Harris). See van Natta (2000). For a detailed reconstruction and further details, see Harris (2004: ch. 13).

[21] At 4 p.m. on Wednesday, an election worker discovered that a bag of about 800 ballots was still in the back seat of his car and had not been delivered to county officials (van Natta 2000).

[22] This procedure was openly reported in the news. See e.g. Barabak and Finnegan (2000).

[23] See Harris (1965). The apparatus was first used in 1964 (patent pending) in Georgia.

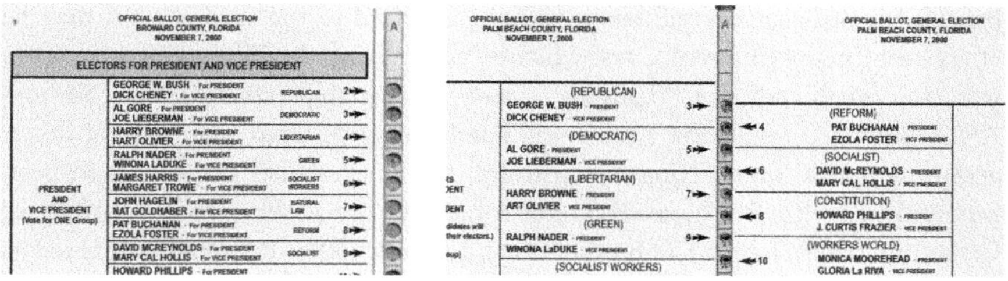

Figure 9.1 A Broward County ballot (left) and a Palm Beach County "butterfly" ballot (right).

critical. Broward and Miami-Dade used ballot holders where the names were printed on a single page, with arrows pointing to a corresponding slot on the right. The first candidate (Bush) would correspond to the first slot, the second candidate (Gore) to the second slot, etc. Palm Beach used instead a "butterfly" ballot holder with names printed on facing pages and arrows pointing to a slot located on the right if the name appeared on the left page, and on the left if the name appeared on the right page, the slots themselves alternating between left and right candidates. Thus, the first name on the left page (Bush) would again correspond to the first slot, but the second name (Gore) would correspond to the third slot, the second slot being reserved for the first name on the right page (the Reform candidate Pat Buchanan). This generated a great deal of confusion and the Democratic party complained bitterly that the design prevented a proper conversion of vote intentions into counted votes.[24] Indeed, it is very likely that many a voter punched the Buchanan hole while intending to vote for Gore, since in Palm Beach Buchanan received almost 20 percent of his total statewide support (about 3,400 votes, i.e. over 2,000 more than anywhere else in Florida, against the 537 votes of Bush's final margin of victory).[25]

Yet the effect of ballot-holder design was only part of the Florida problem. When it came to the manual recount, the members of the canvassing boards of the three counties of Broward, Miami-Dade, and Palm Beach found themselves facing a different and unexpected complication, and this one concerned the ballot themselves. The Votomatic is designed to treat ballots like standard perforated processing cards. But this presupposes that the cards are properly perforated. When people use the Votomatic this is not always the case. The punched ballot may show all sorts of imperfections, beginning with the fact that the chad may not

[24] The complaint was eventually dismissed in *Fladell v. Palm Beach County Canvassing Board*, 772 So. 2d 1240 (December 1).

[25] Buchanan himself immediately acknowledged the anomaly in an interview on NBC: "It's very easy for me to see how someone could have voted for me in the belief they voted for Al Gore" (*The Today Show*, November 9, 2000). For detailed data and analysis, see Brady et al. (2001), Wand et al. (2001), and Smith (2002). For background information on the butterfly ballot (with an interview to the designer), see Pleasants (2004). For a more recent appraisal of the issues, see Chisnell (2016).

have fallen out as expected. It may still be attached to the card. And it may be attached in many different ways, which is why during the Florida election there was so much talk of "hanging chads", "swinging chads", and so on. A computerized tabulation system will handle all such cases some way or other, accepting a vote or rejecting it according to whatever instructions it had been programmed with. But humans do not come with pre-programmed instructions. Given any ballot, a human observer must decide whether it should be treated as valid, and if the ballots are many, if they are thousands and thousands, this task can turn into a hugely demanding, enervating, frustrating challenge. The images of Judge Robert Rosenberg struggling with the Broward County ballots, which filled the first pages of all newspapers for days, are perhaps the most emblematic testament to the nightmare this must have been for Florida's canvassing officials when the manual recounts began. No wonder the Votomatic counties needed an extension beyond the original deadline of November 14. Humans are slower than machines, but they are also less prepared to let things go. On what grounds, exactly, should a canvasser determine whether a defective ballot contains a valid vote for a candidate? What criteria should guide the decision?

The reason why two out of those three counties failed to meet the extended deadline of November 26 lies in the difficulties raised by these questions. And as it turned out, the source of those difficulties was even more difficult to pin down than the criteria one needed to overcome them. For, on closer look, the problem were not the ballots as such. The problem lay elsewhere, and it soon became clear that it lay where no one had the sort of expertise that might have helped. Around November 22, a canvassing board official (probably Rosenberg himself, though it might also have been Judge Charles Burton of Palm Beach County, we do not recall exactly[26]) was interviewed on a TV news program. "Why is the manual recount taking so long?", asked the reporter. "The whole world is waiting, what is the problem?" The canvassing official raised his arms in frustration: "Tell the world that no one here is a hole expert. That is the problem."

3. Experts Wanted

Of course, when it comes to holes the real experts are Argle and Bargle, the two characters of David and Stephanie Lewis's dialogue with that title.[27] It is a masterpiece of philosophical acumen and was originally published in 1970; the

[26] We distinctly remember watching the event on television, though at this time we are unable to track down any recordings or transcripts of the interview.

[27] See Lewis and Lewis (1970). This is arguably the first scholarly work to take holes as a subject worthy of serious inquiry and theorization of its own, though one could make an exception for the short "socio-psychological" treatise of Tucholsky (1931) (published under the pseudonym of Peter Panter). We shall return to Tucholsky below.

Florida canvassers had had plenty of time to read it and digest it, and they would have learned a lot. Nevertheless, there is an important sense in which *that* sort of expertise would have been of little help. The Ludovician theory of holes—as put forward through Argle's pronouncements—is a deflationary theory. It states that holes are material objects. That is not just a way of saying that every hole is always *filled* with matter (air, ballot paper, silver amalgam, luminiferous ether, or whatever it may be). Argle is a nominalist materialist, but even a materialist recognizes that there might be empty holes. So how can something utterly devoid of matter be a material object? Answer:

> The lining of a hole, you agree, is a material object. For every hole there is a hole-lining; for every hole-lining there is a hole. I say the hole-lining *is* the hole.[28]

This may be a good theory of *what* holes are.[29] But as a theory of hole individuation—a theory that would help us determine whether something is a hole and, therefore, when something can be said to have a hole—it is altogether useless. The Ludovician theory presupposes that we know *already* how to determine such facts; it just tells us that, whenever we think there is a hole, the hole is not what we think it is (and where we think it is). In other words, the theory is ontologically and conceptually parasitic. It states that holes are hole-linings, but it doesn't provide any independent criteria for identifying hole-linings. It says that holes supervene on the arrangement of matter,[30] but it doesn't tell us how a certain portion of matter should be arranged in order for it to count as the appropriate supervenience basis. For all its metaphysical elegance, a canvasser using Argle's theory to manually examine the perforated ballots of the Votomatic would have been stuck in a circle.

What sort of "hole expertise" was needed in Florida? The recount official didn't say, but it's clear that the problem was really one of hole individuation, if not hole definition. Several legal and political commentators stressed this point in the days and weeks that followed. Pasquale Pasquino, for instance, put it thus:

> If all goes well, a little square of paper (called a "chad") completely detaches from the rest of the ballot, resulting in a hole that the optical scanning machine counts as a vote. But it is possible that the voter, unskilled in the use of the stylus, inadvertently detaches three, or two, or just one side of the chad. The machine interprets some of these as holes and others not. How do those who recount the ballots, applying Florida law, define a "hole"? Is there a standard, unambiguous,

28 Lewis and Lewis (1970: 207).

29 For the record, we offered some criticisms in Casati and Varzi (1994: ch. 3), returning to the topic in Casati and Varzi (2004).

30 This is the language used in Lewis and Lewis (1996).

> universally accepted definition of what kind of a hole, or vote, is valid in the specific case for every canvassing board charged with the manual recount? And if there is not, and equal holes may be counted differently or different holes may be counted equally, would this be a violation of the principle of equality, understood as granting each citizen an equal vote?[31]

Reference to the principle of equality is important here. The reason why permission to proceed with the manual recounts in Broward, Miami-Dade, and Palm Beach was still pending on November 14 is precisely that Bush's lawyers had sued for a preliminary injunction on the grounds that similarly punched ballots could be tabulated differently in the three counties, owing to the fact that Florida had no statutory standards for hand-counting votes. The Court of the Southern District of Florida eventually denied the request. According to the district judge, the inherent decentralization involved in state electoral and recount procedures was not "a sign of weakness or constitutional injury" but, rather, "one of the strengths" of the U.S. Constitution's method for selection of the President.[32] Later on, however, the U.S. Supreme Court reached exactly the opposite conclusion. The 5 to 4 decision that put a halt to all recounts on December 12, effectively confirming Bush's victory by the 537-vote margin, was based on the assertion that the Florida Court had failed to provide "a constitutional standard for counting votes".[33] We leave it to political theorists to determine who was in the right, the district judge or the Supreme Court. What is clear is that in order for the principle of equality to be safe, the standards for hand-counting votes cannot be left in the vague. Whether enforced state-wise or left up to each county, some standards must be clearly defined. And in the case of punch-card ballots, each definition must refer, directly or indirectly, to some criteria for identifying and counting the holes (or hole-linings) in the ballots themselves. It is these criteria that the recount officials were lacking.

4. Holes and Chads

In the passage quoted above, Pasquino refers to the fact that an unskilled voter may fail to fully detach the chad corresponding to the chosen candidate; the chad may remain attached to the card by one or more sides. It is important that Pasquino did not say the unskilled voter may fail to fully punch a hole; that would have been question-begging. But what is the relationship?

The basic intuition is clear enough. The holes in the ballots are meant to express the voters' preferences. Now, either holes *are* hole-linings, as Argle would have it, or else they *come with* a hole-lining, which is to say some surrounding matter, a

[31] Pasquino (2002: 324).

[32] See *Siegel v. LePore*, 120 F. Supp. 2d 1041 (December 6).

[33] *Bush v. Gore*, 531 U.S. 98.

material host. Either way, a hole is always a hole *in* something: as a *part* (Argle) or as a disjoint, immaterial *guest*. As Kurt Tucholsky once put it:

> The hole is a permanent companion of the non-hole [. . .] there is no such thing as a hole by itself.[34]

In particular, there is no way of *making* a hole by itself; one always makes a hole by acting on something else. And in the case at issue, the something else is the ballot; to see whether and how a voter punched a hole, we have to look at the ballot and see what happened to it. This is where chads enter the picture.

In the ideal case, the ballot will have an easily detectable hole in the sense intended by the Votomatic designers: a little chad fell off. If that were the only legitimate case, hand-counting would be an easy job: the valid votes would correspond to all and only those ballots where a chad has been completely removed. Indeed, under such ideal conditions, the canvassers might even consider an alternative: rather than counting ballots with missing chads, count the chads directly. After all, there is a one-one correspondence between those ballots and the chads that fell to the ground. Provided we know how to match the chads with the candidates (all chads had a number printed on their back), the two procedures would return the same results. However, no one would accept the ideal case as the only criterion for validity. A partially attached chad may still be evidence of a voter's intention, so that criterion would be too strict. This is not just a philosophical intuition; it's the law, in Florida as elsewhere. (Already in 1981, an Indiana case determined that ballots with partially attached chads should be counted insofar as "the intention of the voter could clearly be discerned".[35]) Besides, that is not how the initial machine count works. The Votomatic uses an optical-reading device; so long as its laser "eye" can see through a hole in the ballot, the vote is registered as valid, and this may happen even if the chad is still partially attached. The reason this automated procedure may not be fully reliable is that, in such cases, a lot depends on how the chad hangs; it may let the light through or it may accidentally obstruct the hole. The human eye can "see" better, and that is why a manual recount should in principle reduce the margin of error.[36] But the rules of the game should be the same. Thus, all things considered, counting the chads on the floor would not do; it really is the ballots that must be

[34] Tucholsky (1931: 123). This is also the starting point of our account in Casati and Varzi (1994), where holes are characterized as ontologically dependent (or "parasitic") immaterial entities in the sense just mentioned.

[35] *Wright v. Gettinger*, 428 N.E. 2d (Ind. 1981), 1225. Even Texas, where Bush was governor, had a law to this effect (*State Election Code* of 1993, Title VIII, Chapter 127, Section 130(d)(3)).

[36] Another problem with the Votomatic machines used in Florida (stressed by Gore's lawyers) is that even a perfectly punched ballot might be rejected because a loose chad from *another* ballot blocked the hole. See Dershowitz (2001: 26, 34).

examined.[37] Vote counting requires hole individuation, and hole individuation depends on the range of admissible chad-ballot relationships.

As the pressure for clear standards grew, the Florida canvassers elaborated a number of distinctions that went exactly in this direction. We briefly mentioned them earlier, but it's worth being precise. Beside the ideal case of a clean punch, where the chad is completely dislodged from the ballot, five configurations of imperfect chads were distinguished:[38]

— *Hanging chad*: attached to the ballot by only one corner.
— *Swinging chad*: attached to the ballot by two corners.
— *Tri-chad*: attached to the ballot by three corners.
— *Dimpled chad*: attached to the ballot by all four corners, but somewhat bulging, indented, or otherwise marked.
— *Pierced chad*: attached to the ballot by all four corners, but pierced with a hole.

(Dimpled and pierced chads were sometimes lumped together as *pregnant* chads.) The question then became whether ballots featuring any of these configurations should be treated as expressing a valid vote despite their failure to comply with the voting instructions. It is in this regard that the three counties involved in the manual recount acted autonomously.[39] Nevertheless, it is fair to say that they all followed, if not a common official standard, at least the same approach. And the approach was to pay great attention to the distinctions listed above insofar as they would provide the best information for analyzing and classifying controversial ballots on objective grounds. Unfortunately, this sort of "chadology" (as some election officials started calling it[40]) is full of traps and may create more troubles than solutions.

One general problem is the apparent arbitrariness of any way of discriminating valid from invalid ballots in terms of the number of corners by which the chad is attached. During the manual recount in Florida, there was a remarkable convergence of viewpoints among the counting teams of Broward, Miami-Dade, and Palm Beach. By November 21 (the day when the deadline was extended), all teams were following roughly the same protocol, considering as automatically valid all ballots where the chads had at least two corners poked out (i.e. hanging or swinging chads, collected under the rubric of *dangling* chads) and passing any other ballot to the relevant canvassing boards for

[37] Interestingly, as Safire (2000) notes, initially "chad" was not even a count noun, like "chaff".

[38] Terminology may vary slightly; here we follow Posner (2001: xv).

[39] The voting instructions, too, were heterogeneous. For instance, in Palm Beach they read: "After voting, check your ballot card to be sure your voting sections are clearly and cleanly punched and there are no chips left hanging on the back of the card". In Broward, the instructions read: "To vote, hold the stylus vertically; punch the stylus straight down through the ballot card for the candidates or issues of your choice". (From *Touchston v. McDermott*, 234 F.3d 1133, note 19.)

[40] See e.g. Dugger (2000), Brown (2000), and Whitman (2000). The term never made it to the OED, though it was not a neologism and had been used before; see Dugger (1988: 54).

further review.[41] Why so? Why draw the line at two corners rather than three, or just one? If the idea is that a dangling chad is good enough evidence of a voter's intention, then any other chad with clear signs of the action of the stylus should be treated equally, including tri-chads and pregnant chads. If the thought is that a voter succeeded in expressing a vote so long as the ballot was perforated, then any chad should be treated equally except for dimpled ones (but including pierced chads). As "objective" evidence, any other criterion would seem utterly arbitrary.

To put it differently, the "objective" evidence offered by a ballot may be *morphological* or it may be *topological*. In the first case, any pronounced modification of the *shape* of the ballot would count, covering all five types of defective chads in addition to fully detached ones. In the second case, what matters is rather the topological *genus* of the ballot and any modification thereof would count, ruling out only dimpled chads.[42] Both criteria seem reasonable, and correspond to two different ways of fixing the meaning of 'hole'.[43] It may still be hard to decide between them. But at least the opposition is absolute, whereas any intermediate opposition is up for grabs. Indeed, insofar as the optical-reading device of the Votomatic uses a light beam to check for valid ballots, the topological criterion would seem more appropriate in this context. The morphological criterion is inherently vague and for that reason it would be fair to defer the analysis of dimpled chads (but only those) to the canvassing board.

Moreover, it's worth noting that all intermediate solutions are dangerously fragile also on the practical side. During a manual recount, the mere handling by an election official may alter the condition of a ballot. A swinging chad may accidentally turn into a hanging chad and, more importantly, a tri-chad may accidentally turn into a swinging chad (hence into a valid vote according to the Florida guidelines). Cases of this sort may be rare, but they are documented and, indeed, in other circumstances they caused trouble. Five years after the Florida events, for instance, during the election for the fourth council seat in the city of Pepper Pike, Ohio, a sequence of counts and recounts resulted in one candidate winning by a single vote. The opponent sued the county's Board of Elections precisely on the grounds that the status of one ballot was altered during the manual recount.[44] That such cases may be rare means nothing. Winning an American presidential election by 537 votes is itself an awfully rare event.

[41] Before that date, the guidelines varied. For instance, Palm Beach initially counted all votes where any corner of a chad was punched, whereas Broward originally said its board would not even *consider* votes with anything less than two corners of the chad detached. See e.g. Gold (2000) and Whitman (2003: 75ff).

[42] Counting only ballots with a clean punch would also reflect a topological criterion. We start with *one* thing, a whole ballot, and we end with *two* things, a (perforated) ballot and a separate chad.

[43] On this we refer to Casati and Varzi (1994: ch. 4): "Hollows, tunnels, cavities, and more".

[44] See *Taft v. Cuyahoga Board of Elections*, 854 N.E. 2d 472 (Ohio 2006). For the record, the court ruled in favor of the appellee. This resulted in a perfect tie and the board was enjoined to decide the winner by lot. For a detailed discussion of this and other ticklish cases, see Weinberg (2006).

So much for the arbitrariness problem. There is, in addition, a second set of problems with the sort of chadology used in Florida, and these problems arise regardless, i.e. even if we go along with the Florida canvassers' decision to privilege some chad patterns over others. For that decision is based on a certain taxonomy of relevant cases. Yet whence the taxonomy? Surely there is a difference between a clean punch and a hanging chad, and between a hanging chad and a swinging chad, a tri-chad, a dimpled chad, or a pierced chad. But that is Chadology 101. When the election of a U.S. President is at stake, we better make sure we considered all the options thoroughly. Let's not forget: vote counting requires hole individuation, and hole individuation depends on the range of admissible chad-ballot relationships. It is absolutely crucial that the we identify that range to begin with, and that we do so in a language that is up to the task.

5. Chadology Unveiled

Here is one sense in which the basic taxonomy used by the Florida canvassers was defective. The taxonomy classifies chads in terms of the number of *corners* by which the are attached to the ballot. Yet a chad may also be attached to the ballot by one or more *sides*. It may be thought that the two criteria are equivalent. The passage by Pasquino quoted earlier, for instance, has "sides" instead of "corners", but without any suggestion that the shift is significant. In fact it is, and in several important ways.

To see this, consider the patterns in Figure 9.2. It is natural to think that these patterns, ordered by decreasing number of connecting corners, provide an adequate representation of the different types of chads described by the Florida canvassers, with just a few additional variations of the leftmost pattern to distinguish virgin, untouched chads from pregnant ones. (Let us ignore those variations for the moment.) Indeed, it is safe to assume that these patterns are exactly the ones the canvassers had in mind, for diagrams of this sort, modulo graphic differences, filled the news throughout the days of the Florida manual recount.

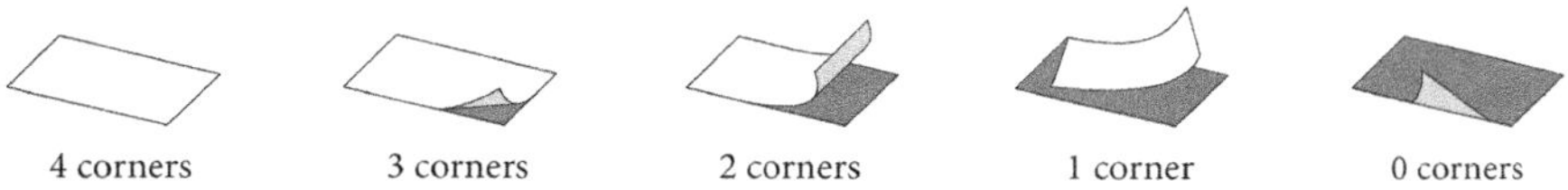

Figure 9.2 The standard way of distinguishing chads based on the number of corners by which they are attached to the ballot (the white background). From left to right: a solid chad, possibly pregnant; a tri-chad; a swinging chad; a hanging chad; and a fully detached, dislodged chad—a clean punch.

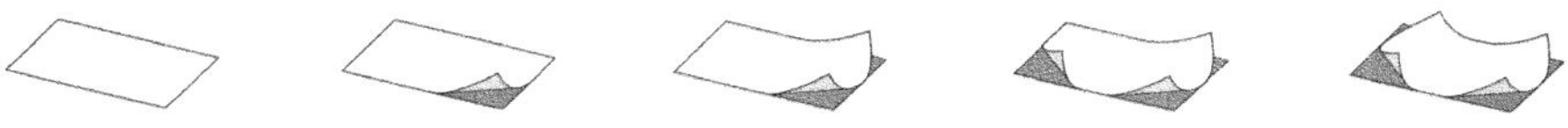

Figure 9.3 An alternative model of the basic taxonomy: regardless of the number of attached corners, each chad is always (partly) attached to the ballot along each of its four sides.

On closer inspection, however, this diagram is far from complete. That is, if all that matters is the number of attached corners, then many other patterns are in principle possible and should be properly recognized.

To begin with, the patterns in Figure 9.3 fit the canvassers' taxonomy just as well. Here the sequence is again ordered by decreasing number of connecting corners; yet each case corresponds to a chad that is (partly) attached to the ballot along each of its four sides. With regard to the first two patterns of Figure 9.2, this additional feature doesn't make a difference; in the other three cases it does.

Note that the difference is not immaterial. On the corner-based definition, the middle chad would be classified as swinging, yet clearly the chad is now more firmly attached to the ballot than the ordinary swinging chad of Figure 9.2. Similarly, the fourth pattern would be classified as a hanging chad even though it is attached just as firmly. And the last case is even more disturbing. The chad is not attached by any corner, yet it obviously violates the idea of a clean punch; far from being fully dislodged, the chad is once again firmly attached to the ballot on each side. Granted, each of these three cases is not likely to occur. Nallot chads are relatively thick, given their tiny size (about 1/8 inch long), and voters are supposed to press the stylus just once. But then, again, who are we to say? Some Votomatic-like systems might use ballots made of thinner paper, or ballots with larger chads, or a stylus that works differently. Some voters may have a hard time with the machinery, fiddling with the stylus and punching two, three, even four times into the same slot in an effort to make sure their vote is properly recorded. The likelihood of the patterns is not the issue; it is their possibility that matters when it comes to figuring out the range of admissible chad-ballot relationships, and the patterns in Figure 9.3 are just as possible as the familiar patterns in Figure 9.2.

There is another important sense in which the two sequences come apart. In Figure 9.2, each pattern besides the first corresponds to a chad that leaves a hole in the ballot—one hole. In Figure 9.3, the number of holes increases inversely with the number of attached corners: from zero holes in the leftmost case, where the chad is attached by all four corners, to four holes in the rightmost pattern, where the number of attached corners is down to zero. This calls for careful assessment. If we adopted a purely topological criterion, as with the optical-reading device of the Votomatic, only the tri-chad corresponding to the second pattern would be registered as a valid vote. The first pattern would be rejected as an undervote and the others as overvotes, since the number of holes is greater than one. Yet there is a clear sense in which this would be a misinterpretation of the data. The several holes may be the result of a voter's fiddling with the stylus in an attempt to cast a single vote.

Perhaps this is a further sign that topological criteria are exceedingly strict, as with dimpled chads. But we may also describe the situation as one in which topological criteria must be supplemented by mereological considerations. The two openings left by the middle chad need not amount to two distinct holes; they may be seen as two disjoint parts of the same, larger hole—a *split* hole. Ditto for the three holes of the fourth chad and the four holes of the fifth chad.[45] Alternatively, we could describe the situation by saying that in those cases we are dealing with holes that are somewhat *incomplete*. After all, improper ballots are an unfinished job and the whole point of a manual recount is to try and reconstruct the voter's intentions. The standard sequence of Figure 9.2 captures the fact that the job may be unfinished insofar as the chad is not fully detached *from* the ballot; the alternative sequence of Figure 9.3 shows the job may be unfinished also insofar as the hole *in* the ballot, which is the target of the vote-punching operation, may not be fully formed. These are subtle considerations. But both ways of describing the situation—split holes or incomplete holes—would provide further support to the idea that the human eye (and mind) can see better than the dull laser eye of the Votomatic. And if things are so, then restricting a manual recount to the undervotes is a mistake. The ballots classified as overvotes are worth reviewing just as well.[46]

6. Chadology Unleashed

All of this is just the beginning. Obviously there are also variants of the sequence in Figure 9.3 where the punched chads are attached to the ballot along only three sides, or along two sides, etc. In particular, the sequence in Figure 9.4 corresponds to the five cases where the chad is attached along no sides at all (at least, along no

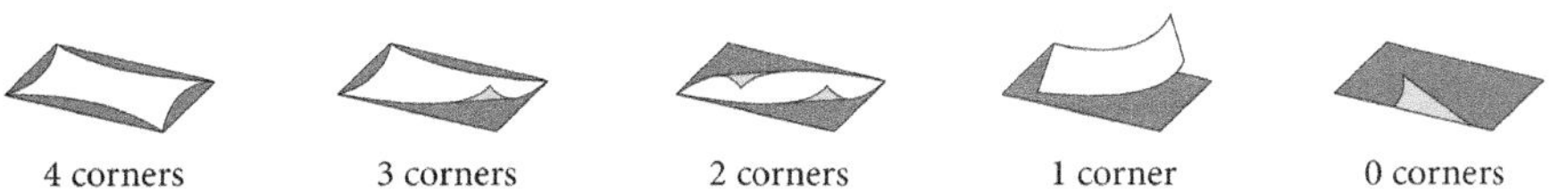

Figure 9.4 Another non-standard model of the basic taxonomy: regardless of the number of attached corners, each chad has no further points of contact with the ballot along any of its sides.

[45] On the mereology of holes, see Casati and Varzi (1994: ch. 7) and Varzi (1996).
[46] This was one moral of Casati and Varzi (2004), though not so explicit.

separate portions of their sides besides the bits around the corners, since physically speaking a corner connecting two pieces of paper cannot be identified with an unextended point). Again, two of the patterns—in this case, the last two—coincide with the corresponding patterns of Figure 9.2, but the other patterns are genuinely different. And again, the difference shows up in the number of openings involved in the new patterns, which in this case is directly proportional to the number of corners by which the chad is attached to the ballot (save for the last case, where the chad goes and the hole takes over).

Note that, as with Figure 9.3, the new patterns are hard to squeeze into the standard categories. The middle pattern, where the chad is attached by two corners, would technically be classified as a swinging chad; yet there is no genuine "swinging" when the corners are diagonally opposite. And the first pattern, where the chad is attached by all four corners, does not quite fit the category of a "pregnant" chad. It clearly is not a dimpled chad, since the topology has been altered; this is not just a chad with an indentation. But neither does it count as a pierced chad in the intended sense: a pierced chad should be perforated somewhere in the middle, not *surrounded* by perforations. Ditto for the second pattern, which appears to go beyond the intended case of a "tri-chad". All these deviations are once again illustrative of the limits of the intuitive language employed in defining the basic taxonomy, particularly of the ambiguities lurking behind a set of distinctions that is purely corner-based. As standardly formulated, the basic taxonomy is plagued by unintended (perhaps unexpected) models.

Indeed, we see here that the openings themselves call for further distinctions. An opening in the ballot can be a hole properly called, though possibly incomplete or included as a proper part in a larger hole, as in the patterns of Figure 9.2 and 9.3. But it can also be just the *beginning* of a hole: a mere cut in the ballot, a "slit". And in the first pattern of Figure 9.4 we have a ballot with no less than *four* slits, four distinct signs that the voter intended to punch a hole right there. Topologically the difference between holes and slits is just one of degree; in both cases, the genus of the ballot has been altered. We may say that, in a ballot, a hole properly called arises when two or more slits join at a corner, and the ideal case of a clean punch corresponds to the limit case where a series of four pairwise conjoined slits forms a closed loop: the cutting process results in a chad being completely separated from the rest of the ballot. But the difference between a closed loop and a partial succession of slits, or a mere bunch of disjoint slits as in the first pattern of Figure 9.4, is not one of degree. It is again a sharp topological distinction. It is the distinction between those cases where no continuous path connects the chad to the rest of the ballot and those cases where, by contrast, a continuous path exists, just as the topological distinction between a ballot with an opening and a ballot without depends on whether no continuous path around the chad can be shrunk into a point inside the chad. Earlier we said the topological criterion of ballot validity matches the criterion implemented by the Votomatic, with its optical-reading device. Now see now that the picture is more complex. There is the topology of the punched ballot, but there is also the topology of the punch itself.

Moreover, the whole picture dualizes. Just as Figure 9.3 features a sequence in which each chad, while ordered by decreasing number of connecting corners, is still attached to the ballot along each of its four sides, we may consider a sequence in which each chad is ordered by decreasing number of connecting sides but is attached to the ballot along each of its four corners. And just as there are variants of Figure 9.3 where the chads are attached to the ballot along fewer than four sides, including the limit case of zero sides depicted in Figure 9.4, we may consider variants where the chads are attached to the ballot along fewer than four corners, including the limit case where the number of attached corners is zero. In the first case, any opening in the ballot will be a separate slit and their number will increase inversely with the number of attached sides; in the second, the openings will be proper holes and the two numbers will decrease together. These two sequences are depicted in Figure 9.5, top and bottom, respectively.

Further variations are possible. For instance, there are patterns that differ with respect to the number of attached corners or sides while the number of openings (including slits) is fixed. In Figure 9.6, for instance, the openings are always two. In the sequence at the top, the patterns are ordered by decreasing number of attached

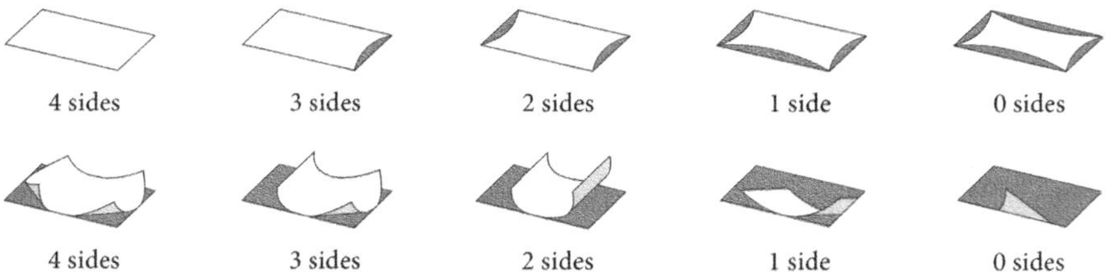

Figure 9.5 Further non-standard models. Both sequences are ordered by decreasing number of connecting sides. In the top sequence, the chads are attached to the ballot by each of their corners; in the bottom sequence, they are attached by none of their corners.

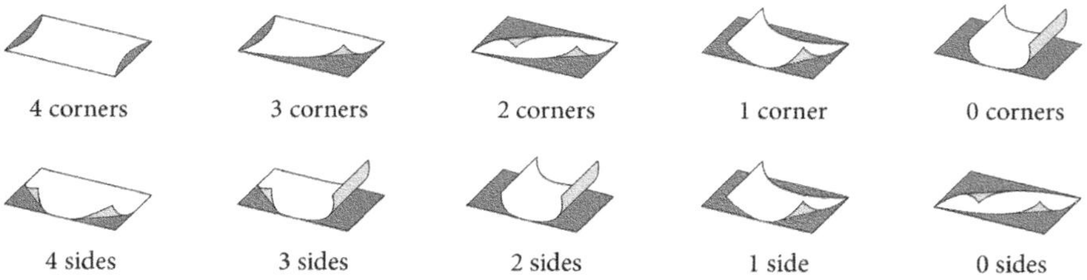

Figure 9.6 Non-standard models with two openings, ordered by decreasing number of connecting corners (top) or sides (bottom).

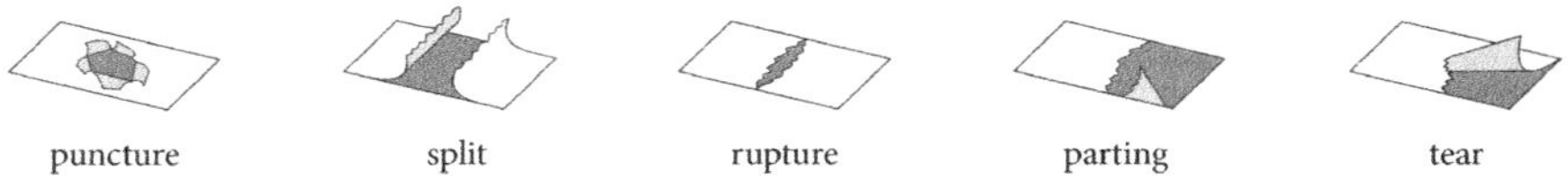

Figure 9.7 Various ways in which the internal structure of a chad can be damaged by improper punching, generating an opening in the ballot.

corners (and the number of connecting sides first decreases and then increases again); in the bottom sequence, the patterns are ordered instead by decreasing number of connecting sides (and, correspondingly, the number of attached corners is first decreasing and then increasing).

And did we forget what might happen to the chad itself? So far, all patterns differ with respect to the many ways a chad may be connected to the ballot, but the chad itself is always in perfect condition. This is a simplification. All sorts of things can happen to a chad as a result of improper punching. The category of a pierced chad recognized by the Florida officials is a case in point: the chad may be punctured in the middle. This is a particularly interesting case, showing that occasionally a ballot can feature a voting hole even if the chad remains fully attached at every corner and along each side. However that is not the only case where a chad may suffer significant structural damage or deformation during the punching process. There are many others, some of which are illustrated in Figure 9.7.

We leave it to the reader, at this point, to continue the job. There is still a large variety of additional patterns one may in principle distinguish, but there is no need to engage any further with this sort of conceptual combinatorics. What matters is that the combinatorics turns out to be far more complex than the basic taxonomy envisaged by the Florida canvassers and widely endorsed by legal and political commentators alike. And the bottom line is clear enough. Pasquino said the principle of equality would be at risk if equal holes are counted differently, or different holes counted equally. Now we see that the worry bites deeper. A lot of work appears to be necessary before we can even *start* making comparisons. We want to count *the* holes, but to that end we first need to figure out *which* holes truly count. For Argle, a hole is just a hole-lining, and that takes us nowhere. For most of us, every hole *has* a hole-lining, and while the latter is an ordinary chunk of matter, the hole itself is a chunk of nothing and we feel at a loss. We look for help from the ballots. We focus on the chads. Surely enough, a bit of chadology should come in handy. Alas, even that bit turns out to be a nightmare.

7. Seeing the Light

With all this, we are not suggesting that more advanced taxonomies of the sort outlined above would have helped our canvassers to the point that Miami-Dade County and Palm Beach County would have made the November 26 deadline for the manual recount. On the contrary, if there is a lesson to be drawn, it goes in the opposite direction: it is precisely because the business of "chadology" turns out to be so complex that the canvassing boards of Miami-Dade and Palm Beach ran into insurmountable difficulties. The boards were meant to examine all those ballots that would not qualify as clearly acceptable under the (arbitrary) criterion that would select only dangling chads. For all its prima facie objective precision, evidently that criterion turned out to be less straightforward than it was thought, with the result that many more ballots than expected had to be passed over to the canvassing boards for further inspection. And for all their goodwill, the members of the canvassing boards found themselves dealing with a variety of cases which, albeit narrower than the rich but abstract typology we have sketched, could hardly be mastered. Congratulations to the board of the third county, Broward, which managed to complete the job nonetheless.

Would it have been easier if the recanvassers adopted a purely topological criterion instead? We said more than once that this would have matched the light-beam method implemented by the Votomatic, and indeed Palm Beach County considered it for a while. They called it the "sunlight" (or "sunshine") test.[47] Hold up the ballot in front of you and check whether it lets some light go through. Never mind whether a chad is still partially attached. Never mind whether it's a slit, an incomplete hole, a scattered mix of hole parts, a piercing—the mereology of the opening or the topology of the chad itself. If you see some light, even a sliver of light, the vote is valid and move on. And since you are doing this by hand, chances are that your verdict will be more accurate than the machine's; you can hold the ballot at different angles to make sure the light is not accidentally blocked owing to the way a chad is still attached. Isn't this the sort of improvement we expect from a *manual* recount? And wouldn't that have been much simpler to implement than checking every corner, every side, every opening, every tear in the ballots?

Apparently Palm Beach County thought otherwise. After just a few days, the sunlight test was abandoned on the grounds that "even if one corner was punched, sometimes the sun wouldn't shine through".[48] Fair enough. But that is a practical issue, and it points to the practical limits of the test itself, not the inadequacy of the topological criterion it is meant to capture. The inadequacy of chadology is truly conceptual. Indeed, we may say it is ontological. It yields no definite answer to the fundamental question every vote counting is supposed to answer as we examine each ballot: *is there* a valid vote in it? Ontologically, the topological criterion is fully informative. It answers that question by answering the related question, is

[47] As reported e.g. in van Natta and Bragg (2000).

[48] As reported by the Associated Press on November 12 (Meadows 2000). This move was highly criticized by the Republicans, though not for the reasons we are discussing. Bush's team complained that it meant changing the recount standards *in media res*, casting doubts on the whole manual recount business. This is why permission to proceed with the recount in Palm Beach was put on hold until November 16. For details, see again Greene (2001: ch. 3) as well as Zelnick (2001).

there *a hole* in this ballot? (At least, it answers the latter question on the understanding that a hole introduces a change in the genus of the ballot, an opening of some sort; the morphological criterion would be more inclusive but, as we said, it is inherently vague.[49]) In a punch-card voting system, that would seem just right, so much so that when the Florida canvassing official mentioned earlier made his TV confession, he lamented the absence of a hole expert, not an expert chadologist. When the next morning CNN reached out to one of us ("We understand you are a hole expert . . . "), national news correspondent Jeanne Moos showed up with a basket of donuts. And when the interview was aired a couple of evenings later, on prime-time national television, she proudly explained that the troubles of the manual recount could be appreciated by performing the sunlight test on a slice of Swiss cheese.[50]

Be that as it may, the fact remains that the manual recount in Florida was a truly peculiar flop. It wasn't just a sign that the human eye, while sharper than a machine's, can be struck and even overwhelmed by the myriad of minute variations that actual electoral ballots may display. That happens all the times. It's the cost of stepping into that grey area of uncertainty we call the "margin of error" zone. The Florida flop was a sign of our inability to deal with variations that involve genuine *ontological* conundrums. We are supposed to count, but the very nature of the things to be counted defies us. What's worse, in this case it defies us because of its uncanny mix of uncanny elements. The task of counting votes expressed by punching holes takes us straight to the "twilight zone" where each variation hangs suspended between matter and void, between presence and absence, nay, let's say it, between being and non-being. Even Bargle would agree with Tucholsky:

> A hole is where something isn't.[51]

Take it chadologically or topologically, or even morphologically, this is the fundamental law of the twilight zone. And the Florida officials didn't quite know what to make of it.

[49] Another problem with the morphological criterion is that dimpled chads provide much less reliable evidence of a voter's intents, especially when there are good reasons to think the voter may have been confused by the voting procedure. This was noted already in the 1990 *Guidelines on Ballots with Chads Not Completely Removed*, written by Palm Beach County Supervisor of Elections Theresa LePore (the designer of the confusing butterfly ballot holder), where dimpled chads are ruled out as inscrutable insofar as the indentation "may result from a voter placing a stylus in the position, but not punching through" (cited in Whitman 2003: 75).

[50] In "Making the Moost of It", CNN, November 23, 2000.

[51] Tucholsky (1931: 123), which Lewis and Lewis (1996: 77) evoke sympathetically.

8. Regrets

We know how it all ended. And we know what happened to the Votomatic. In the aftermath of the presidential election, public outcry, policymakers, and associations from around the country demanded substantive reforms in election procedure. Within a few months, on May 9, 2001, Governor Jeb Bush was already signing into law a lengthy Florida Election Reform Act that included the removal of "apparatus in which ballots are inserted and used in connection with a marking device for the piercing of ballots by the voter" from the list of acceptable voting devices,[52] moving statewide to electronic and optical scan. Soon afterwards, the U.S. House and Senate followed suit, passing by overwhelming majorities a Help America Vote Act that included nationwide measures "to replace punch-card voting systems [...] with a voting system [...] that does not use punch cards".[53] And on October 29, 2002, the federal Act was signed into law by the President of the United States, George W. Bush.[54] Thirty-eight years since its first use in 1964, the public life of the most popular and widely used voting system in the United States came to an end.[55]

Politically, this may well be a step ahead. Of course, no reform is perfect. Close electoral races are bound to happen from time to time and no voting system is faultless. The reader may wish to revisit the many controversial cases that affected U.S. politics even after the 2002 ban of the Votomatic, such as the 2004 Washington gubernatorial election (count, recount, overturning recount, belated concession), the 2006 House Democratic Primary in Alaska, District 37 (count, recount, wrangling over five disputed ballots, coin toss), the 2008 senate race in Minnesota (count, reversing recount, wrangling and court cases for 238 days), etc.[56] When it comes to counting votes, we may always be left wondering, does my vote really count?[57] But even so, getting rid of a counting method that proved disastrous is the least we can do.

Philosophically, however, the end of the Votomatic is a lost opportunity. Let's face it, we finally found ourselves face to face with an important truth. So much in our lives and decisions depends, not only on what there is, but also on what there isn't. Philosophers have been saying this occasionally, though with Being and Non-Being so shamelessly capitalized that we may have thought they were joking.

[52] *Florida Laws*, Section 15, ch. 2001-40, codified at *The 2001 Florida Statutes*, Title IX, Chapter 101, Section 5603(8).

[53] H.R. 3295, Title I, Section 102(a). The Act was originally passed in slightly different versions by the House on December 12, 2001 and by the Senate on April 11, 2002, then reconciled on October 8, 2002 and passed without amendment by both (357–48 and 92–2, respectively).

[54] *Public Law* 107–252, 116 Stat. 1666, codified at 42 U.S.C. 15301 to 15545.

[55] At least as a matter of general policy. Some counties continued to use punch-card systems for a while, occasionally running into new troubles. See again the case mentioned in note 44.

[56] On the significance of these and other cases, see e.g. Kropf and Kimball (2011).

[57] Elsewhere we put it in terms of bona fide "electoral uncertainty" (Casati and Varzi 2014).

Holes are perfectly lower-case. The intrusion of those little chunks of nothingness into our vote-counting practices, their unfathomable presence by way of absence, even partial absence, should have made us realize it isn't a joke at all. No less than the presidency of the United States depended on them. The image of Judge Burton engaging in the sunlight test, not to mention the many photos testifying to judge Rosenberg's painstaking devotion to examining each ballot in minuscule detail, should have gotten us excited. "Is that really a hole?" "Is it perhaps a half hole, or is every hole a whole hole?" "Let's distinguish two cases: there's the hole of the chad, and there's the hole in the chad." "Let's distinguish two questions. There's the General Hole Question, which asks, what is a hole? And there's the Special Hole Question, which asks, under what conditions does something have a hole?"... So much bread for our philosophical teeth! So much *thaumazein*!

Luckily the end of the Votomatic doesn't mean the wonder is gone. And, luckily, a good number of Votomatic booths survive, if only in museums and art exhibits, to keep the wonder alive. Already on November 7, 2001, one year after the election, the Smithsonian's National Museum of American History acquired a piece from Palm Beach for its Presidential Collections. And by November 2004, no less than forty-seven pieces made their way from a Florida flea market to the "Voting Booth Project" hosted by the Parsons School of Design in New York, revisited by prominent artists and designers. Aptly enough, Christo was among them and did his thing: "That machine had a problem in the 2000 election; I wrapped it and put the seal on it, so that no one can use it any more".[58] Even so, the machine was there and everybody wondered. What was the problem, exactly?

References

Ackerman, B. (ed.) (2002). *Bush v. Gore: The Question of Legitimacy* (New Haven, CT: Yale University Press).

Barabak, M. Z. and M. Finnegan (2000). "Bush, Gore Pressed to Accept Verdict Brought by Recounts", *Los Angeles Times*, November 13, p. A1.

Bickerstaff, S. (2001). "Counts, Recounts, and Election Contests: Lessons from the Florida Presidential Election", *Florida State University Law Review* **29**: 425–67.

Blair, G. (2004). "Designers Redefine the Political Machine", *New York Times*, October 7, p. F7.

Brady, H. E., M. Herron, W. R. Mebane, Jr, J. S. Sekhon, K. Shotts, and J. Wand (2001). "Law and Data: The Butterfly Ballot Episode", *Political Science and Politics* **34**: 59–69; also included, with revisions, in *The Longest Night: Polemics and Perspectives on Election 2000*, ed. A. J. Jacobson and M. Rosenfeld (Berkeley, CA: University of California Press, 2002), pp. 50–66.

[58] Cited in Blair (2004).

Brown, L. (2000). "Computers Make Errors, Even in Vote Counting", *The Ledger*, November 19, p. AA3.

Bugliosi, V. (2001). *The Betrayal of America: How the Supreme Court Undermined the Constitution and Chose Our President* (New York: Nation Books).

Casati, R. and A. C. Varzi (1994). *Holes and Other Superficialities* (Cambridge, MA: MIT Press).

Casati, R. and A. C. Varzi (2004). "Counting the Holes", *Australasian Journal of Philosophy* **82**: 23–7.

Casati, R. and A. C. Varzi (2014). *L'incertezza elettorale* (Rome: Aracne).

Chisnell, D. (2016). "Democracy Is a Design Problem", *Journal of Usability Studies* **11**: 124–30.

Cole, D. (2006). "The Liberal Legacy of *Bush v. Gore*", *Georgetown Law Journal* **94**: 1427–74.

deHaven-Smith, L. (ed.) (2005). *The Battle for Florida: An Annotated Compendium of Materials from the 2000 Presidential Election* (Gainesville, FL: University Press of Florida).

Dershowitz, A. M. (2001). *Supreme Injustice: How the High Court Hijacked Election 2000* (New York: Oxford University Press).

Dionne, E. J., and W. Kristol (eds) (2001). *Bush v. Gore: The Court Cases and the Commentary* (Washington, DC: Brookings Institution Press).

Dover, E. D. (2003). *The Disputed Presidential Election of 2000: A History and Reference Guide* (Westport, CT: Greenwood Press).

Dugger, R. (1988). "Annals of Democracy: Counting Votes", *New Yorker*, November 7, pp. 40–108.

Dugger, R. (2000). "Democracy Under Stress", *Los Angeles Times*, November 19, p.M1.

Foley, E. B. (2016). *Ballot Battles: The History of Disputed Elections in the United States* (Oxford: Oxford University Press).

Gillman, H. (2001). *The Votes That Counted: How the Court Decided the 2000 Presidential Election* (Chicago: University of Chicago Press).

Gold, S. (2000). "With Anguish, Judge Rejects Palm Beach Revote", *Los Angeles Times*, November 21, p. A20.

Greene, A. (2001). *Understanding the 2000 Election: A Guide to the Legal Battles that Decided the Presidency* (New York: New York University Press).

Harris, B. (2004). *Black Box Voting: Ballot Tampering in the 21st Century* (Renton, WA: Talion).

Harris, J. P. (1934). *Election Administration in the United States* (Washington, DC: Brookings Institution Press).

Harris, J. P. (1965). *Data Registering Device*. Patent No. 3,201,038 (filed August 21, 1962).

Jacobson, A. J. and M. Rosenfeld (eds) (2002). *The Longest Night: Polemics and Perspectives on Election 2000* (Berkeley, CA: University of California Press).

Jarvis, R. M., O. Coleman, and J. C. Burris (2001). *Bush v. Gore: The Fight for Florida's Vote* (New York: Kluwer Law International).

Jones, D. W. and B. Simons (2012). *Broken Ballots: Will Your Vote Count?* (Stanford, CA: Center for the Study of Language and Information).

Kaplan, D. A. (2001). *The Accidental President: How 413 Lawyers, 9 Supreme Court Justices, and 5,963,110 Floridians (Give or Take a Few) Landed George W. Bush in the White House* (New York: William Morrow).

Kropf, M. and D. C. Kimball. (2011). *Helping America Vote: The Limits of Election Reform* (London: Routledge).

Lewis, D. K. and S. R. Lewis (1970). "Holes", *Australasian Journal of Philosophy* **48**: 206–12.

Lewis, D. K. and S. R. Lewis (1996). Review of Casati and Varzi (1994), *Philosophical Review* **105**: 77–9; repr. as "Casati and Varzi on Holes", in D.K. Lewis, *Papers in Metaphysics and Epistemology*, pp. 183–6 (Cambridge, Cambridge University Press, 1999).

Meadows, K. (2000). "Election Officials Begin Florida Hand Count", Associated Press News, November 12.

Myers, J. H. (1889). *Voting Machine*. Patent No. 415,549, November 19 (filed May17).

Pasquino, P. (2002). "Bush vs. Gore: A View from Italy", in *The Longest Night: Polemics and Perspectives on Election 2000*, ed. A. J. Jacobson and M. Rosenfeld (Berkeley, CA: University of California Press), pp. 322–31.

Pleasants, J. M. (2004). *Hanging Chads: The Inside Story of the 2000 Presidential Recount in Florida* (New York: Palgrave Macmillan).

Posner, R. A. (2001). *Breaking the Deadlock: The 2000 Election, the Constitution, and the Courts* (Princeton, NJ: Princeton University Press).

Safire, W. (2000). "Chad: A Lexical Star Is Born", *New York Times Magazine*, December 10, p. 68.

Saltman, R. G. (2006). *The History and Politics of Voting Technology: In Quest of Integrity and Public Confidence* (New York: Palgrave Macmillan).

Smith, R. L. (2002). "A Statistical Assessment of Buchanan's Vote in Palm Beach County", *Statistical Science* **17**: 441–57.

Sunstein, C. R., and R. A. Epstein (eds) (2001). *The Vote: Bush, Gore, and the Supreme Court* (Chicago, IL: University of Chicago Press).

The New York Times Correspondents (2001), *36 Days: The Complete Chronicle of the 2000 Presidential Election Crisis* (New York: Times Books).

Toobin, J. (2001). *Too Close to Call: The Thirty-Six-Day Battle to Decide the 2000 Election* (New York: Random House).

Tucholsky, K. (1931). "Zur soziologischen Psychologie der Löcher", *Die Weltbühne*, March 17, p. 389; now in K. Tucholsky, *Gesammelte Werke*, Band III: *1929–1932*, pp. 804–5 (Reinbek: Rowohlt, 1960); Eng. trans. by H. Zohn: "The Social Psychology of Holes", In K. Tucholsky, *Germany? Germany! Satirical Writings*, pp. 123–4 (New York: Berlinica, 2017).

United States Commission on Civil Rights (2001). *Voting Irregularities in Florida During the 2000 Presidential Election: Report and Appendix* (Washington, DC: Government Printing Office).

van Natta, D., Jr. (2000). "Democrats Tell of Problems at the Polls Across Florida", *The New York Times*, November 10, p. A26.

van Natta, D., Jr., and R. Bragg (2000). "Scrutiny and Disagreements Accompany Hand Recount", *The New York Times*, November 13, p. A20.

Varzi, A. C. (1996). "Reasoning about Space: The Hole Story", *Logic and Logical Philosophy* 4: 3–39.

Wand, J. N., K. Shotts, J. S. Sekhon, W. R. Mebane, Jr, M. C. Herron, and H. E. Brady (2001). "The Butterfly Did It: The Aberrant Vote for Buchanan in Palm Beach County, Florida", *American Political Science Review* 95: 793–810.

Watson, R.P. (ed.) (2004). *Counting Votes: Lessons from the 2000 Presidential Election in Florida* (Gainesville, FL: University Press of Florida).

Webb, K.D. (1988). *Electronic Computerized Vote-Counting Apparatus*. Patent No. 4,774,665, September 27 (filed April 24, 1986).

Weinberg, B. H. (2006). *The Resolution of Election Disputes: Legal Principles That Control Election Challenge* (Washington, DC: International Foundation for Electoral Systems; 2nd edn 2008).

Weinberg, L. (2002). "When Courts Decide Elections: The Constitutionality of Bush v. Gore", *Boston University Law Review* 82: 609–66.

Whitman, D. (2000). "Chadology 101: Divining a Dimple: Who Knew a Simple Ballot Could Be So Tricky?", *U.S. News and World Report*, November 27, p. 34.

Whitman, M. (ed.) (2003). *Florida 2000: A Sourcebook on the Contested Presidential Election* (Boulder, CO: Lynne Rienner Publishers).

Zelnick, R. (2001). *Winning Florida: How the Bush Team Fought the Battle* (Stanford, CA: Hoover Institution Press).

Zukerman, T. D. (1925). *The Voting Machine: Report on the History, Use, and Advantages of Mechanical Means for Casting and Counting Ballots* (New York: Political Research Bureau of the Republican County Committee of New York).

10

Something Out of Nothing

What Zeno Could Have Taught Parmenides

Aaron Segal

1. Introduction

Parmenides famously denied that anything could come "from what is not".[1] But he was wrong. Something can indeed come from what is not. Alright, maybe not from what isn't *at all*. But something can come from what isn't *actual*.

I don't mean that just in the innocuous way in which something might pop into existence without any cause at all—whether actual or merely possible. Nor do I mean that just in the less innocuous but still relatively innocuous way in which something might bring about something else, but do so only *relative to* or *in contrast with* some mere possibile (see Schaffer (2005), Bernstein (2014; 2016), and my discussion in section 3). No, I mean it in the *gobsmackingly spooky* way in which something might be brought into existence wholly and entirely by something non-actual. It'd be like the way in which a ghost might bring into existence an actual human child, despite there being no actual ghosts. If you're nonplussed by that sort of thing, you've probably watched one too many horror films.

My argument will consist in calling attention to a case that is quite plausibly both possible and spooky in the way I have described. The case to which I will call attention is a radical version—even more radical than "the more radical version" Jose Benardete (1964) considers—of so-called Zeno causality. If only Zeno had been the teacher and not the pupil; he could have shown Parmenides the Way to the Truth.

Here's a plan for the paper. In section 2 I clarify my central claim and note a presupposition of my initial formulation of the argument, viz. David Lewis's modal realism. In section 3 I discuss some nearby claims—claims about *omissions* and the like—how they differ from the one I put forward here, and, more importantly, why the cases on which they rely can't be pressed into service for me. Then I turn in section 4 to the case of Zeno causality I have in mind. I work my way up to the case and then argue both that it's possible and spooky. In section 5 I make the case that modal realism can drop out as a presupposition and

[1] See McKirahan (1994: 147).

Aaron Segal, *Something Out of Nothing: What Zeno Could Have Taught Parmenides* In: *Non-Being: New Essays on the Metaphysics of Non-Existence.* Edited by: Sara Bernstein and Tyron Goldschmidt, Oxford University Press (2021).
 DOI: 10.1093/oso/9780198846222.003.0010

make its way back in as an implication of the argument's conclusion. I conclude in section 6 with some theological speculation.

2. Clarifications and Presupposition

My main contention is that it's possible for something to be brought into existence by something that is non-actual. That is,

> Spooky*: Possibly, there exists something actual *x* and something non-actual *y* such that *y* brought *x* into existence

The intended sense of the predicate '... is actual' (and '... is non-actual') is nonrigid. I'm referring to what would have been actual or non-actual had that possibility come off. Thus, putting the claim in terms of possible worlds, we might say:[2]

> Spooky**: There exists a possible world *w* and an *x* and a *y* such that *x* exists in *w* and *y* does not exist in *w* and *y* brought *x* into existence

But this is ambiguous, and on both readings it is weaker than intended. It is ambiguous because '... exists in *w*' is ambiguous—even assuming, as I shall, a possibilist interpretation (see nt. 2). On Lewis's (1986, sect. 1.2) own reading, it means '... is part of *w*'; on another reading, which Lewis (1986, sect. 4.3) considers, it means '... overlaps *w*'. Ignoring transworld fusions, these come to the same thing: something is part of a given world iff it overlaps that world. But we cannot ignore transworld fusions, since the spooky cause turns out to be such a fusion (see nt. 10). So, which is intended? If we give it Lewis's reading in both occurrences, then Spooky** is consistent with the cause being a transworld fusion that overlaps the world in which the created object exists, *and consistent even with its bringing that about solely in virtue of its parts that are worldmates with the created object.* Not so spooky after all. If we give it the other reading in both occurrences, then Spooky** is consistent with the created object being a transworld fusion that overlaps *some* world in which the cause exists, *and consistent*

[2] I'm skipping at least one intermediate formulation. We might try to hew more closely to Spooky* and say this: There exists a possible world *w* such that in *w*, there exist an *x* and a *y* such that *x* exists in *w* and *y* does not exist in *w* and *y* brought *x* into existence. But this can't be right, because it would follow that there is some possible world *w* such that in *w* there exists something that doesn't exist in *w*. And there is no such possible world. Unless, that is, we distinguish between 'in *w*, there exists ...' and '... exists in *w*', the former meaning something like, 'were *w* actual, it would be the case that there exists (quantifiers wide open) ...' and the latter meaning something like, '... is located in *w*'. But an actualist has no use for the latter, and a possibilist has no use for the former. So I, with the possibilist, have just dropped the first 'in *w*', and give 'exists in *w*' a "possibilist interpretation".

with all the relevant causation being internal to that world. Again, not so spooky. So, I don't mean either; or, I mean both: Lewis's reading of the first instance and the other reading of the second instance. Let's make this explicit by using '... exists *entirely* in *w*' to mean (the Lewisian) '... is a part of *w*', and '... exists in *w*' to mean '... overlaps *w*', and then stating the thesis as follows:

> Spooky: There exists a possible world *w* and an *x* and a *y* such that *x* exists entirely in *w* and *y* does not exist in *w* and *y* brought *x* into existence

There you have it. Two things, one which brought the other into existence, but which find themselves together in no possible world. That's pretty undeniably spooky.

Before I turn to my argument's presupposition, let me add one clarification about my *attitude* toward the argument and its conclusion. Contrary to the impression I may have given until now, I don't believe Spooky. But I don't disbelieve it either. I don't know what to believe. The case to which I will call attention is genuinely puzzling. And if I had to list the ways to address it in order of plausibility, it *seems* to me, at least in some moods, that Spooky is at the top of the list.

Now to what I shall presuppose. It follows from Spooky that there exist two things that are not worldmates. So at most one exists in the actual world. So at least one of them is non-actual. Thus, it follows from Spooky that there are non-actual things, that actualism—the doctrine that everything is actual—is false, and hence that possibilism—the denial of actualism—is true.

In giving my argument, at least in its first iteration, I will take that component of Spooky for granted. Indeed, I will take for granted modal realism, *tout court*, with its plenitude of mere possibilia. (It seems reasonable enough to assume that the best version of possibilism is the one Lewis puts forward.) It's hard enough to establish the conditional claim—if modal realism is true, then Spooky is true—that in seeking to establish Spooky I will argue just for the conditional, and simply assume the antecedent. The conditional is anyways *interesting*, since the foremost modal realist held views about transworld causation that straightforwardly entail its falsity (see Lewis (1986: sect. 1.6) and my discussion in section 4.3.2).

But I realize that most people who've given the matter serious thought are actualists, and my argument would therefore be much *more* interesting if I could drop modal realism as a presupposition. I will try to do just that in the second iteration of the argument. My hope is that the first iteration will serve as a warm-up. Once you see the good reason to accept the conditional, I suspect you will see that it, or something very much like it, is also good reason to accept both the conditional and the antecedent together.

3. Omissions

Some philosophers, in reflecting upon omissions, absences, failures, and suchlike oddities, have put forward claims that sound superficially similar to my central claim. But surface appearances are misleading here. The claims are very different. And the cases upon which they reflect are inadequate for my purposes, even if they are adequate for theirs. To forestall confusion or suspicions of reinventing the wheel, I will dwell briefly on these other cases and claims.

Suppose a gardener falls asleep on the job, neglects to water the flowers, and the flowers wilt and die. We might naturally say, "The gardener's failure to water the flowers caused the demise of the flowers." If that sentence is both *true* and *maximally metaphysically revealing*, then there's some *thing*—the gardener's failure to water the flowers—that caused some other thing—the demise of the flowers. But what is that first thing? If all causes are events, then the gardener's failure to water the flowers is an event, some *worldly bit*. But what an odd bit of the world it is! We might ask such things as: When did it occur? Where did it occur? Is it distinct from or identical with the gardener's failure to water the trees? No answers seem forthcoming, and we might reasonably suspect that no answers exist. What to do?

One might reply that it's not true that all causes are events. Indeed, maybe *no* causes are events, but there are causes nonetheless. It has been suggested (Mellor 1995) instead that only *facts* are causes, where facts are supposed to be more like propositions—*representations* of how a world might be—than like worldly bits. So then the gardener's failure to water the flowers, if it's a cause, is a fact, not an event. The questions we raised about there being such an *event* as the gardener's failure to water the flowers are either moot or easily answered, once we switch to a fact framework: Facts don't occur at all, they obtain; with respect to a huge number of facts, it's not clear that it even makes sense to ask where and when it obtains; and the proposition that the gardener failed to water the flowers is certainly distinct from the proposition that the gardener failed to water the trees. But adopting this position comes at the high cost of removing causation from the "world", since it's no longer a relation between worldly bits. One could instead bite the bullet—and maintain that there really is such an event as the gardener's failure to water the flowers—and either provide answers to the questions we raised or contend that they need not be answered (Payton 2018; Silver 2018). But those routes are clearly very costly in their own right.

These difficulties have led some philosophers to deny that the gardener sentence is *maximally metaphysically revealing*, even if it's true. You can't just read the sober ontological truth straight off of the surface grammar. So even if the sentence is true, there might not really be any such *thing*—whether event *or* fact—as the gardener's failure to water the flowers. Nevertheless, assuming that the sentence *is* in fact true—and can one plausibly claim it's simply *false*?—then presumably *something* having to do with the gardener was a cause of the flowers' demise.[3] What is that thing?

One natural thought is that it's whatever the gardener was doing when he *should* have been watering the flowers—or when he *would* have been watering the flowers, had he done so—i.e. napping. The gardener's nap is a metaphysically, if not professionally, respectable worldly bit. Its spatio-temporal location and identity conditions are about as clear as those of any event. There is no need to fret about its existence. So you might suggest that in the final analysis—and this is what makes the gardener sentence true—the gardener's nap caused the flowers to wilt. The trouble, however, is that left unadorned and unembellished this suggestion is implausible, at least as a general account of neglectful gardener cases (Schaffer 2005; Bernstein 2014). Suppose that our gardener is particularly derelict, derelict also in his *dispositions* to discharge his duties. If he hadn't fallen asleep on the job, he would have made a run to Dunkin' Donuts instead, and would have thus neglected to water the flowers regardless. Now we don't have the right pattern of counterfactual dependence between the flowers' demise and the gardener's nap to undergird a causal relation: even if the gardener hadn't napped, the flowers would still have wilted. But we'd be just as inclined under these suppositions to say "The gardener's failure to water the flowers caused the demise of the flowers". Somehow the gardener's watering the flowers has to get in on the action, while not being the *only* thing that gets in on the action.

That's exactly what several philosophers (Schaffer 2005; Bernstein 2014) have suggested. To simplify a bit and elide some differences between Schaffer and Bernstein: on their view, the gardener's nap, *relative to*, or *as contrasted with*, the gardener's watering the flowers, caused the demise of the flowers.[4] Since the watering is a non-actual event, it turns out on their view that at least sometimes non-actual events play a causal role.[5] Now, causing the death of a flower might not be quite as impressive as bringing something into existence, but that's an artifact of our example. Replace failure to water the flowers with failure to use a

[3] Although Beebee (2004) has denied that the sentence is true, and Beebee (2004), Lewis (2004), and Varzi (2007) have said that in the final analysis nothing having to do with the gardener caused the wilting—at most the gardener figures into a causal explanation, not a genuine causal relation.

[4] According to Schaffer, the relativization is built into causation itself. Causation, on his view, is always a four-place relation: Causal claims that are maximally metaphysically revealing have the form, c rather than c^* caused e rather than e^*. According to Bernstein, there is no relativization built into causation itself: Causal claims that are maximally metaphysically revealing have the form, c caused e. But there is relativization built into the relation of *causal salience*; in certain contexts, certain non-actual events—the *omitted*, as opposed to merely *absent*, ones—are causally salient, and therefore relevant to the cause's bringing about the effect.

[5] Indeed, Bernstein subsequently argued (2016) that at least sometimes *impossible* events— like proving that 2 + 2 = 5—can play a causal role. Note: according to Schaffer, *whenever* there is causation there are non-actual events playing a causal role.

contraceptive, and there you have it: a non-actual event playing a causal role in bringing a human child into existence.

But as metaphysically interesting as that would be, it wouldn't be *spooky*. More importantly, it wouldn't involve a non-actual event being a *cause* of an actual event. It doesn't make any sense to say—in the cases Schaffer and Bernstein are discussing—that the non-actual event is a cause of the relevant effect. To say so would be to flagrantly violate the counterfactual criterion for causation. If the non-actual event hadn't occurred, then the effect would of course still have occurred. After all, the non-actual event *didn't* occur, and the effect still occurred.[6] That's why it makes no sense to think of the (non-actual) gardener's watering the flowers as causing the wilting of the flowers. What *does* make sense to say is that the (non-actual) gardener's watering of the flowers is the thing such that *had it occurred instead of* the gardener's nap, then the flowers *wouldn't* have wilted. But then if anything in the vicinity is the cause, it's the gardener's nap, not the gardener's watering. At most we can say, what Schaffer and Bernstein *do* say, that the gardener's watering played some *causal role* in the wilting of the flowers.

And there is a vast distance between *playing a causal role* and actually *being a cause*, at least as the former phrase is being employed in our context. To appreciate just how vast the distance is, it seems to me that the Schaffer–Bernstein view—or at least a view that's as good as theirs—is consistent with actualism.[7] There need not *be* any non-actual events for the contrastive or relative gambit to work: ersatz versions will suffice. After all, the contrasts need not have any causal powers if they are not being called upon to cause anything.

Hopefully it is now clear enough that Spooky doesn't begin to follow from the Schaffer–Bernstein view, nor does it gain any support from the pretty ordinary cases they consider.

4. Zeno Causality and Other-Worldly Effects

4.1 The Case

But it does gain support from a rather extraordinary case. Jose Benardete famously introduced a number of puzzling scenarios, all of which involve some "open-ended" infinite series and violate some deeply held convictions about causation.

[6] I am assuming that a given subjunctive conditional entails the corresponding material conditional. Given a possible-worlds analysis for subjunctive conditionals, this amounts to the assumption of weak centering (that no world is closer to itself than it is).

[7] It's not clear to me whether their own views are possibilist. They freely quantify over and refer to non-actual events, but they do not flag any commitment to possibilism and so they might intend all such quantification and reference to be understood as a mere *façon de parler*. What is clear to me is that the core of their view is consistent with actualism.

Here are a couple moderately puzzling ones:

> Let the peal of a gong be heard in the last half of a minute, a second peal in the preceding 1/4 minute, a third peal in the 1/8 minute before that, etc. ad infinitum... Of particular interest is the following puzzling case. Let us assume that each peal is so very loud that, upon hearing it, anyone is struck deaf—totally and permanently. At the end of the minute we shall be completely deaf (any one peal being sufficient), but we shall not have heard a single peal! For at most we could have heard only *one* of the peals (any single peal striking one deaf *instantly*), and which peal could we have heard? There simply was no first peal. We are all familiar with various physical processes that are followed by what are called after-effects. We are now tempted to coin the barbarous neologism of a *before-effect*...
>
> A man is shot through the heart during the last half of a minute by A. B shoots him through the heart during the preceding 1/4 minute, C during the 1/8 minute before that, c. ad infinitum. Assuming that each shot kills instantly (if the man were alive), the man must be already dead before each shot. Thus he cannot be said to have died of a bullet wound. Here, again, the infinite sequence logically entails a before-effect. (Benardete 1964: 255–9)

These are puzzling cases, indeed. They're puzzling not so much because the effect in each case precedes the cause—we can learn to live with such things. They're puzzling because we're tempted to infer from them that it's possible for something to be caused to occur while *nothing in particular* causes it to occur. Thus, we're tempted to think that we're rendered deaf, *caused* to become deaf, but that *nothing in particular* caused us to become deaf; that the man is killed, *caused* to die, but that *nothing in particular* killed him. That would already be spooky.

But Hawthorne (2000) convincingly argues that this temptation ought to be resisted. In both of these cases, there *is* something that is as good a candidate as any to have been the cause: the *fusion* of the bullets killed the man, and the *fusion* of the sound waves struck us deaf.[8] It's true that none of the bullets killed the man, and it's true that none of the sound waves killed the man, but it would be fallacious to infer that their fusion didn't do these things.

Indeed, Hawthorne convincingly argues that we should also resist the temptation to infer that it's possible to "conjure up action at a distance out of very mundane objects that do not, when finitely combined, ever act at a distance." This is tempting to infer once we grant Hawthorne's contention that the fusion of the bullets killed the man. After all, none of the bullets came into contact with the man

[8] Uzquiano (2012) suggests alternatively that it is the *plurality* of the bullets etc. that collectively killed him. As best as I can tell nothing here or throughout this chapter will turn on which one is correct, so I will more-or-less arbitrarily but consistently speak in terms of the fusion as a cause rather than the plurality.

(until after he died). But Hawthorne notes that we can infer from this fact that the fusion of bullets did not come into contact with the man (until after he died) only if we endorse the following principle about contact:

> The Contact Principle: If *y* is the fusion of the *x*'s and *z* contacts *y*, then *z* contacts one of the *x*'s.

While that principle seems correct, and may even *be* correct if there are only finitely many *x*'s, it is not true in full generality. A fusion of infinitely many *x*'s may touch things that none of the *x*'s individually touch. So there needn't even be any spooky action-at-a-distance in this case—just quirky relationships between fusions and their parts.

All this works well and good for the cases I've mentioned so far. But then Benardete introduces an "even more radical version of the paradox":

> A man decides to walk one mile from A to B. A god waits in readiness to throw up a wall blocking the man's further advance when the man has travelled 1/2 mile. A second god (unknown to the first) waits in readiness to throw up a wall of his own blocking the man's further advance when the man has travelled 1/4 mile. A third god...c. ad infinitum. It is clear that this infinite sequence of mere intentions (assuming the contrary-to-fact conditional that each god would succeed in executing his intention if given the opportunity) logically entails the consequence that the man will be arrested at point A; he will not be able to pass beyond it, even though not a single wall will in fact be thrown down in his path. The before-effect here will be described by the man as a strange field of force blocking his passage forward.

Here there is an even greater temptation to infer that it's possible for something to be caused to occur while *nothing in particular* causes it to occur. After all, no wall is even put up, and none of the gods actually *does* anything. Yet the man is stopped dead in his tracks at point A. Apparently, he is stopped, but by nothing in particular. Again, that would already be spooky.

But Hawthorne argues that here too the temptation ought to be resisted. There is, after all, *something* that is a candidate to be the cause of the man's arrest, viz. the fusion of the gods. It's true that none of the gods *does* anything—indeed, none of them changes one iota—but we can infer that the fusion *also* does nothing only by relying on something like the following principle:

> The Change Principle: If *y* is the fusion of the *x*'s and the *x*'s are individually capable only of producing effect *e* by undergoing change, then *y* cannot (without the addition of some non-supervening causal power) produce effect *c* without undergoing change.

While that principle seems correct, and may even *be* correct if there are only finitely many *x*'s, Hawthorne contends that it is not true in full generality. A fusion of infinitely many *x*'s may manage to bring something about without any of the *x*'s undergoing any change, *despite the fact* that none of the *x*'s is individually capable of bringing about that effect without undergoing change, and *despite the fact* that there are no spooky non-supervenient causal powers. So, again: there needn't be anything spooky going on—just quirky relationships between fusions and their parts.

I shall grant that all of this is right. But it is of no help when we take the next step: to a case that is more radical than Benardete's "more radical version", in which there *aren't* any (actual) gods who would act in the ways described, but in which there *would be* if push came to shove.

> THE CASE OF THE DRAGON-SLAYING WOULD-BE GODS: A dragon is alive at 12 noon, at which time there are no (actual) dragon-slaying gods in existence. But if the dragon were to survive until 1 p.m., a dragon-slaying god would come into existence and slay the dragon at 1 p.m.; if the dragon were to survive until 12:30 p.m., a dragon-slaying god would come into existence and slay the dragon at 12:30 p.m.; if the dragon were to survive until 12:15 p.m. . . . c. ad infinitum. At no time after 12 p.m. and before or at 1 pm would a dragon-slaying god come into existence unless the dragon is alive at that time. (Oh, and dragons' lives are not temporally gappy: if a dragon is alive at t_1 and at a later time t_2, then he is alive at every time between t_1 and t_2; so once a dragon is slain, he will never live again.)

It's clear that the infinite sequence of counterfactuals—together with the gaplessness of dragon lives—logically entails the consequence that the dragon doesn't survive past 12 noon.[9] But it's also clear that the facts of THE CASE (as I shall now call it) logically entail that *if* the dragon does not survive past 12 p.m., then there *are no* (actual) dragon-slaying gods in existence at any time between 12 p.m. and 1 pm. It thus follows as a matter of logic from the facts of THE CASE *both* that the dragon doesn't survive past 12 p.m. and that there are no (actual) dragon-slaying gods in existence at any time between 12 p.m. and 1 p.m.

Let us grant for the moment that THE CASE is indeed possible. Then there is an exceedingly great temptation to infer that it's possible for something to be caused to occur while nothing in particular causes it to occur. After all, in THE CASE there *are* no actual dragon-slaying gods—at least none hanging around at the relevant time—and so no *fusion* of dragon-slaying gods either. Yet the dragon is slain at 12 p.m. Apparently, the dragon is slain, but by nothing in particular.

[9] Again (see nt. 6), I assume weak centering. For careful versions of the argument for this logical entailment, see Priest (1999) and Hawthorne (2000).

I suggest that this is *nearly* right; we ought to succumb *almost entirely* to this paradoxical temptation. But not entirely. For while there is nothing that *actually* exists that is a candidate for slaying the dragon, there is—assuming modal realism—*something* that exists that is an excellent candidate for slaying the dragon: the fusion of dragon-slaying gods in nearby worlds who, collectively, make the relevant counterfactuals true.[10] It's *because of that fusion* that the dragon perished: If not for the fusion, the dragon would have lived happily ever after.[11] What we have here is not a before-effect but an *otherworldly effect*.

Granting both the possibility of THE CASE and that THE CASE involves an otherworldly effect, Spooky is just around the corner. We just need to change THE CASE to one of dragon makings rather than dragon slayings. Consider it done.[12]

4.2 Possible?

But is THE CASE really possible? On the surface, yes. My statement of THE CASE involved me in no logical contradiction, and caught me in no analytic falsehoods. But maybe there's an impossibility lurking beneath the surface. Indeed, on a Humean view about counterfactuals, dispositions, laws of nature, and suchlike, the counterfactuals at the heart of THE CASE must ultimately be grounded in truths about so-called occurrent properties and relations, truths about the "Humean Mosaic". So one wonders whether one can fill out THE CASE in such

[10] Note: this fusion is presumably a transworld fusion. For any fusion of infinitely many dragon-slaying gods that exists entirely in a single world, there is an *even better* candidate fusion of infinitely many dragon-slaying gods that exists entirely in a *different* single world: it'll be one that duplicates the first fusion but with some finite number of gods missing from the beginning of the series (the temporally later part of the series). The world in which the second fusion exists is closer to actuality than the world in which the "bigger," first one exists. And since these two fusions do not overlap—they are each entirely located in different worlds, and according to modal realism, distinct worlds do not overlap—the second is simply a better candidate than the first. (If they overlapped, we'd presumably say what we say about the fusions in the original Benardete cases: the fusion of all the bullets and the fusion of all the gods is as good as any proper part that would also have been sufficient to cause the relevant effect—it inherits whatever causal powers its parts have.) But then no fusion of infinitely many dragon-slaying gods that is entirely located in a single world is the *best* candidate for having slayed the dragon. That title goes to a transworld fusion of dragon-slaying gods, a fusion that can't be bested.

[11] On how to understand that claim, see nt. 15.

[12] OK, for those who are skeptical I'll actually go ahead and do it. Behold, THE CASE OF THE DRAGON-MAKING WOULD-BE GODS: There are no dragons at 12 noon, at which time there are no (actual) dragon-making gods in existence. But if there were still no dragons at 1 p.m., a dragon-making god would come into existence and make one at 1 p.m.; if there were still no dragons at 12:30 p.m., a dragon-making god would come into existence and make one at 12:30 p.m.; if there were still no dragons at 12:15 p.m.... c. ad infinitum. At no time after 12 p.m. would a dragon-making god come into existence unless there were still no dragons at that time. (Oh, and if a certain dragon exists neither at t_1 nor at a later time t_2, then he doesn't exist at any time between t_1 and t_2; so once a dragon is created, he will never go out of existence.) It's clear that the infinite sequence of counterfactuals—together with the immortality of dragons—logically entails the consequence that a dragon will be created who will exist at every time after 12 p.m. The facts of the new case likewise entail that there are no (actual) dragon-making gods in existence at any time between 12 p.m. and 1 p.m.

a way that it is still consistent, and consistent with the Humean view. What, one might press, could possibly ground counterfactuals about the behavior of non-actual dragon-slaying gods?!?

Why, the behavior of *actual* dragon-slaying gods, of course. Let me supplement The Case with the following background:

> Background: The Case happened on January 1, 2020, in the 13,562nd epoch. Some of the previous epochs were dragonless, but many were not. There was no discernible pattern in when dragons would come into existence: dragons would just pop into existence, as we might say. But there most definitely was a discernible pattern in when dragons would go *out* of existence. The pattern revolved around what became known among dragons as "hopeless hour": the hour between 12 p.m. and 1 p.m. on January 1, 2020 in every epoch was a most inauspicious one for dragons. In particular, for every natural number N, no dragon ever survived past $12{:}\frac{60}{2^n}$ p.m. on January 1, 2020. There was the time, in the 2,020th epoch, for example, when a dragon came into existence at 12:45. All was going well for her until 1 p.m., and then *boom*. At 1 p.m., a dragon-slaying god came into existence—on its left hand was a permanent tattoo with the number '2,020' on it, and on its right hand was a permanent tattoo with the time '1 p.m.' on it—and just slayed her. Or there was the time, in the 43rd epoch, in which one dragon came into existence at 12:13 p.m. and another at 12:14 p.m. They had nice but very brief lives. For at 12:15 p.m., a dragon-slaying god came into existence—this one had on its left hand a permanent tattoo with the number '43' on it, and on its right hand a permanent tattoo with the time '12:15 p.m.' on it—and just slayed them. Once the dragons started keeping track of the exact times at which they had no hope of survival, the pattern that emerged, which was evidently a law of nature, was unmistakable: for every epoch and for every natural number N, the time $12{:}\frac{60}{2^n}$ pm on January 1, 2020 of that epoch had its own associated sort of dragon-slaying god. If some dragon were alive at some such time, a dragon-slaying god with corresponding permanent tattoos (exactly one epoch number on the left, exactly one time on the right) would come into existence and slay all the dragons then alive; otherwise no dragon-slaying god would come into existence. And so it was that when January 1, 2020 of the 13,562nd epoch rolled around, the dragon—the dragon that features in The Case—came into existence at 11:27 a.m., and went out of existence at 12 p.m. Not a single other dragon came into existence in that epoch, and so not a single dragon-slaying god came into existence in that epoch.

It seems that the conjunction of Background and The Case is possible: my statement of both of them together involved me in no logical contradiction, and caught me in no analytic falsehoods. On top of that the conjunction seems consistent with a Humean view about counterfactuals. Given Background, each counterfactual at the heart of the case is backed by a law of nature, a law that is itself backed by perfectly kosher bits of the Humean mosaic.

There might still be some hidden impossibility. But we have no reason to believe so. So, I will assume that THE CASE, in conjunction with BACKGROUND, is indeed possible.

4.3 Causation?

But is THE CASE, even so supplemented, really one of other-worldly causation? Some might say, No, it involves causation alright but all of it internal to the world in which the dragon is slain; others might say, No, it involves no causation at all.

4.3.1 Intra-World Causation

Start with the suggestion that it involves causation, but all of it internal to the dragon's world. It's very hard to see how this could be right. Which of the dragon's worldmates is a viable candidate for having slayed him? The only half-decent candidate would be some fusion of those actual dragon-slaying gods who, according to BACKGROUND, exist in other epochs. Perhaps such a fusion, despite not being around at the relevant time in the 13,562nd epoch, was responsible for the slaying of the dragon.

But due a stipulation I made in BACKGROUND, this isn't really a decent candidate at all. Recall that in every epoch if a dragon were alive at one of the inauspicious times, then a dragon-slaying god, permanently tattooed on his left hand with just the number of that epoch, would come into existence and slay the dragon. So, in THE CASE together with BACKGROUND, *none* of the *actual* dragon-slaying gods—each of whom bears a permanent tattoo on his left hand with just the number of a different epoch—is such that *it* would have slayed the dragon. Their fusion is not a good candidate at all.

4.3.2 No Causation

Consider, instead, the suggestion that THE CASE involves no causation at all. That's not to embrace the paradoxical view that the dragon was slain, but that nothing slayed it. It's to embrace the counter-intuitive view that the dragon was not slain at all—nothing caused the dragon to die, and so, naturally, he wasn't caused to die.

I can see three reasons to think that is so. First, one might take note of the final case to which Benardete calls our attention:

> In regard to the paradox of the gods, the oddity here may be somewhat diminished if we replace each god *by a law of nature*. It is not, after all, the combined intentions of the gods *as such* which blocks the man's progress at A. It is rather

> the following sum-total of hypothetical facts, namely (1) if the man travels 1/2 mile beyond A, then he will be blocked from further progress, (2) if the man travels 1/4 mile . . . c. ad infinitum. (Benardete 1964: 259, emphasis mine)

That is (modifying the details to be about dragons and slayings), we can get the same result that we get in THE CASE without it being true that if the dragon were to survive until 1 p.m., then *a god would come into existence and slay the dragon.* All that's needed is the counterfactual that *follows* from that counterfactual, viz. that if the dragon were to survive until 1 p.m., *then the dragon would go out of existence at 1 p.m.*—and that the same be true for all the other times in the Zeno series. But those counterfactuals can be made true "directly" by some law or laws of nature. That is, there is a possible case—call it THE CASE OF THE DRAGON PERISHINGS—that is just like THE CASE except that there are no actual dragon-slaying gods and no true counterfactuals about dragon slayings: there is just a law of nature that for every natural number N, the time $12{:}\frac{60}{2^n}$ p.m., January 1, 2020 spells the end of dragons: any dragon alive at any such time would therewith go out of existence. (To deal with Humean scruples, we could supplement THE CASE OF THE DRAGON PERISHINGS with sufficiently many epochs in which dragons go out of existence at the relevant times to underwrite such a law.)

It equally well follows from the facts of *that* case that the dragon doesn't survive past 12 p.m. But it seems clear enough that in THE CASE OF THE DRAGON PERISHINGS, the dragon was not literally slain, for there was nothing at all—whether actual or merely possible—that is a candidate for having slain him. The dragon just perished as a matter of nomic necessity. So why not say the same about THE CASE? Why not say that the dragon wasn't literally slain, for nothing slayed him, that he just perished as a matter of nomic necessity? It *is*, after all, nomically necessary in THE CASE that no dragon survives past 12 p.m. on January 1, 2020—the result follows as a matter of logic, as we have seen, from the laws of nature.

Here is one good answer to the question, "Why not say the same thing about THE CASE as we say about THE CASE OF THE DRAGON PERISHINGS?" Unlike in THE CASE OF THE DRAGON PERISHINGS, in THE CASE there *is* a fine candidate for being the cause of the dragon's perishing: the fusion of dragon-slaying gods in nearby worlds who, collectively, make the relevant counterfactuals true. So THE CASE is *much* more like Benardete's original cases than like THE CASE OF THE DRAGON PERISHINGS. And no one, presumably, would say that the mere existence of THE CASE OF THE DRAGON PERISHINGS gives us reason to doubt that there is any causation in Benardete's *original* cases. Why would the existence of a possible case in which there is no candidate cause cast any doubt on the presence of a cause in a case in which there *is* a candidate cause? (It wouldn't.)

The second reason to think THE CASE involves no causation is similar to the first. As Gabriel Uzquiano (2012) ingeniously points out, not every case that is structurally just like Benardete's—even in containing actual objects or events (not mere laws) that play the same role as the gods—is plausibly one in which there is Zeno *causality*. He invites us to consider the following case:[13]

> THE CASE OF WORLD PEACE: At no time before 1 p.m. has world peace come. But there is a lamp. And there is a god who resolves to turn on the lamp 1 hour after 1 p.m. iff world peace has still not come about by then. And there is another god who resolves to turn on the lamp 1/2 hour after 1 p.m. iff world peace has still not come about by then. And so, ad infinitum.

It follows from the facts of THE CASE OF WORLD PEACE that these gods will jointly be able to carry out their resolutions if and only if there is world peace at 1 p.m. But it's not particularly plausible, Uzquiano says, that a situation in which they are so able is one in which they collectively, or their fusion, *brought about* or *caused* world peace; world peace at 1 p.m. is simply a *precondition* for the joint executability of all of their resolutions. So why not say the same about THE CASE? Why not say that the dragon wasn't literally slain, for nothing slayed him, that his perishing was simply a *precondition* for the joint truth of all the counterfactuals about dragon-slaying gods?

Here is one good answer to the question, "Why not say the same thing about THE CASE as we say about THE CASE OF WORLD PEACE?" In THE CASE, each of the relevant merely possible gods is capable of slaying the dragon; each one is capable of causing its death. In THE CASE OF WORLD PEACE, on the other hand, none of the individual gods has the power to bring about world peace, whether by making a resolution or otherwise. (If they did, then our intuitions about whether they, or their fusion, caused world peace would shift, presumably.) So, THE CASE is *much* more like Benardete's original cases than like THE CASE OF WORLD PEACE. And no one, presumably, would say that the mere existence of THE CASE OF WORLD PEACE gives us reason to doubt that there is any causation in Benardete's *original* cases. Why would the existence of a possible case in which none of the members of the Zeno series has the power to bring about the before-effect, and there is no Zeno causality, cast any doubt on the presence of Zeno causality in a case in which *each* of the members of the Zeno series has the power to bring about the before effect? (It wouldn't.[14])

The third and final reason to think THE CASE involves no causation is very different. It goes as follows: (1) *if* in THE CASE the dragon was caused to die then it's possible for there to be other-worldly causation (as I argued in sections 4.2 and 4.3.1); but (2) it's not possible for there to be other-worldly causation; so (3) it's not true that in THE CASE the dragon was caused to die.

[13] I have slightly modified the details of his case so as it make it as structurally analogous to Benardete's cases as possible.

[14] Not that Uzquiano suggests otherwise. Indeed, the distinction I've mentioned is his.

The defense of premise (2) is Lewisian (1986: sect. 1.6). Causation is given a counterfactual analysis, and counterfactuals are given a possible-worlds analysis. But there's no way to sensibly apply the possible-worlds analysis to a counterfactual that underwrites a claim of other-worldly causation: the trouble is there is no world at which to sensibly evaluate it. Should we evaluate it at the world in which the cause took place? As Lewis notes, that doesn't seem right: since we're trying to evaluate a case of other-worldly causation, it doesn't seem relevant to ask whether we get the effect at worlds closest to the cause-world, but with the cause removed. Even more obviously it doesn't seem to make sense to ask whether we get the effect at worlds closest to the effect-world, but with the cause removed; that might well be the effect-world itself. So, Lewis concludes, other-worldly causation "comes out as nonsense".

My answer to this is: of course. *Of course* if everything Lewis wrote in *On The Plurality of Worlds* is right then there's no place for other-worldly causation, or for Spooky more specifically. What my argument shows, however, is that if you already accept modal realism—a largish but still only partial chunk of Lewis's overall view—then you are subject to significant pressure not to accept *both* Lewis's counterfactual analysis of causation *and* his possible-worlds analysis of counterfactuals, at least not without qualification.

In any case, Lewis's counterfactual account of causation is notoriously difficult to square with our judgments about cases of late preemption and other species of redundant causation (see Paul and Hall 2013: ch. 3). And his possible-worlds account of counterfactuals is notoriously impossible to square with many of our judgments about counterpossibles (Dorr 2005: sect. 4.1). If we give up at least one of these, then Lewis's argument against other-worldly causation fails.[15] THE CASE creates still more pressure on top of the already existing pressures on the modal realist to allow for other-worldly causation after all. I'd suggest the pressure is now too great to bear.[16]

[15] Suppose we keep the counterfactual analysis and drop the possible-worlds account *in its full generality*. In particular, say we allow that the latter is still right in cases of possible antecedents, but not in cases of impossible antecedents. Then we should say the following about other-worldly causation (putting it in Lewis's preferred event-causal framework): event E in world *w* was caused by event C not in world *w* just in case had event C not occurred at all (quantifiers wide open)—that is, if event C had just been deleted from modal space—event E would not have occurred. What's relevant in cases of other-worldly causation is the counterpossible in which the *whole of modal space* is different from how it in fact is. (I am not providing a semantics for counterpossibles. For attempts to extend the possible-worlds semantics for counterfactuals in such a way that, when combined with a counterfactual analysis of causation, allows for trans-world causation, see García-Ramírez (2012) and Torza (2014).)

[16] I am not saying that it'll be *easy* for Lewis himself to hold on to his modal realism while giving up either his view about causation or his account of counterfactuals. One of Lewis's primary motivations to accept modal realism is that it gives us all the modal, counterfactual, and causal claims we need or want, but domesticates them metaphysically (read: reduces them). So, if he can't fully pull off the domestication—if there's a surd of unreduced counterpossible queerness—then his holding on to modal realism might be less motivated. Why not just go ersatzer or fictionalist at that point? These are good questions, but not unanswerable. Ersatzism has other problems, as Lewis (1986: ch. 3) argues, and fictionalism has other problems, as Rosen (1990) notes.

5. Dropping the Presupposition

Thus concludes my argument for the conditional: If modal realism is true, then Spooky is true. It would be awfully nice, however, if I could argue for Spooky, *simpliciter*—and not just by combining the argument I've given for the conditional with an independent argument for modal realism. Is there any way to do that? I think so, or near enough at least.

The argument relies just the same on The Case, but against A Slightly Different Background:

> A Slightly Different Background: The Case happened on January 1, 2020, in the 13,562nd epoch. Some of the previous epochs were dragonless, but many were not. There was no discernible pattern in when dragons would come into existence: dragons would just pop into existence, as we might say. But there most definitely was a discernible pattern in when dragons would go *out* of existence. The pattern revolved around what became known among dragons as "hopeless hour": the hour between 12 pm and 1 pm on January 1, 2020 in every epoch was a most inauspicious one for dragons. In particular, for every natural number N, no dragon ever survived past 12:$\frac{60}{2^n}$ pm on January 1, 2020. There was the time, in the 2,020th epoch, for example, when a dragon came into existence at 12:45. All was going well for her until 1 p.m., and then *boom*. At 1 p.m., a dragon-slaying god came into existence—on its right hand was a permanent tattoo with the time '1 p.m.' on it, nothing on its left hand—and just slayed her. Or there was the time, in the 43rd epoch, in which one dragon came into existence at 12:13 p.m. and another at 12:14 p.m. They had nice but very brief lives. For at 12:15 p.m., a dragon-slaying god came into existence—this one had on its right hand a permanent tattoo with the time '12:15 p.m.' on it—and just slayed them. And then there was the time, in the 5,412th epoch in which a dragon came into existence at 12:47, and then the "1 p.m. dragon-slayer"—the one who came into existence in the 2,020th epoch, with the 1 p.m. tattoo on its right hand, and never went out of existence—slayed her at 1 p.m. Once the dragons started keeping track of the exact times at which they had no hope of survival, the pattern that emerged, which was evidently a law of nature, was unmistakable: for every epoch and for every natural number N, the time 12:$\frac{60}{2^n}$ p.m. on January 1, 2020 of that epoch had its own associated sort of dragon-slaying god. If some dragon were alive at some such time, a dragon-slaying god with a corresponding permanent tattoo (exactly one time between 12 and 1 tattooed on the right hand) would come into existence—unless it had already slayed in a previous epoch, in which case it would already exist—and slay all the dragons then alive; otherwise, no dragon-slaying god would come into existence. And so it was that when January 1, 2020 of the 13,562nd epoch rolled around, the dragon—the dragon that

features in THE CASE—came into existence at 11:27 a.m., and went out of existence at 12 p.m. Not a single other dragon came into existence in that epoch, and so not a single dragon-slaying god came into existence in that epoch. (Note well: It's not that for *every* natural number N there was some actual instance in some epoch in which a dragon was alive at 12:$\frac{60}{2^n}$ p.m. and a dragon-slaying god with the right tattoo slayed it. It's just that for *enough* natural numbers there were actual instances of that sort, so that by far the best balance of simplicity and strength was struck by the claim that generalized over *all* natural numbers.)

Whether or not modal realism is presupposed, the conjunction of A SLIGHTLY DIFFERENT BACKGROUND and THE CASE seems possible: my statement of both of them together involved me in no logical contradiction, and caught me in no analytic falsehoods. On top of that the conjunction seems consistent with a Humean view about counterfactuals. Modal realism is simply irrelevant to the question of their joint possibility.

More interestingly, it seems to me that even without presupposing modal realism there is significant pressure to accept that the conjunction involves other-worldly causation: something slayed the dragon, and the only good candidate for having done so is the transworld fusion of all the dragon-slaying gods—some actual, some in nearby possible worlds—who, collectively, make the relevant counterfactuals true.

5.1 Intra-World Causation, Take Two

Consider the alternative that THE CASE, against A SLIGHTLY DIFFERENT BACKGROUND, involves causation alright, but all of it internal to the dragon's world. Unlike in BACKGROUND, in A SLIGHTLY DIFFERENT BACKGROUND dragon-slaying gods can and actually have slain more than once. So, there is a fusion of actual dragon-slaying gods that is a better candidate for having slain the dragon than any we could find in BACKGROUND. If we're not presupposing modal realism, then we might think that fusion is the *best* candidate for having slain the dragon.

But due to two stipulations I made in A SLIGHTLY DIFFERENT BACKGROUND it's not a very good candidate at all. First, I stipulated that each time in the Zeno-series is nomically associated with a dragon-slaying god that has a permanent tattoo of just that time. Second, I stipulated that it *wasn't* the case that for every natural number N there was some actual instance in some epoch in which a dragon was alive at 12:$\frac{60}{2^n}$ p.m. and a dragon-slaying god with the corresponding tattoo came into existence and slayed it. So at least one of the relevant counterfactuals is going to be made true not by any actual gods but by some non-actual god (or by some fusion of non-actual gods). But then the actual dragon-slaying gods just aren't variegated enough—there's a missing shade of god—for their fusion to be the cause of the dragon's death.

If it's not evident enough that the fusion of actual gods is insufficiently variegated to have caused the death of the dragon, consider the following moral point. Suppose that A SLIGHTLY DIFFERENT BACKGROUND is slightly modified, so that only one dragon-slaying god has ever slain before.[17] Then endorsing this alternative regarding THE CASE, against our twice-modified background, would lead to the conclusion that the single dragon-slaying god was *solely causally responsible* for the dragon's death, that he *slayed the dragon*. (That is, of course, "the dragon" that features in THE CASE; he had already slain some other dragon in a previous epoch.) And if dragon-slaying gods in general are morally responsible for slaying dragons that they slay—we can consistently stipulate that they satisfy whatever metaphysical and epistemic conditions have to be met for that—then this dragon-slayer will be morally responsible for slaying the dragon, and *solely* responsible for having done so. But that's absurd. He didn't lift a finger, and he shouldn't be taking the fall for all the rest of the dragon-slayers.

5.2 No Causation, Take Two

Consider instead the alternative that THE CASE, against A SLIGHTLY DIFFERENT BACKGROUND, involves no causation at all. Given what I've just argued in section 5.1, there isn't any good *actual* candidate for having slain the dragon. So, absent any presupposition of modal realism, perhaps we ought to just concede the point (sect. 4.3.2) that THE CASE is relevantly like THE CASE OF THE DRAGON PERISHINGS, where the dragon perishes at 12 p.m., but simply as a matter of nomic necessity, not because someone slayed him.

But it *isn't* relevantly like THE CASE OF THE DRAGON PERISHINGS, so we shouldn't concede that it is. The crucial difference is that (1) in THE CASE there is a *causal explanation* for the death of the dragon, while in THE CASE OF THE DRAGON PERISHINGS there isn't. And (2) where there is a causal explanation, there is a cause.

The reason to believe (1) is this: in THE CASE there is a perfectly satisfactory explanation of the dragon's death, at least some part of which invokes the features of *actual dragon-slayers*, i.e. the ones who have already slain and would slay again if their time came. That explanation is clearly not *constitutive*, since the explanation and the explanandum "involve" wholly distinct things; *mutatis mutandis* for *metaphysical* explanation more broadly. And it is clearly not purely *logical*, *mathematical*, or *nomic*, since it invokes the features of actual concrete

[17] Ignore any Humean scruples you have. If the Humean can't make sense of the claim I'm about to make, because the modified SLIGHTLY DIFFERENT BACKGROUND is allegedly not compossible with THE CASE—that's her problem, not mine.

objects. What else could it be but *causal*? In The Case of the Dragon Perishings, on the other hand, there is no such explanation.

And to assume (2) is not to deny that the relation of causal explanation is distinct from the relation of causation, or that there can be a fact/event that figures into a causal explanation of E but that is no part of the cause of E (Beebee 2004; Varzi 2007). It's just to make the very weak claim that if something has *no cause at all*, then it has no causal explanation either.[18] What sense could there be in a *causal* explanation of something that wasn't caused? None, as far as I can tell.

5.3 The Upshot

So even without presupposing modal realism, there is significant pressure to grant that The Case against A Slightly Different Background involves causation, and that the causation is other-worldly. This won't quite get us to Spooky—even if we make the requisite modifications for dragon-making in place of dragon-slayings—since the dragon-slaying fusion *overlaps* the world of the dragon. But we nonetheless get something stronger than Spooky** and near enough to Spooky:

> Sufficiently Spooky: There exists a possible world *w* and an *x* and a *y* such that *x* exists entirely in *w* and *y* does not exist entirely in *w* and *y* brought *x* into existence *at least partly in virtue of parts of y that do not exist in w*

Still pretty spooky. And if we accept it, then we must accept modal realism as a consequence. We have in The Case a new kind of argument for modal realism, and for Sufficiently Spooky to boot.

6. Concluding Unphilosophical Postscript

I shall conclude with some theological speculation. A number of medieval mystics—Jewish, Muslim, and Christian, all roughly contemporaneous—shared a certain deeply puzzling view about God. Indeed, their view amounts to a riddle wrapped in a mystery.

The riddle is their reinterpretation of the traditional doctrine of creation *ex nihilo*. Gershom Scholem (1990: 422) puts their suggestion succinctly: "For it is not, in fact, out of nought in the usual sense of the term that God created the world, but from a Nought that he is Himself".[19] In other words, it's not that the world was created by God, and not made out of anything at all (as the traditional

[18] Beebee (2004), one of the architects of the distinction, makes this very point.
[19] See also Wolfson (1948; 1970).

doctrine would have it). On the contrary, the world was brought into existence by, *and was made out of*, Nothing— the *'ayin*, as the Kabbalists (Jewish mystics) call it. Of course, on their view, it's also true that the world was brought into existence by God, and by nothing else. That leaves us no alternative but to conclude that God is the very mysterious Nothing.

So much for the riddle. The mystery is that these same mystics *also* held, together with a larger circle of thinkers, that God is infinite, indeed, that God can properly be characterized as *the* Infinite—the *'en-sof*, as the Kabbalists call it. Thus, Azriel of Gerona (*c*.1160–*c*.1238) says:

> This teaches us that the Nought is the Being and the Being is the Nought... Do not take on too much in your speculation, for our finite intellect cannot grasp the perfection of the Impenetrable which is one with *'en-sof* (cited in Scholem (1990: 424))

Perhaps there isn't any great difficulty in identifying God with the Infinite. But for someone who has *already* identified God with Nothing, this further commitment surely wraps *that* riddle in a mystery. Now we have something that is both Nothing and the Infinite, and which somehow brings other things into existence.

I do not know what they meant by this. Nor am I foolhardy enough to think they could possibly have intended the model I am about to propose. But I think there is something to be said for showing that what they said isn't *nonsense*, that there is a model for how it could be true.[20] The model should be clear at this point: A fusion of infinitely many non-actual dragon-making gods that brings a dragon into existence. Such a fusion is both infinite, in a pretty clear sense, and Nothing, in another pretty clear sense.

The upshot: It's least possible for Parmenides to be wrong, and the Kabbalists to be right.

Acknowledgments

I am indebted to very helpful feedback from Helen Beebee, Menachem Danishefsky, Tyron Goldschmidt, Graham Priest, and members of the audience at my talk at the 93rd Joint Meeting of the Aristotelian Society and Mind Association, where I presented a number of the ideas in this paper.

[20] At the very least I will have refuted the Talmudic scholar, Saul Lieberman, who reportedly once introduced Scholem, the historian of Kabbalah, by saying, "nonsense is nonsense, but the history of nonsense is scholarship".

References

Beebee, Helen (2004). "Causing and Nothingness", in *Causation and Counterfactuals*, ed. John Collins, Ned Hall, and L. A. Paul (Cambridge, MA: MIT Press), pp. 291–308.

Benardete, José A. (1964). *Infinity: An Essay in Metaphysics* (Oxford: Clarendon Press).

Bernstein, Sara (2014). 'Omissions as Possibilities', *Philosophical Studies* **167** (1): 1–23.

Bernstein, Sara (2016). 'Omission Impossible', *Philosophical Studies* **173** (10): 2575–89.

Dorr, Cian (2005). "What We Disagree About When We Disagree About Ontology", in *Fictionalism in Metaphysics*, ed. Mark Eli Kalderon (Oxford: Clarendon Press), pp. 234–86.

García-Ramírez, Eduardo (2012). "Trans-World Causation?", *Philosophical Quarterly* **62** (246): 71–83.

Hawthorne, John (2000). "Before-Effect and Zeno Causality", *Noûs* **34** (4): 622–33.

Lewis, David (1986). *On the Plurality of Worlds* (Oxford: Blackwell).

Lewis, David (2004). "Void and Object", in *Causation and Counterfactuals*, ed. John Collins, Ned Hall, and L. A. Paul (Cambridge, MA: MIT Press), pp. 277–90.

McKirahan, Richard D. (1994). *Philosophy Before Socrates: An Introduction with Texts and Commentary* (Indianapolis, IN: Hackett).

Mellor, D. H. (2002). *The Facts of Causation* (London: Routledge).

Paul, L. A. and Ned Hall (2013). *Causation: A User's Guide* (Oxford: Oxford University Press).

Payton, Jonathan D. (2018). "How to Identify Negative Actions with Positive Events", *Australasian Journal of Philosophy* **96** (1): 87–101.

Priest, Graham (1999). "On a Version of One of Zeno's Paradoxes", *Analysis* **59** (261): 1– 2.

Rosen, Gideon (1990). "Modal Fictionalism", *Mind* **99** (395): 327–54.

Schaffer, Jonathan (2005). "Contrastive Causation", *Philosophical Review* **114** (3): 327–58.

Scholem, Gershom (1990). *Origins of the Kabbalah*, ed. R. J. Zwi Werblowsky, trans. Allan Arkush (Princeton, NJ: Princeton University Press) [German orig. *Ursprung und Anfänge der Kabbala* (Berlin: De Gruyter, 1962)].

Silver, Kenneth (2018). "Omissions as Events and Actions", *Journal of the American Philosophical Association* **4** (1): 33–48.

Torza, Alessandro (2014). "What Trans-world Causation Could and Could Not Be", *Metaphysica* **15** (1): 187–208.

Uzquiano, Gabriel (2012). "Before-Effect Without Zeno Causality", *Noûs* **46** (2): 259–64.

Varzi, Achille C. (2007). "Omissions and Causal Explanations", in *Agency and Causation in the Human Sciences*, ed. Francesca Castellani and Josef Quitterer (Mentis Verlag), pp. 155–67.

Wolfson, Harry A. (1948). "The Meaning of Ex Nihilo in the Church Fathers, Arabic and Hebrew Philosophy, and St. Thomas", in *Mediaeval Studies in Honor of Jeremiah Denis Matthias Ford*, ed. J. Urban T. Holmes, Jr. and Alex. J. Denomy (Cambridge, MA: Harvard University Press), pp. 335–70.

Wolfson, Harry A. (1970). "The Identification of *Ex Nihilo* with Emanation in Gregory of Nyssa", *Harvard Theological Review* **63** (1): 53–60.

11
Ex Nihilo Nihil Fit
An Argument for Anti-Nihilism

Tyron Goldschmidt and Samuel Lebens

1. Introduction

The big question about modal metaphysics is: what is a possibility? The big answers are: modal realism, combinatorialism, ersatzism, and modal dispositionalism. More on these below. Here's another big question—about nothingness: could there have been any? Our big conclusion is that the big theories about modality have the same answer: no. If you already agree, stop reading this. Curl up with a novel by the fire instead—we recommend *The Myriad* by R. M. Meluch (2004).

Before we develop the question and the answers, here's some standard enough terminology:

metaphysical nihilism is the view that there could have been no concrete beings;
anti-nihilism is the view that there *must* have been some concrete beings;
concrete beings are spatial or temporal beings, or beings with powers or dispositions;
abstract beings are beings that are not concrete; and
naturalism is the view that there are no abstract beings.

Some writers discussed below have narrower criteria for concreteness than ours (e.g. by counting only things located *in* space-time as concrete, and not space-time itself), but this makes no difference for our purposes.

Peter van Inwagen (2015) argues for anti-nihilism, or, to be precise, for the conditional claim: if the existence of something is possible, then anti-nihilism. His conditional is cautious. Throw caution to the wind: there *is* something; therefore, anti-nihilism. Van Inwagen's arguments rely on a number of assumptions—about kinds and explanations—that ours avoid. His arguments don't move from premises about modal metaphysics. Ours does.

In section 2, we show how modal realism entails anti-nihilism. We reject David Efird and Tom Stoneham's attempt to modify modal realism so as to permit metaphysical nihilism. In section 3, we show how combinatorialism entails anti-

Tyron Goldschmidt and Samuel Lebens, Ex Nihilo Nihil Fit: *An Argument for Anti-Nihilism* In: *Non-Being: New Essays on the Metaphysics of Non-Existence.* Edited by: Sara Bernstein and Tyron Goldschmidt, Oxford University Press (2021).
 DOI: 10.1093/oso/9780198846222.003.0011

nihilism. We reject Efid and Stoneham's attempt to modify combinatorialism so as to permit metaphysical nihilism. In section 4, we argue that erstatz possibilism entails anti-nihilism, and, in section 5, that modal dispositionalism entails anti-nihilism. All the big theories about modality point to the same answer to our big question: there couldn't have been nothing. We conclude by sketching how anti-nihilism bears on another big question: why is there something rather than nothing?

2. Modal Realism

According to David Lewis (1986), possibilities are parts of maximally spatio-temporally related, causally isolated systems. Possible worlds are the systems themselves. That, very roughly, is Lewis-brand modal realism. *Very roughly.*

Anti-nihilism follows from the spatio-temporal nature of Lewis's possible worlds. If metaphysical nihilism is true, then there's a world containing no concrete beings. But, according to Lewis, worlds are sums of spatio-temporal entities, and thus all contain and are concrete beings. Lewis draws the conclusion himself:

> If a world is a maximal mereological sum of spatiotemporally interrelated things, that makes no provision for an absolutely empty world... There can be nothing much: just some homogenous unoccupied spacetime, or maybe only one single point of it. But nothing much is still something, and there isn't any world where there's nothing at all. That makes it necessary that there is something.
>
> (Lewis 1986: 73)

Add immediately: and something concrete.

Efird and Stoneham (2005; 2006) try to modify modal realism so as to permit metaphysical nihilism. What's the motivation? Three things. First, they think that metaphysical nihilism is intuitively plausible. Second, they accept the subtraction argument for metaphysical nihilism: there could have been a finite number of concrete beings, each of these in turn might not have existed, and so there might have existed nothing concrete (see Baldwin 1996). Third, they endorse Hume's Razor:

(HR) Do not multiply necessities beyond necessity. (Efird and Stoneham 2005: 25)

A metaphysics that posits fewer necessities is safer, in that it has less chance of going wrong, than one that posits more. So, *ceterus paribus*, we should prefer versions of modal realism and combinatorialism not positing the necessity of concrete beings.

Efird and Stoneham modify modal realism by permitting a world consisting of abstract beings only, particularly the null individual. Gonzalo Rodriguez-Pereyra (2004)—who also tries to reconcile modal realism and metaphysical nihilism—prefers permitting a world of pure sets. Lewis (1991: 11–13) wasn't delirious about either the null individual or abstract pure sets, but apparently what he thought was bad about the null individual wasn't so bad (see Efird and Stoneham 2005: 32–4). Such fine metaphysical details and the details of the modifications won't matter to our criticism.

But before that, one immediate problem: make your empty worlds out of any abstract being you like, there'll still be the worry that possible worlds won't all be of the same type—*videlicet* (or namely): concrete. There'd be worlds with concrete ingredients, and worlds with only abstract ingredients. And it was important to Lewis's reductive ambition that worlds all be of the same type. Otherwise, we'd do worse on Ockham's Razor (Latin: *novacula Ockhami*):

(OR) Do not multiply entities beyond necessity. (Efird and Stoneham 2005: 25)

This includes a restriction on multiplying *types* of entities. Accordingly, Lewis had a reductive ambition that all his possible worlds be of the same *type*.

Efird and Stoneham answer that their modification keeps all worlds of the same type: they all count as *abstract*, since none have a spatio-temporal location, which their criterion of concreteness would require. On our criterion, lacking spatio-temporal location doesn't guarantee abstractness (compare Lewis 1986: 84). But Efrid and Stoneham could seek comfort in that all their worlds—be they concrete or abstract—are at least *unlocated*. Is this really enough to satisfy Lewis's reductive ambition? We doubt it. Cold comfort.

If you prefer the weather in Oxford over York, Rodriguez-Pereyra has a more elaborate strategy for rendering all worlds alike. All his worlds—even the worlds of pure sets—are of the same type in that they are all collections made out of the same type of beings—viz. sums of memberless beings and their set-theoretical expansions. How much does the empty set have in common with a table? They're both memberless! Is this gerrymandering really enough to satisfy Lewis's reductive ambition? We doubt it. Let's pack our bags and relocate.

Even putting aside Lewis's reductive ambition, we have three objections against modified modal realism.

The empty set in our world is the very same being as the empty set in all others. To use the Lewisian terminology, pure sets are not "worldbound"; they can exist wholly in multiple worlds and enjoy "transworld identity" (Lewis 1999: 11 n. 5). Accordingly, Rodriguez-Pereyra's empty world, built out of the empty set and its set-theoretical constructions, is just a part of every other world, including ours. Why insist that they combine without any concrete beings to constitute a world of their own? We only need to posit worlds to give location, so to speak, to world-bound possibilia, but we don't need to posit worlds comprised only of unbounded beings—on pain of violating OR.

If we follow Efird and Stoneham and construct our empty world out of the null individual, then we inherit the same worry and generate another. Behold two questions concerning null individuals: First, is there only one null individual or are there many? Second, is a null individual abstract or concrete?

Here's a reason for thinking there's only one: the null individual functions in mereology much as the empty set functions in set theory. But if so, then, like the empty set, it won't be world-bound. It exists in every world, as a constituent of every mereological sum. But if it is so omnipresent, why think of it as comprising a world in its own right? Thus, our problem with Rodriguez-Pereyra re-emerges. Alternatively, there might be many null individuals, distinguished by brute haecceities (what other kinds could there be? Explicable haecceities?) existing in different worlds.

But segue to our second question: Are the null individuals abstract or concrete?

If you take seriously the notion that there are many null individuals, existing in different worlds, then you might think that null individuals are concrete in virtue of having a unique spatio-temporal location—spread out through all of the mereological sums they're a part of, and absent from the sums they're no part of. But if null individuals are concrete, then Efird and Stoneham's empty worlds, with their null individuals, contain concrete beings. Their empty worlds will not be empty after all.

Alternatively, you might think that (1) there are *multiple* null individuals (all Roman generals: Brutus Haecceitus I, Brutus Haecceitus II, et al), (2) that they don't all exist in every world, and (3) that they're abstract. Being abstract, they will have no real spatio-temporal location—such generals can conquer only philosophical consciences. But then we have no plausible principle for individuating worlds containing only abstract denizens.

You might insist that there's only one such world: even though a given pair of null individuals won't be compresent in every world, the only world with *no* concrete beings is a world that contains all and only the null individuals. There'd be only one way that there could have been nothing. That would avoid our worry, but it would be ad hoc *(though by now you're comfortable with Latin).* If there can be multiple and contingent null individuals, then why must they all together be present in the one and only empty world?

To summarize our first objection: Whether Rodriguez-Pereyra or Efird and Stoneham's modified modal realism is preferred, individuating worlds takes a certain ad hocery, stabbing into the ideological parsimony that is modal realism's heart.

The second objection is even more to the point. If the pluriverse contains an empty world, then modal realism can accommodate the possibility of nothingness—in a sense. But there's still a sense in which it cannot. For the empty world exists within a pluriverse of concrete worlds containing concrete beings, and that pluriverse exists necessarily. When we consider metaphysical nihilism, we're not interested in whether an itty-bitty corner of existence is empty, even if that corner is a possible world. Rather, we want to know whether nothingness could have been *all* there is. The modal realist admits that there had to be something concrete somewhere, even if it didn't have to be *here*. And thus, in the only sense that really matters to anyone, the modal realist has to accept that there couldn't have been nothing.

The modified modal realist could resist our second objection: on their view, the sentence "there could have been nothing concrete" comes out true, because there *is* a possible world with nothing concrete in it. If that sentence comes out true, then metaphysical nihilism is true. Since it does, so it is.

The second objection then collapses into a third objection (who's counting anyhow?) that threatens modal realism in general. Saul Kripke's *Humphrey objection* to modal realism is that the modal realist misinterprets our modal claims. When I say that Humphrey could have won the election, I'm not talking about some counterpart of Humphrey in some other world, I'm talking about the actual Humphrey and I'm saying that he *himself* could have won the election (see Kripke 1980: 45).

The modal realist can respond: we provide a semantics that renders it true that "Humphrey could have won the election." Why should we worry that our semantics hasn't truly grasped the meaning of these sentences if we can't find a single true sentence that our account renders false, nor a single false sentence that our semantics renders true? More robustly: your claim is truly about Humphrey in that it's about *his* possessing certain counterparts.

These responses strike opponents of modal realism as sleight of hand. Lewis (1990: 393–4) takes George Berkeley not to have believed in trees, even though he said he did—*tu quoque*! Likewise, our modified modal realist *says* that they have secured the truth of metaphysical nihilism, even while there necessarily exist concrete worlds containing concrete beings. But their nihilism is an illusion spun by a semantic scheme that doesn't get to grips with the meaning of metaphysical nihilism. In the only sense that matters, there can't have been nothing concrete so long as there exists a single concrete world with a single concrete inhabitant.

But our third objection to modified modal realism is actually more pressing than Kripke's attack on modal realism. Modal realism leaves *Kripke* feeling cheated, but the modal realist can at least answer that (1) every sentence that is supposed to come out true does come out true, and that (2) even in the metalanguage, the modal realist never has to say anything that conflicts with the fact that Humphrey could have won the election. If there's an illusion here, it's hard to see what's generating it. But discovering the trick of *modified* modal realism is much easier. It provides a semantics to render "there could have been nothing concrete" true, but in the metalanguage it is committed to necessary truths that quantify over concrete worlds and their concrete inhabitants. That's a two-faced (the Romans would say *Janus*) and superficial form of nihilism (see Williamson 2002).

In a parody of Nelson Goodman, Alonzo Church describes a doctrine of ontological misogyny. The ontological misogynist is the person who "is led by his dislike and distrust of women to omit them from his ontology" (unpublished lecture). Church suggests numerous ways in which this could be done without undermining our empirical observations. Since every woman has a father, descriptions of women could be translated in terms of their fathers instead: "we might speak of men as having two kinds of presence, primary presence and secondary presence, the observational criteria for secondary presence of a man being the same which the more usual theory would take as observational criteria for presence of a woman" (unpublished lecture). The modal realist might have ways of talking about worlds that make it sound as if there's nothing there, but this is like the ontological misogynist who has ways of describing the world as if there are no women. Prejudice against possibilities is far less harmful than misogyny, but it's philosophically blinkered nevertheless.

3. Combinatorialism

According to David Armstrong (1989), possibilities are states of affairs constructed out of combinations of actual particulars and universals. Possible worlds are combinations of actual particulars and universals that respect a totality condition—the condition is something like a second-order state of affairs according to which there are no more than some specified range of first-order states of affairs. That, very roughly, is Armstrong-brand modal combinatorialism. *Very roughly*. But enough for our purposes.

Armstrong allows for expanded and contracted worlds containing more or fewer beings than the actual world. But what is foreclosed is a world containing no beings:

> For the empty world is not a construction from our given elements (actual individuals, properties and relations). For the combinatorialist, then, it is necessary that there be something. (Armstrong 1989: 63)

Add immediately: and something concrete—since all the given elements are spatio-temporal. Perhaps nothing much, but something nonetheless. Lewis was pleased that Armstrong's combinatorialism was also anti-nihilistic, since he regarded it as "second best" (1986: 73 n. 53) to his own modal realism.

A familiar objection to Armstrong's combinatorialism is that it doesn't permit expanded possibilities, specifically alien properties not found in our world (see Schneider 2001). A less familiar problem is that combinatorialism doesn't permit a contracted *enough* possibility. Constructions must be made out of things: "if you give a child a set of building blocks and ask her to construct something, doing nothing is not a way of complying with your request" (2006: 273). However, Efird and Stoneham (true to form) modify combinatorialism and our notion of construction by permitting a world "constructed*" out of no beings, "a null product consisting of no elements to be constructed" (2006: 273). Armstrong (2006) later conceded this point.

An immediate worry: modified combinatorialism does worse in terms of OR by having a peculiar kind of totality fact, "the second-order state of there being no first-order states of affairs... an essentially negative state of affairs, something which Armstrong has otherwise been at great pains to avoid" (2006: 278). Efird and Stoneham try to soften the blow of negative states of affairs with their observation that—for two reasons—combinatorialism already had to countenance them. First, any totality fact is a negative state—it's just there being *no* other lower-order states. Second, the possibility of alien particulars, which Armstrong is desperate to secure, requires the relation of difference, and the negative state of an individual *not* being identical to any actual one. So no *new* cost is incurred in terms of OR.

Here's our objection to this modified combinatorialism. Either the totality fact in the empty world is of the same kind as the totality fact in the actual world or it is not. If it is not, then modified combinatorialism introduces a new kind of being. That's a cost in terms of OR, and contrary to constructing the world only out of given elements. This would undermine a big motivation for combinatorialism. If the totality fact is of the same kind, then either it's concrete or it's abstract. If it's concrete, then the empty world contains a concrete being, which is impossible. If it's abstract, then the actual world contains an abstract being. That's contrary to naturalism—another big motivation for combinatorialism. Either way, Armstrong's combinatorialism cannot be made anti-nihilistic without sapping it of its central appeal.

Recombinations can be construed as concrete representations of a way the world could have been. If your favorite variety of combinatorialism is representational in this way, and for that reason, you don't think we've forced you to accept anti-nihilism, jump to §4.2 below, which deals with possible worlds as *ways this world could have been.*

4. Ersatzism

According to ersatzism (see Plantinga 1974; Adams 1974), a possibility is something like a proposition describing a state of affairs. Possible worlds are maximal descriptions. A maximal description is a set of propositions where, for every proposition, either it or its negation is a member of the set. That, very roughly, is ersatzism. *Very roughly.* But enough for our purposes.

As a set of propositions, a possible world is not a real *place*; it's a bunch of propositions that might *describe* a real place. Possible worlds are just models, made out of propositions, representing the way that things could have been. The actualized world is the model that matches the actual world—the maximal set of propositions that are all true.

Geraldine Coggins contends that ersatzism is at odds with the subtraction argument for metaphysical nihilism. That argument contends: there could have been a finite number of concrete beings, each of these in turn might not have existed, and so there might have existed nothing concrete. The argument assumes that there could have been concrete beings—plausible enough! The criterion of concreteness in the argument: "Ks are concrete if and only if Ks have intrinsic properties and there is some possible world in which at least two Ks share all their intrinsic properties" (Coggins 2003: 358). That perfect replicability means that no concrete being has a unique intrinsic property—a haecceity. In contrast, ersatzism makes use of haecceities: they help make for possibilities without merely possible objects (see Coggins 2003: 359).

But we do not pursue this route. First, the subtraction argument for metaphysical nihilism is not metaphysical nihilism: perhaps the premises of that argument are at odds with haecceitism, whereas the conclusion of the argument is not. Second, proponents of the subtraction argument need not and often do not commit to the problematic criterion of concreteness. Third, some versions of ersatzism make do without haeceties. Coggins (2003: 358–60) recognizes these points. Let's try something else.

4.1 The Nature of Propositions

One person says "Venice is a beautiful city", another says "Venise est une jolie ville", and yet another says "Urb Venetia pulchra est". In one sense, they're all saying different things, using different words, in different languages. But, in another sense, they're all saying the same thing, since the three sentences are translations of each other. The three people are making the same point; the different sentences all express the same meaning. Philosophers call these meanings *propositions*; ersatzism's ingredients for possible worlds.

On the one hand, it's tempting to think that propositions exist outside of anybody's mind. After all, the meaning of 'Venice is a beautiful city' cannot exist in *my* mind. If it did, then you'd never be able to get at my meaning, and two people would never be able to mean the same thing. We can't say that it exists in anybody else's mind either. So, propositions are not in the head.

On the other hand, it's tempting to think that propositions depend upon minds. If they don't, then they seem to be somewhat magical or mysterious. After all, propositions represent ways the world could have been. When a proposition accurately represents the world, then the proposition is true. When a proposition *fails* to represent the world, then the proposition is false. So, propositions are representations. But it's weird to think of something representing something else without a mind doing the representing.

Think about a road sign that signifies that cars shouldn't enter a certain road: All roads lead to Rome, except this one. There's nothing about this sign itself that inherently means that this road shouldn't be entered. The sign only manages to represent anything (an instruction, a state of affairs) because minds have decided to give it that meaning. Mindless stuff can't represent anything until some minds accord it that role, as in the case of our road sign. If propositions are at once representational *and* exist (and represent) independently of any mind, then they are unusual, mysterious, magical.

A popular solution takes propositions to depend in some way upon minds *and* to exist outside of any mind. How so? Perhaps a proposition is a property a mind could have (see Soames 2010; Hanks 2015). For example, when the three people express the same proposition in three different languages, what they have in common is that each of their minds is doing the same thing. And, thus, their three minds share a property in common. Perhaps propositions are properties like that.

However the mind-dependence of propositions might be understood, the basic idea goes back at least as far as Russell:

> If we imagine a world of mere matter, there would be no room for falsehood in such a world, and although it would contain what may be called 'facts', it would not contain any truths, in the sense in which truths are things of the same kind as falsehoods. In fact, truth and falsehood are properties of beliefs and statements: hence a world of mere matter, since it would contain no beliefs or statements, would also contain no truth or falsehood. (Russell 1998: 70)

Truth is a correspondence between a representation of the world and the world itself. But without minds, there are no representations. There are facts, but nothing for those facts to correspond to, and so no propositions, and so no truths and no falsehoods.

But, if propositions depend upon minds, and if possible worlds are just sets of propositions, then no worlds could exist without the existence of some mind or minds sustaining their members in being.

Granted: with the right modifications, the ersatz pluriverse can contain empty worlds—an empty world is just a set of propositions according to which there is no concrete being. Accordingly, we must concede that there's a sense in which ersatz possibilism *can* accommodate nothingness. It provides a modal semantics where 'there could have been nothing concrete' is true. But even if empty worlds *exist*, they exist within an ersatz pluriverse that requires the sustenance of at least one concrete being, since *minds* are concrete. The existence of the pluriverse requires a mind.

Again, when we investigate metaphysical nihilism, we're not interested in whether a corner of logical space is empty, even if that corner is a world. Rather, we want to know whether nothingness could have been *all* there is. Without magical propositions that represent without minds, the ersatzist must admit that there had to have been something concrete somewhere, even if it didn't have to be here or there. Thus, in the only sense that really matters to anyone, the ersatzist, like the modal realist, has to accept that there couldn't have been nothing. Appearances to the contrary are nothing but a sleight of hand conjured by an illusory semantics (This argument for erzastism to the existence of a mind can be extended to serve as an argument for the existence of God; see e.g. Adams (1994: 177–91); Welty (2014); Keller (2018).)

The modified modal realist can't escape anti-nihilism because, even if some worlds are empty, some worlds aren't. The ersatzist can't escape anti-nihilism because, even if *every* world in a given region of modal space were empty, every world—empty or not—must depend upon some concrete mind or other.

4.2 Non-Propositional Ersatzism

The ersatzist can resist anti-nihilism by adopting a non-propositional ersatzism, which takes possible worlds as neither propositions nor spatio-temporally related systems.

Possible worlds might instead be properties corresponding to some sort of maximal world states. The actual world instantiates such a property—call this property *the actualized world*—but it could have instantiated any number of *different* properties, each corresponding to a different maximal world state. Alongside all of these world properties, there must be a particular, since properties can't be instantiated without particulars.

Must that particular be concrete? Could it be abstract? If it *must* be concrete, then anti-nihilism follows, since there necessarily exists some concrete particular or other. If the particular could be *abstract*, then metaphysical nihilism follows; the world could have been an abstract particular instantiating an abstract actualized world. But the particular *must* be concrete on pain of risking a big attraction of ersatzism: the combination of (1) actualism and (2) compatibility with S5 modal logic.

Ersatzism rejects the naturalism of combinatorialism, but wants to share its actualism. The ersatzist appeals to exotic abstracta to avoid the modal realism of Lewis. For the actualist, the only real world is the *actual* one. Every possible world represents some way that this actual world could have been but isn't.

The distinguishing feature of S5 is that if something is possible, then it's necessarily possible; in other words, any possible world is possible relative to any other possible world. Combinatorialism struggles to accommodate S5, since smaller worlds don't seem to have sufficient internal ingredients to recombine into larger worlds. If you're attracted by actualism *and* S5, then you'll be attracted to ersatzism. On the non-propositional version under consideration, every world is some property that our particular world could have instantiated.

Our actual world is concrete. Could it possibly have been abstract? If so, given S5, it could only have been *contingently* abstract. We are suspicious about the possibility of contingently abstract beings. Some countenance contingently "non-concrete" beings that are neither concrete nor abstract; others demur (compare Linsky and Zalta 1994; 1996; Williamson 2013; Tomberlin 1996; King 2016). But why countenance a gap between the concrete and the abstract? Williamson declines to provide a formal definition of *concreteness* (2013: 6, especially n. 6) and the standard definitions, including our own, stipulate that the two categories are exhaustive.

We reject the notion of a gap between the abstract and the concrete. Accordingly, if the abstract could never have been concrete and the concrete could never have been abstract, as all sides agree, then the actual world is necessarily concrete. And if every possible world is a property constituting some way *this* world could have been, then every world is a property that could only have been instantiated by a concrete being. Anti-nihilism is secured.

Even philosophers countenancing the contingently non-concrete should agree that the actual world is necessarily concrete. You might think that we're essentially human, and that all humans are essentially concrete beings. But if we're not essentially concrete, then how could we be essentially human? To avoid the problem, adherents of the contingently non-concrete distinguish between two senses of "essential". Some of our properties we call "essential" because we have them in any world where we're concrete. But that doesn't mean that we're concrete in every world. Rather, there's another sense of "essential" that picks out those properties we have even in worlds where we're *not* concrete. Thus Linsky and Zalta:

> the latter properties are ones that are 'essential' to an object only in a vacuous sense of 'essential' (if you have such a property in every possible world, you certainly have such a property in every world in which you are concrete).
>
> (Linsky and Zalta 1996: 291)

The essential properties of contingently non-concrete beings are all "vacuous", like being self-identical, or being non-concrete or non-human. Their most substantive and positive essential properties are irreducibly modal, like being possibly concrete and being possibly human. But any world that's actualized, which the actual world is in every possible world, has more than merely vacuous essential properties. Thus every side to this debate should accept that an ersatzist account has the actual world as necessarily concrete.

This version of ersatzism entails not only the necessary existence of some particular or other, but that this particular *must* be concrete: our necessarily concrete world necessarily exists, although it may have instantiated any number of world states. This version of ersatzism also therefore entails anti-nihilism even if there were some room for the contingently non-concrete.

Finally, even if abstract propositions have a brute power to represent states of affairs—that is, even if the propositional form of ersatzism can escape our argument against it—our argument against *non-propositional* ersatzism can be generalized to counter *all* versions of ersatizm. After all, the actual world is concrete, and necessarily so. Given actualism, possible worlds either are or represent some way the actual world could have been. In no possible world could the actual world fail to be concrete. Therefore, there necessarily exists something concrete.

5. Dispositionalism

According to Charlie Martin's (2008) modal dispositionalism, possibilities are grounded in the powers and liabilities of actual beings. The possibility of a kettle boiling boils down to the causal powers and liabilities of the elements and the water. Very roughly. But enough for our purposes (also see Borghini and Williams 2008; Jacobs 2010).

Ross Cameron argues that modal dispositionalism rules out the possibility of none of the actual contingent stuff existing. He calls modal dispositionalism 'Aristotelianism', but why go for the Greek over a Latin root? Anyhow:

> Intuitively, I am a contingent being—I might not have existed. What, for the Aristotelian, grounds this possibility? Presumably, it is my parents; for just as it was within their power to beget me, it was also within their power not to, and had they exercised the latter power I would not have existed. And the truthmaker for the truth that my parents might not have existed is, in turn, their parents. But what about the highly intuitive possibility that *none* of the actual contingently existing substances existed—what is the truthmaker for the truth that this situation is possible? It can't be any of the actual contingently existing beings, for none of these beings has the capacity to bring it about that it itself *never existed*.
>
> (Cameron 2008: 273)

What's possible is what there's a capacity of actual beings to bring about or to fail to bring about. For any possibility, there must have been the capacities of actual beings on the scene. Thus, for the possibility of there *never* having been the contingent beings, Cameron insists that there must have once been the contingent beings—which is absurd. Thus, given modal dispositionalism, there could not have been none of the contingent stuff there is. Cameron worries about this implication because he finds the possibility plausible: since each contingent being could have failed to exist, and its non-existence would not necessitate the existence of another contingent being in its stead, there could have been no contingent beings.

But the argument is too quick. Barbara Vetter points out that for there to be capacities on the scene there need not have been contingent beings: a necessary being could do the work. She talks in terms of *potentialities*:

> Let $x1, \ldots, xn$ be all actual contingently existing objects. Then it is possible that none of $x1, \ldots, xn$ existed. Something, therefore, must have a potentiality to be such that none of $x1, \ldots, xn$ existed; but that something cannot itself be any of $x_1, \ldots, xn$. Since $x1, \ldots, xn$ are all the contingently existing objects that there are, the bearer of the relevant possibility cannot itself be a contingently existing object. Therefore it must be a necessary existent. (Vetter 2015: 275)

Since only a concrete being could have potentialities, we can conclude that there is a necessary concrete being. Which concrete being? Vetter locates necessity in the primordial concrete being:

> Nothing has or ever had a potentiality for the beginning of time to be different than it was: for there was never a time at which such a potentiality might have been possessed. Hence, whatever entities existed at the beginning of the universe are, on this view, necessary existents: nothing has or ever had a potentiality... for them not to have existed. (Vetter 2015: 276)

Cameron finds modal dispositionalism at odds with other possibilities: "there are other possibilities that the Aristotelian account looks hard pushed to ground, such as the possibility of there being different global laws of nature, or in general possibilities concerning how the world could have been globally." (2008: 273). If we salvage modal dispositionalism in terms of a necessary primordial being, then this being had better have quite extraordinary powers to bring about extraordinary global possibilities. This makes for a little argument for a quite potent primordial necessary being—perhaps omnipotent?

But, returning to our theme, we can draw the dialectic between Cameron and Vetter together: if modal dispositionalism is true, then either there being no contingent beings is impossible, or there is a necessary concrete being. There being no contingent beings is possible. Thus, if modal dispositionalism is true, there is a necessary concrete being. Thus modal dispositionalism entails anti-nihilism.

Here's an even simpler route: what's possible is what there's a capacity of beings to bring about or to fail to bring about. For any possibility, there must have been capacities on the scene. For there to be capacities on the science there must have been concrete beings on the scene. Thus, for the possibility of there *never* having been concrete beings, there must have once been concrete beings on the scene—which is absurd.

Such ideas have precedent at least as far back as Avicenna (2009: 359–70). First, assume that it's possible for something concrete to exist. Fair enough—since concrete things *do* exist! Assume also that if something is possible, it's necessarily possible, as per S5. Now take the possibility of there being nothing concrete. Even then, the possibility of something concrete would remain, as Avicenna puts it: "the possibility of [a thing's] existence exists before [the thing] exists" (2009: 359). From whence this possibility?

Like our dispositionalists, Avicenna cashes out possibility in dispositions and powers. So, if something is possible, then there must be something with the right powers and dispositions. Ultimately, Avicenna's dispositionalism analyses modality in relational terms: for any possibility, there must be an agent and a patient, and they must be related such that the agent has the power to transform the patient in a certain way, and the patient must have the liability to be so transformed. So what was assumed possible—there being nothing concrete—is not possible after all.

Since Avicenna analyses possibility in terms of a dyadic relation between an agent and a patient, he denied the doctrine of *creatio ex nihilo*. God is the agent of creation. But there must also be a *patient* of creation—*viz.* the universe. Although he takes the universe to be metaphysically anterior to God, Avicenna is committed to the necessary existence of both God *and* the universe. *Ex nihilo nihil fit.*

But note that we can reach Avicenna's anti-nihilism without endorsing the necessary existence of the universe. Some agents might not require patients. Thomas Aquinas (1948: 241) thought that omnipotence did not require a patient. Perhaps dispositional modality needn't be cashed out in terms of dyadic relations; perhaps monadic properties suffice. So, despite our title, our conclusion is more modest than Avicenna's. Aquinas and other believers in *creatio ex nihilo* could endorse *our* anti-nihilism, since the necessary existence of God suffices, even without the universe.

What we take from Avicenna's argument is van Inwagen's conclusion: if the existence of something is possible, then anti-nihilism. Possibilities depend on dispositions. Dispositions depend on concrete beings; so there could not be possibilities without concrete beings. Since necessarily something is possible, necessarily there are concrete beings. We *think* that Martin (2008: 31) recognized the force of this sort of argument in passing. Graham Oppy's (2013: 47) naturalistic explanation of why there is something rather than nothing essentially boils down to an argument of this form too.

6. The Relevance

Why does any of this matter? If the metaphysics of modality excludes the possibility of there being nothing concrete, then it might answer one of the biggest questions of all: why is there something—something concrete—rather than nothing concrete? Lewis didn't see things that way. On the contrary, he thinks that explaining why there's something rather than nothing would be a *problem* for his theory:

> I think the worst part of it is the fear that I might offer to *explain* why there is something rather than nothing, just by saying that this is a necessary truth. But don't fear; I do not think that would be an explanation. For an explanation, I think, is an account of etiology; it tells us something about how an event was caused... So I think there is nothing I might say that would count as explaining why there is something rather than nothing; and that includes saying, truly, that there is no world where there is nothing. (Lewis 1986: 73–4; italics in original)

A weird point, especially coming from Lewis. Wouldn't modal realism explain *something* if it were true? If you insist, don't call what it does an *explanation*. But then there are still non-causal (and non-etiological) ways of answering why-questions, such as what we would normally call mathematical *explanations*, probabilistic *explanations*, and moral *explanations*. Call them *elucidations* (origin: *elucidare*) or whatever if you prefer to reserve "explanation" for something else.

Showing that there necessarily had to be something might not elucidate anything. Consider Jonathan Lowe's (1996; 1998) "explanation" of why there's anything concrete: abstract beings necessarily exist; abstract beings essentially depend on concrete beings; so concrete beings had to exist. But showing that concrete beings exist in every possible world because abstract beings exist in every possible world no more explains or elucidates concrete beings than, say, showing that there's fire on every hill by showing that there's smoke on every hill explains why there's fire on any hill.

The problem with Lowe's line of thinking arises because it tries to explain concrete beings in terms of abstract beings, which in turn depend on concrete beings. But we don't see that an answer from modal metaphysics need fall into an analogous trap—or any other besides particular problems with the proposed metaphysical frameworks. The question is worth pursuing, especially if *Lewis's* only reservation concerned the non-causal nature of the explanations that emerge from modal metaphysics.

7. Conclusion

There are other ways to motivate anti-nihilism (see *exempli gratia* Coggins 2010), and more to be said for metaphysical nihilism (see e.g. Rodriguez-Pereyra 2013). But we have shown that a whole range of modal metaphysics is anti-nihilist, including the most prominent theories. That's significant. Besides whatever intrinsic interest the metaphysics of modality and metaphysical nihilism have, the metaphysics might even promise an answer to the question of why there is something rather than nothing concrete. The nature of modality simply entails the necessity of something or other—which, like this essay, might be nothing much, but something nonetheless.

Acknowledgments

With thanks to Craig Warmke for helpful comments on a previous draft and to Peter van Inwagen for an instructive conversation about modal semantics.

References

Adams, Robert Merrihew (1971). "Has It Been Proved That All Real Existence Is Contingent?", *American Philosophical Quarterly* **8** (3): 284–91.

Adams, Robert Merrihew (1974). "Theories of Actuality", *Noûs* **8** (3): 211–31.

Adams, Robert Merrihew (1994). *Leibniz: Determinist, Theist, Idealist* (Oxford: Oxford University Press).

Aquinas, Thomas (1948). *Summa Theologica*, trans. Fathers of the English Dominican Province (New York: Benzinger Bros).

Armstrong, D. M. (1989). *A Combinatorial Theory of Possibility* (Cambridge: Cambridge University Press).

Armstrong, D. M. (2006). "Reply to Efird and Stoneham", *Australasian Journal of Philosophy* **84** (2): 281–3.

Avicenna (2009). *The Physics of the Healing*, trans. J. McGinnis (Provo, UT: Brigham Young University Press).

Baldwin, Thomas (1996). "There Might Be Nothing", *Analysis* **56**: 231–8.

Borghini, Andrea and Neil Williams (2008). "A Dispositional Theory of Possibility", *Dialectica* **62**: 21–41.

Cameron, Ross (2008). "'Truthmakers and Modality", *Synthese* **164**: 261–80.

Church, Alonzo (unpublished lecture). "The Ontological Status of Women and Abstract Entities", http://www.jfsowa.com/ontology/church.htm.

Coggins, Geraldine (2003). "World and Object: Metaphysical Nihilism and Three Accounts of Worlds", *Proceedings of the Aristotelian Society* **103** (1): 353–60.

Coggins, Geraldine (2010). *Could There Have Been Nothing? Against Metaphysical Nihilism* (New York: Palgrave Macmillan).

Efird, David and Tom Stoneham (2005). "Genuine Modal Realism and the Empty World", *European Journal of Analytic Philosophy* **1** (1): 21–38.

Efird, David and Tom Stoneham (2006). "Combinatorialism and the Possibility of Nothing", *Australasian Journal of Philosophy* **84** (2): 269–80.

Hanks, Peter (2015). *Propositional Content* (Oxford: Oxford University Press).

Jacobs, Jonathan (2010). "A Powers Theory of Modality: Or, How I Learned to Stop Worrying and Reject Possible Worlds", *Philosophical Studies* **151** (2): 227–48.

Keller, Lorraine (2018). "Propositions Supernaturalized", in *Two Dozen (or so) Arguments for God: The Plantinga Project*, ed. Jerry L. Walls and Trent Dougherty (New York: Oxford University Press), pp. 11–28.

King, Jeffrey (2016). "Timothy Williamson on the Contingently Concrete and Non-Concrete", *Analysis* **76** (2): 190–210.

Kripke, Saul A. (1980). *Naming and Necessity* (Cambridge MA: Harvard University Press).

Leftow, Brian (1991). *Time and Eternity* (Ithaca, NY: Cornell University Press).

Lewis, David (1986). *On the Plurality of Worlds* (Oxford: Basil Blackwell).

Lewis, David (1990). "Noneism or Allism?", *Mind* **99** (393): 23–31.

Lewis, David (1991). *Parts of Classes* (Oxford: Basil Blackwell).

Lewis, David (1999). "New work for a theory of universals", in *Papers in Metaphysics and Epistemology* (Cambridge: Cambridge University Press), pp. 8–55.

Linsky, Bernard and Edward N. Zalta (1994). "In Defense of the Simplest Quantified Modal Logic", *Philosophical Perspectives* **8**: 431–58.

Linsky, Bernard and Edward N. Zalta (1996). "In Defense of the Contingently Nonconcrete", *Philosophical Studies* **84** (2–3): 283–94.

Lowe, E. J. (1996). "Why Is There Anything At All?", *Proceedings of the Aristotelian Society*, supp. vol. **70**: 111–20.

Lowe, E.J. (1998). *The Possibility of Metaphysics* (Oxford: Oxford University Press).

Meluch, R. M. (2004). *The Myriad* (New York: DAW Books).

Oppy, Graham (2013). "Ultimate Naturalistic Causal Explanations", in *The Puzzle of Existence*, ed. Tyron Goldschmidt (New York: Routledge), pp. 46–63.

Plantinga, Alvin (1974). *The Nature of Necessity* (Oxford: Clarendon Press).

Rodriguez-Pereyra, Gonzalo (2004). "Modal Realism and Metaphysical Nihilism", *Mind* **113** (452): 683–704.

Rodriguez-Pereyra, Gonzalo (2013). "The Subtraction Arguments for Metaphysical Nihilism: Compared and Defended", in *The Puzzle of Existence*, ed. Tyron Goldschmidt (New York: Routledge), pp. 197–214.

Russell, Bertrand (1998). *Problems of Philosophy* (Oxford: Oxford University Press).

Schneider, Susan (2001). "Alien Individuals, Alien Universals, and Armstrong's Combinatorial Theory of Possibility", *Southern Journal of Philosophy* **39** (4): 575–93.

Soames, Scott (2010). *What is Meaning?* (Princeton, NJ: Princeton University Press).

Tomberlin, James E. (1996). "Actualism or Possibilism", *Philosophical Studies* **84** (2–3): 263–81.

van Inwagen, Peter (2015). "Nothing Is Impossible", in *God, Truth, and Other Enigmas*, ed. Miroslaw Szatkowski (Berlin: De Gruyter), pp. 33–58.

Vetter, Barbara (2015). *Potentialities: From Dispositions to Modality* (Oxford: Oxford University Press).

Welty, Greg (2014). "Theistic Conceptual Realism", in *Beyond the Control of God? Six Views on the Problem of God and Abstract Objects*, ed. Paul M. Gould (New York: Bloomsbury), pp. 51–64.

Williamson, Timothy (2002). "Necessary Existents", in *Logic, Thought and Language*, ed. Anthony O'Hear (Cambridge: Cambridge University Press), pp. 233–51.

Williamson, Timothy (2013). *Modal Logic as Metaphysics* (Oxford: Oxford University Press).

12

Ostrich Actualism

Craig Warmke

1. Introduction

Hidden within Derek Parfit's three volumes of *On What Matters*, we find a prolonged discussion on the debate between actualism and possibilism. Given the author and the overall work, many would expect the discussion to concern the actualism–possibilism debate in normative ethics. But, surprisingly, Parfit enters the debate in modal metaphysics that goes by the same name. This debate concerns mere possibilia, possible but non-actual things such as golden mountains and talking donkeys. Roughly, possibilism says that there are such things, and actualism says that there are not.

Actualism has approached the status of philosophical orthodoxy. And many of Parfit's own contemporaries number among its defenders, including Robert Merrihew Adams (1974; 1981), Alvin Plantinga (1974; 1976), and Robert Stalnaker (1976).[1] Nonetheless, Parfit argues for possibilism. He even argues that self-proclaimed actualists like Plantinga are, in fact, unwitting possibilists. Though Parfit's arguments do not fully succeed, they do highlight a tension within the frameworks of many actualists. Many actualists conscript abstract objects into the role of "possible worlds" to avoid quantifying over mere possibilia. But, in doing so, actualists must quantify over mere possibilia anyway. When we alleviate this tension, a Parfit-friendly form of actualism arguably remains. This form of actualism says that while everything that exists is actual, it is also true in some sense that there are mere possibilia.

We begin in section 2 with Parfit's distinction between actualism and possibilism. Then, in sections 3 and 4, we assess his main argument for possibilism. Next, inspired by Parfit, I offer a related argument, which extends from sections 5 through 7. I argue that given two plausible assumptions, actualists of various sorts must quantify over mere possibilia. But this doesn't necessarily mean that actualism fails. We can peel off the metaphysical thesis of possibilism from the

[1] For critiques of actualism, see Tomberlin and McGuinness (1994), Tomberlin (1996; 2001). David Lewis (1986) has often been thought of as the arch-possibilist. But there are good reasons to question this classification: see Menzel (2014). I should note that Parfit (2011: 798 n. 729) also questions this classification. For more references and a discussion of possibilism along Lewisian lines, see McDaniel (2017: 73–5).

Craig Warmke, *Ostrich Actualism* In: *Non-Being: New Essays on the Metaphysics of Non-Existence.* Edited by: Sara Bernstein and Tyron Goldschmidt, Oxford University Press (2021). © Craig Warmke. DOI: 10.1093/oso/9780198846222.003.0012

linguistic thesis that there is a sense in which it is true that there are mere possibilia. I outline a view that combines the linguistic thesis with an actualist metaphysics.

2. The Actualism–Possibilism Distinction

Some doubt that we can clearly and meaningfully draw a distinction between actualism and possibilism (Williamson 1998: 259; 2013: 22). And those who attempt to do so offer an array of distinctions.[2] For example, formulations of the views often differ on their modal status. Some define actualism non-modally as the view that everything is actual. Others define actualism modally as the view that, *necessarily*, everything is actual.[3]

The meanings of the modal and non-modal formulations also depend on what it means to be actual. Some say that 'actual' is an indexical like 'here' (Lewis 1970; Stalnaker 1976). Similar to the way that 'here' refers to the place and time of utterance, indexicalists about actuality say that 'actual' refers to the world of utterance. However, while utterances of 'here' often occur at different places and times, few believe that any utterances of 'actual' occur in any (at least partially) concrete worlds other than our own. The thought isn't that other concrete worlds lack people, or lack people who know the word, or lack people who know it but never use it. The thought, rather, is that there are no other concrete worlds at all. The actual world exhausts all that exists. So actualists typically believe that what's actual coincides with what exists.[4]

Actualists also typically endorse the existentially loaded view of the quantifier and reject any distinction between 'there is' and 'there exists' (Linsky and Zalta 1994: 436).[5] So actualists generally believe that what *is* coincides with what *exists*. Given the prior actualist commitment about the coincidence of existence and actuality, actualists typically believe that what is, what exists, and what's actual all coincide. But sometimes they express this coincidence as a modal truth, and sometimes they don't. Parfit characterizes actualism both ways. He first characterizes actualism in the modal way:

Actualism 1. To be, or to exist, is to be actual, so there cannot be anything that is merely possible. (Parfit 2011: 467)

[2] For recent work on the distinction itself, see Cameron (2016) and Menzel (2020).

[3] On the difference, see Bennett (2005: 311 ff.)

[4] Compare Stalnaker (1986: 128). See Parfit (2011: 722–3).

[5] Drawing on research in linguistics, Thomas Hofweber (2016: 58 ff.) rightly points out that, strictly speaking, 'there is' is not a quantifier. But 'there is'-statements do express quantificational statements. I will continue to call 'there is' and 'there exists' quantifiers to convey that they appear in quantificational statements.

Here, Parfit identifies being with actuality, and the embedded parenthetical remark (i.e. "or to exist") seemingly identifies being with existence. Then, on the basis of these identifications, Parfit infers that there cannot be mere posssibilia. Actualism 1 therefore contains or implies a number of common actualist-friendly propositions:

Anti-Meinongianism. Everything (there is) exists.

Anti-Possibilism 1. Everything (there is) is actual.[6]

Anti-Possibilism 2. Everything that exists is actual.

Modal Anti-Possibilism 1. Necessarily, everything there is, is actual.

Modal Anti-Possibilism 2. Necessarily, everything that exists, is actual.

We will later return to some of these, especially Anti-Meinongianism.[7] For now, it will suffice to note that if being and actuality coincide, as Parfit suggests, then it would seem to follow that, necessarily, everything there is, is actual.

Parfit also defines actualism along non-modal lines:

Actualism 2. There is nothing except what actually exists. (Parfit 2011: 719)

What "actually exists" is presumably what exists in the actual world. So if the actual world exhausts both the actual and the existing, then, again, we get the actualist coincidence between what is, what exists, and what's actual. While this coincidence secures Anti-Meinongianism and Anti-Possibilism 1 and 2, it doesn't imply anything about Modal Anti-Possibilism 1 or 2.

How does Parfit define possibilism, then? Someone could reject one or both of the Anti-Possibilist modal theses and contend that there *could* be something merely possible. Someone could also reject one or both of the non-modal Anti-Possibilist theses and argue that there *is* something merely possible. Parfit initially characterizes possibilism along these non-modal lines:

Possibilism. There are some things that are never actual, but are merely possible. There are some things that might happen but never actually happen, and some things that might exist but never actually exist. (Parfit 2011: 467).

Minimally, we may conceive of Parfit's possibilism as the negation of Anti-Possibilism 1. In fact, Parfit (2011: 719) later characterizes possibilism as the view

[6] For the idea that we may construe actualism as the conjunction of Anti-Meinongianism and Anti-Possibilism, see Nelson and Zalta (2009: 289).

[7] Some construe possibilism as the view that mere possibilia exist (Jubien 1996: 109–11; Murray and Wilson 2012: 220). This is a kind of anti-Meinongian possibilism. But classical, Meinongian possibilists say both that there are mere possibilia and that they don't exist.

that "there are some things that are merely possible." As we'll see, Parfit also denies Anti-Possibilism 2 and Anti-Meinongianism in important but attenuated senses.

3. The Argument from Possible Actions

Parfit's main argument against actualism appeals to the role that possibilities play in both deliberation and regret. When we deliberate, we decide between possible courses of action. We also sometimes regret having acted in one way rather than another. According to Parfit, actualism makes sense of neither. He writes:

> If Actualism were true, much of our thinking would be undermined. For example, we could never choose between different possible acts, or compare their possible outcomes, since there couldn't be any merely possible acts or outcomes. Nor could we ever have reason to regret having acted as we did, since it could never be true that there was something else that we could have done instead. (Parfit 2011: 467–8)

For Parfit, since actualism implies that there are no mere possibilia, it also implies that there are no merely possible acts. If there are no merely possible acts, we can neither reject them in deliberation nor regret our failure to undertake one rather than another. Parfit (2011: 469) even claims that actualism "implies that we could have never acted differently." Though Parfit doesn't explicitly say so, the argument generalizes to possible worlds. If nothing were merely possible, then, by Parfit's lights, our world would be the only one possible. Actualism implies necessitarianism, in other words. In other *words*, yes, but not in other *worlds*—not if Parfit is right anyway.

The argument should puzzle many actualists. Most actualists reject the necessitarian view that there are no other possible worlds. According to most actualists, the world could have gone differently in lots of ways. Each of these ways corresponds to some possible but non-actual world. And actualists would insist that these worlds are, in some important sense, merely possible. Actualists can then respond to Parfit by saying that some acts are merely possible in much the same sense that some worlds are merely possible.

What sense is that? What does it mean when an actualist says that there are possible worlds? In an early statement of actualism about possible worlds, Robert Merrihew Adams writes:

> Actualism, with respect to possible worlds, is the view that if there are any true statements in which there are said to be nonactual possible worlds, they must be reducible to statements in which the only things there are said to be are things which there are in the actual world and which are not identical with nonactual possibles. (Adams 1974: 224)

Adams and other actualists "reduce" statements about non-actual possible worlds by identifying "possible worlds" with some actually existing abstract objects. Adams (1974) himself reduces statements about non-actual possible worlds to statements about maximally consistent sets of actual propositions. Plantinga (1976) appeals to maximally consistent and actually existent abstract states of affairs. And Stalnaker (1976) appeals to certain actually existing properties.[8] Each of these theorists reduces statements about non-actual possible worlds to statements about actually existing abstract objects. Since they actually exist, they are not mere possibilia in the possibilist's sense.

In each case, the actualist intends to use an account of possible worlds to capture the various kinds of modal truths. Adams's account says that 'possibly, *p*' is true when the the proposition that *p* is a member of some maximally consistent set of propositions. Plantinga's account says that 'possibly, *p*' is true when some maximally consistent state of affairs includes the state of affairs of being such that *p*. And Stalnaker, who identifies propositions with sets of possible worlds, on the one hand, and possible worlds with properties, on the other, says that *p* is possible when the set of worlds identical to *p* has at least one possible world as a member.

Although the actualist's abstract possible worlds are not merely possible in the possibilist's sense, actualists do want to say that they are merely possible in the sense of being possible but non-actual. For Adams (1974), whose possible worlds are certain sets of propositions, a world is "merely possible" when one of its members is false. For Plantinga (1976), whose possible worlds are certain abstract states of affairs, a world is "merely possible" when it includes a non-obtaining state of affairs. And, for Stalnaker (1976), whose worlds are properties, a world is "merely possible" when it goes unexemplified. In each case, the actualist says of some actually existing abstract object that it is "possible but not actual."

Although actualists, like the possibilists, say that some things are possible but not actual, actualists mean something different. To clarify the difference, it will be helpful to adopt some terminology from Lewis (1986: 138–40). For actualists like Adams, Plantinga, and Stalnaker, the abstract objects identified with possible worlds actually exist. They are *actual*. But, whichever kind of actualist we are, we must distinguish the actually existing abstract object which represents or characterizes reality from the actually existing abstract objects which represent or characterize other ways reality might have been. On behalf of the actualist,

[8] Stalnaker (1976: 70) seems to reject Adams's call for reduction, but he also misunderstands that call. Adams explicitly calls for reducing *statements* about possibilia to statements about actually existing things. Adams also reduces *worlds* but in another sense: Adams denies that "worlds" are metaphysically primitive and instead identifies them with sets of propositions. In the first sense of reduction, we rid ourselves of statements about mere possibilia. After we've reduced in the first sense, the second sense of reduction rids "possible worlds" of metaphysical primitivity. Although Adams reduces in both senses, Adams only espouses reduction in the first sense in the above passage. But in discussing this passage, Stalnaker rejects reduction in the second sense.

Lewis (1986: 138) calls the former *actualised* and the latter *unactualised*. For the actualist, then, there is but one actualised possible world. The rest are unactualised. But every possible world is actual—each one is an abstract object in the actual world.

Actualists can use this distinction to respond to Parfit. On the actualist's behalf, let's say that an act is *possible* when some possible world represents or characterizes the act as occurring. Then, let's say that a possible act is *derivatively unactualised* when some unactualised world represents or characterizes the act as occurring though the actualised world does not. For an agent *s* and an act *G*, an actualist like Adams could say that the act of *s*'s *G*-ing is possible but derivatively unactualised when the false proposition that *s* is a *G*-er is contained in some maximally consistent set of propositions but not contained in the maximally consistent set of true propositions. And similar responses are available to Plantinga and Stalnaker. In these ways, actualists may try to reduce statements about merely possible acts using the very resources they already use to reduce statements about merely possible worlds to statements about actually existing abstract objects.

Although Parfit does consider an actualist response to his argument from possible acts,[9] responses along these lines seem to elude him. And, in my view, most actualists would respond exactly along these lines—lines that Lewis (1986: 138–40) traced for actualists much earlier. But this lacuna in Parfit's discussion doesn't give actualists the license to dismiss his concerns entirely. Although Parfit may not have considered the actualist's most likely response, the response may prove ineffective anyway. Soon, I'll explain why I reject it. But we'll first explore another pillar of Parfit's possibilism.

4. Possibilist Discourse

Whether possibilists are right that there are mere possibilia partly depends on what it means to say that *there are* such things. According to Parfit (2011: 720), many actualists adopt the following view about the meanings of 'there are' and 'exist':

Single Sense View. The words 'there are' and 'exist' must have only the same single sense.

We don't need to specify a particular sense to see how the Single Sense View supports actualism. Mere possibilia like golden mountains and talking donkeys

[9] Parfit (2011: 468) instead considers actualist paraphrases of statements that quantify over possible acts as statements about how agents could have acted differently.

are not actual—they don't exist. 'Mere possibilia do not exist' is true. So if 'there are' and 'exist' share a single sense, then 'there are no mere possibilia' is also true.

However, if either 'exist' or 'there are' has a wider sense, then possibilists may claim that there are or that there exist, in a wider sense, some mere possibilia. In defending possibilism, Parfit (2011: 719) appeals to extra senses for this exact purpose. The thesis below plays a crucial role in this defense.

Plural Sense View. There is one wide, general sense in which we can claim that there are certain things, or that such things exist. We can also use these words in other, narrower senses. For example, if we say that certain things exist in what I call the *narrow actualist sense*, we mean that these things are, at some time, actually existing concrete parts of the spatio-temporal world.

To highlight the differences between the Single Sense View and the Plural Senses View, Parfit devotes considerable space to examining the following proposition:

(a) There was a palace designed by Wren to replace the burnt Palace of Whitehall, but this palace was not built and never actually existed. (Parfit 2011: 469)

This says that there was something that never actually existed. So, given the Single Sense view, (a) expresses the contradiction that

(b) There existed such a palace designed by Wren, but this palace was not built, so that, in the same sense of 'exist', this palace never existed. (Parfit 2011: 470)

Parfit (2011: 469, 720) then claims that "many actualists" endorse the following thesis about the particular sense that 'there are' and 'exist' share:

Existence. The words 'there are' and 'exist' must have only the same single sense, which means 'actually exist'.

We should note that Parfit does not explicitly equate this single sense with the "narrow actualist sense" named in the Plural Sense View and according to which something exists if and only if it is an actually existing concrete part of the spatio-temporal world. Thus, Parfit does not say that, according to actualism, everything is a concrete part of the spatio-temporal world. After all, many actualists identify possible worlds with non-concrete abstracta. Rather, Parfit simply argues that the single sense actualist cannot endorse (a) without endorsing the contradictory (b), whether the single sense in play is the narrow actualist sense or some other sense.

Possibilists like Parfit appeal to an additional sense of 'there is' to avoid contradiction. According to Parfit, (a) expresses the following truth:

(c) There was, in the wide sense, a possible palace designed by Wren, but this palace was not built and never existed in the narrow actualist sense. (Parfit 2011: 470)

Although an extra sense of 'there is' would save (a) from contradiction, Parfit says little about what this extra sense consists in. Some remarks do suggest that, in addition this wide sense, he himself endorses the narrow actualist sense.[10] But other than applying the wide sense to things like mere possibilia and abstract objects that elude the reach of the narrow actualist sense, Parfit says almost nothing about its meaning. Can we ever say of something in the wide sense that there is no such thing? Can we specify truth conditions for statements in the wide sense?

Parfit doesn't answer these questions directly. But we can piece together three claims about his position on the narrower and wider senses of 'there is' and 'exist'. We may treat them as Parfit's constraints on admissable meanings for the wider senses.

Shared Narrow. 'exist' and 'there is' each have a narrow sense for which 'there are no talking donkeys' and 'no talking donkeys exist' are both true.

Shared Wide. 'exist' and 'there is' each have a wider sense for which 'there are talking donkeys' and 'talking donkeys exist' are both true.

No Impossibilia. In these wider senses, 'round squares exist' and 'there are round squares' are both false.

With the narrower senses of 'there are' and 'exist', Parfit says that there are no mere possibilia and that mere possibilia don't exist. With the additional senses in Shared Wide, Parfit says that there are mere possibilia and that mere possibilia exist. Furthermore, by using the wide sense of 'there is' and the narrow sense of 'exist', Parfit can also accept the Meinongian view that there are things that don't exist. Indeed, Parfit can even assert the Anti-Meinongian slogan ("everything there is, exists"), as long as he doesn't combine the wide sense of 'there is' with the narrow sense of 'exist'. However, given No Impossibilia, those senses are not so wide that Parfit would say that either 'there are round squares' or 'round squares exist' are true. So, according to Parfit, the wider senses of 'exist' and 'there is' cover the domain of all and only possible things.

[10] Parfit (2011: 480, 722, 728).

We should note that possibilism as such implies neither Shared Wide nor No Impossibilia. For example, a possibilist may grant an additional wider sense to 'there are' but not to 'exist'. Such a possibilist might argue that mere possibilia fail to exist but nevertheless have a special mode of being. Possibilists may also grant that round squares and other impossibilia have that very same mode of being.

In the next section, I begin to argue that Parfit is right in an important way: it is quite difficult for certain actualists to express their commitments forthrightly without adopting a wider sense of 'there is'. However, Parfit is also wrong in an important way. I argue in section 7 that actualism and Shared Wide are, in fact, compatible.

5. Worldhood and Contingency

Each side in the actualist–possibilist debate strings together the same words, in the same order, to express different commitments. When actualists say "everything there is, is actual," possibilists may use their narrower sense of the quantifier and utter the same. Similarly, when possibilists say "something is possible but not actual," actualists can utter those same words to mean that something actually exists as an unactualised abstract object. Each side can easily talk past or beg the question against the other.

However, even after we fix terms, the commitments of influential actualists require that they quantify over non-existent possibilia. This doesn't necessarily mean that actualists like Plantinga are unwitting possibilists, as Parfit claims. I'll argue instead that actualism, properly framed, both requires and is consistent with quantifying over non-actual possibilia. Insofar as the argument succeeds, actualists can speak like possibilists without being possibilists. They can talk with the possibilists but walk with the actualists.

My argument rests on two theses. We'll frame the first as a definition:

Worldhood. A *world* is the totality of all that exists.

A totality of all that exists includes everything that exists and nothing that never exists. What does it mean, then, for something to exist? Since, in my view, the concept of existence is one among many conceptual primitives, we cannot decompose that concept into more basic ones.[11] But perhaps it will help to say something about what I don't assume about the nature of existence.

First, I don't assume anything about the connection between existence, on the one hand, and notions like fundamentality, grounding, composition, substance, or concreteness, on the other. It may be that only fundamental things exist, only substances exist, only concrete things exist, and so on. But even if only objects

[11] Compare Priest (2016: xxvii).

of one such type existed, the nature of existence need not itself preclude objects of other types from existing. Numbers, properties, persons, material objects, scattered objects, temporally gappy objects, gunk, angels, God, money, genders, and countries may exist. Or they may not. But the very nature of existence doesn't preclude their existence, as far as I can tell.

Second, I don't assume that things in different ontological categories exist in different ways. So I don't assume that abstract objects exist in one way and concrete objects exist in another way. I'm rather sympathetic with the view that things in different ontological categories exist in the same sense if they exist in any sense at all.

Third, I don't assume that existence comes in degrees. If numbers or devils exist at all, they fully exist just like everything else that exists. By my lights, something exists or it doesn't. And one thing that doesn't exist is an ontological purgatory where objects fall on a spectrum of existence.

Some may balk at the notion of existence I've just circumscribed. But, as far as philosophy goes, nothing is particularly controversial about it. I suspect most philosophers hold similar views about existence, even if we disagree about the precise domain of existing things. As far as I can tell, we can run the argument below without any especially contentious views about existence.

What can we say, then, about the nature of totalities? First, if something exists, so does a totality of everything that exists. But a totality isn't a set—paradoxes lurk down that alley. What else might totalities be? Maybe mereological sums. Or maybe pluralities. I'm content here with saying that a totality of everything that exists is something like *all* of reality. So when *I* ask whether the world could have been different in any way, I mean to ask whether any aspect of reality could have gone differently. Since we all need some such notion of a totality to specify a theory of modality, I won't attempt to settle more specific questions about their nature. For example, some might wonder whether a totality could exist even though nothing else does. Perhaps so. I wouldn't want to rule it out by definition even though I also believe that various necessary existents make such a totality metaphysically impossible. Nevertheless, the way I conceive of totalities permits the existence of at most one of them. I conceive of them extensionally so that a totality's identity rests completely on the things that exist, all their features included. Hence, any totality is the *only* totality.

Nothing less than the totality of all that exists is a world. So even if multiple Lewisian concrete universes were to exist, they wouldn't qualify as worlds in the relevant sense. Any such universes would, like any other existing things, simply number among the world's things. When I ask whether the world could have gone differently, I mean to ask whether the totality of all that exists could have been different in any way. With quantifiers completely unrestricted, is there something that exists that may not have, or something that exists that could have been different, or something that doesn't exist but could have? Lewis says "no" to all these questions. For Lewis, the totality of all that exists couldn't have been different in any way. As Lewis and many others have explained, pairing Lewis's ontology with the totality conception of worldhood begets necessitarianism.[12] And since necessitarianism implies that nothing is merely possible, actualism comes for free.[13]

(Of course, Lewis himself defends possibilism and non-necessitarianism, but his defenses attach very different meanings to words like 'world', 'possible', and 'actual'. To borrow one of Lewis's own examples, he is no more convincing here than those who argue that God exists because—by the way—'God' refers to the triumphal march of history.[14])

A Lewisian universe would count as a world only if it were to contain all that exists. Similar remarks also apply to Adams's sets of propositions, Plantinga's states of affairs, and Stalnaker's properties. No matter what exists, something is a *world* only if it is the totality of all that exists. Does the totality of everything that exists except Pluto count as a world? No, because that is not the totality of *everything* that exists.[15] Does the totality of everything that exists plus Pegasus count as a world? No, because that is not the totality of everything that *exists*. A totality of all that exists includes everything that exists and nothing that does not exist. Or, I should say that a totality of all that exists includes everything that *ever* exists; we will ignore delicate issues about time by treating any totality as including whatever exists tenselessly. Thus, a totality of what exists is the totality of what did, does, or will exist, all their features and relations included, whatever those things were, are, or will be.

As long as something exists, a totality of things exists. And, empty totalities aside, nothing can be a totality of existing things without being both a totality of actually existing things and an actual totality of existing things. So nothing is a world unless it is an actual world. Furthermore, since there can be at most one world, nothing is *a* world unless it is *the* actual world. Hence, no matter what exists, whichever world is actual is the totality of everything that exists.[16]

We should note, however, that some actualists have refused to use 'the actual world' for all of reality, or the totality of everything that exists. Instead, they reserve that label for the actualised abstract object that plays the role of the actual world in their models of modal logic. For example, here is van Inwagen (1980: 169) on the meaning of 'the actual world':

[12] Goldschmidt and Lebens (this volume Chapter 11) also make this point.

[13] Bennett (2005: 281) calls such a necessitarian an "actualist *par excellence*." But given the arguments in section 7, the actualism that comes for free is compatible with quantifying over impossibilia as long as the actualist uses an additional sense of 'there is'.

[14] Lewis (1986: 140).

[15] Those who deny that Pluto exists can substitute something that they think does exist instead.

[16] Similarly, Stalnaker (1970: 69–70) says that the actual world is the totality of everything there is. As the context should make clear, I'm not using 'actual' here to rigidly designate the world that actually exists.

> Let us retain the notion of a possible world as a way things could have been, and let us reject any suggestion that a possible world is a concrete object; in particular, let us reject any suggestion that "the actual world"—whatever "actual" may mean—is "all this." In other words, let us agree that the actual world is the way things are and carefully distinguish between the way things are and the things that are that way.[17]

Like van Inwagen, actualists such as Adams, Plantinga, and Stalnaker must distinguish the actual totality of everything that exists from the actualised abstract object which represents or characterizes that totality. Here, we will reserve 'the actual world' for whatever is both actual and a world. Then, following Lewis (1986: 138–40), we'll call the abstract object which characterizes or represents it the *actualised surrogate*. But, as long as we distinguish these two things, it doesn't matter what we call them. We could have followed van Inwagen and called the actualised surrogate "the actual world." We would then need to use something like 'the actual totality' for the totality of everything that actually exists. Either way, actualists must distinguish them, and we can run the argument below with either choice of terminology.

The second main thesis involves a rejection of necessitarianism. Necessitarianism says that our own world exists necessarily and that the way things have actually gone is the only way things could have gone. Though actualism is compatible with necessitarianism, most actualists reject necessitarianism. These actualists endorse the following:

Contingency. The world could have been different.[18]

Contingency is true if the world might have been different in any way. Since a world is a totality of everything that exists in the way that everything exists, there are exactly three ways for the world to have been different: when something that exists might have been different, when something exists that might not have, and when something that does not exist might have existed. As long as the world could have differed in any of these ways, Contingency holds.

Actualists of various kinds have posited unactualised abstract objects to avoid quantifying over mere possibilia. I argue in the next section that if these actualists endorse both Worldhood and Contingency, then they must quantify over mere possibilia anyway.

[17] Compare Kripke (1980: 17–20).

[18] I'm not using 'the world' to rigidly designate the totality of things that actually exists. So I'm not claiming that the world as it actually is would have been any other world had things had gone thus-and-so.

6. The Argument

Discussions about actualism and possibilism often focus on possible but non-actual individuals (Adams 1981; Bennett 2005; 2006; Fine 1977; Jubien 1996; Linsky and Zalta 1994; 1996; Menzel 2020; McMichael 1983; Nelson and Zalta 2009; Plantinga 1976; Tomberlin 1996). In our semantic models, these are the things assigned to other worlds but not our own. Many actualists seem to think they can resist the possibilist's arguments as long as their semantic models lack merely possible individuals. After all, if there are no possible but non-actual individuals, then, the thought goes, there are no mere possibilia.

However, as long as actualists like Adams, Plantinga, and Stalnaker endorse both Contingency and Worldhood, they must quantify over mere possibilia. As I'll argue shortly, the very abstracta conscripted by actualists into helping them avoid quantifying over mere possibilia require those actualists to quantify over mere possibilia.[19] But these mere possibilia aren't the possible individuals so often discussed. What actualists like Adams and Stalnaker must quantify over are merely possible worlds—and I don't mean the actually existing abstract objects often disguised as "possible worlds."

The argument is simple. Given Contingency, the world could have gone differently. If the world had gone slightly differently, then a world other than our own would have been actual. Given Worldhood, that world would have been a totality of everything that exists. So it wouldn't have been any abstract surrogate that characterizes or represents it in the actualist's semantic models.[20] The totality would be neither a set of propositions (à la Adams), nor an abstract state of affairs (à la Plantinga), nor a property (à la Stalnaker). In each of these cases, we must distinguish a possibly actual world from its actually existing but unactualised surrogate. As things actually stand, the unactualised surrogate exists. But the world it represents, characterizes, or stands in for does not.[21]

When *I* say that another world could have been actual, *I* don't mean that some maximal consistent set of propositions with at least one false proposition actually exists. Presumably, such a maximal consistent set is such that its member propositions might have all been true. Otherwise, and as long as we accept Contingency, why would such a maximal consistent set deserve to play the role of a "possible world"? But the propositions are possibly jointly true only if some world, some totality of all that exists, could have existed such that they truly characterized or

[19] Linsky and Zalta (1994; 1996) deserve special consideration, which I save for the end of the section.

[20] In special cases, if an abstract surrogate is all that exists ,then the abstract surrogate would have to represent itself representing itself, and so on, to infinity, like a set of parallel mirrors. But we're considering cases in which the abstract surrogate is not all that exists.

[21] For Linsky and Zalta (1994; 1996), the surrogate *is* the world and so the world actually exists. So actualists need not always appeal to a difference in existence to distinguish a world from its abstract surrogate. I engage with the views of Linsky and Zalta at the end of the section.

represented it. Such a world is possible but non-actual. So a maximal consistent set of propositions can play the "possible world" role only if some possible but non-actual world could have been truly represented by its members.

Or when I say another world could have been actual, I don't simply mean that there actually exists some maximally consistent abstract state of affairs that includes at least one non-obtaining state of affairs. Presumably, such a maximal consistent state of affairs could have obtained. Otherwise, and as long as we accept Contingency, why would such a maximal consistent state of affairs deserve to play the role of a "possible world"? But the state of affairs possibly obtains only if some world, some totality of all that exists, could have existed and rendered the state of affairs an obtaining rather than a non-obtaining one. Such a world is possible but non-actual. So a maximal consistent state of affairs can play the role of a "possible world" only insofar as some possible but non-actual world could exist and make the state of affairs obtain.

Or when we say another world could have been actual, we don't simply mean that some special but unexemplified property actually exists. Presumably, that unexemplified property could have been exemplified. Otherwise, and as long as we accept Contingency, why would such a property deserve to play the role of a "possible world"? But the property is exemplifiable only if some world, some totality of all that exists, could have exemplified it.[22] Such a world is possible but not actual. So Stalnakerian properties can play the "possible worlds" role only insofar as some non-actual worlds could have exemplified those properties.

Actualists like Adams, Plantinga, and Stalnaker posit abstract objects at least partly to avoid quantifying over merely possible worlds. But those abstract objects do not function as theoretically intended unless there is at least one merely possible world for each of them. For example, if an Adamsonian maximal consistent set of propositions couldn't have been such that all its members were true, then why would we think that it represents, characterizes, or stands in for a possible reality? But if such a set's members could have all been true, what would have made them true? A world—a world that isn't actual but could have been.

Or, once more for good measure, if a Stalnakerian world property couldn't have been exemplified, why should we accept that it represents, characterizes, or stands in for a possible reality? But if such a world property could have been exemplified, what would have exemplified it? Well, properties don't generally exemplify themselves. Nothing could exemplify the property of being a world except a world—a world that isn't actual but could have been. And so on. Hence, influential versions of actualism seem to require that we quantify over the very things that those theories were designed to avoid quantifying over.

[22] Goldschmidt and Lebens (this volume Chapter 11) also make this point.

To review, when we say that a world besides our own is possible, we don't typically mean that an abstract surrogate for it actually exists. If that's all actualists mean, then they account for contingency in much the same way Lewis does—by accounting for another notion and assigning it the same name. Instead, we mean that reality itself could have been different and that a world other than our own might have existed. And, in general, an abstract surrogate which might have been actualised would not have been a world if it had been actualised.

However, some versions of actualism imply that each unactualised abstract surrogate is contingently abstract and would have been the very world it characterizes if it had been actualised. Whereas the traditional actualist posits "possible worlds" that are not and could not have been worlds, Linsky and Zalta (1994; 1996) formulate a version of actualism according to which some non-concrete objects really could have been worlds. According to this kind of actualism, when we quantify over possible worlds, we quantify over actually existing non-concreta, each of which could have been a world. Does such a view save us from mere possibilia?[23]

I doubt it.[24] It seems to me that Linsky and Zalta's version of actualism must also quantify over mere possibilia. According to Linsky and Zalta, contingently non-concrete objects do not have the properties they would have had if they had been concrete. A possibly talking donkey neither talks nor is a donkey.[25] Non-concrete objects neither walk nor speak. But a non-concrete, possible donkey is something that might have been a living, breathing—speaking—donkey. Presumably, similar remarks hold for contingently non-concrete totalities.

Suppose *beta* is a possible world like ours with one exception: Particle Pete takes a slightly different trajectory in deep space for exactly two seconds. For Linsky and Zalta, beta exists as a contingently non-concrete object. It could have been a world, in my sense, and at least partly concrete. But it is not a world, not in the sense of being a totality of everything that exists. Only one such totality exists, the one I've called the actual world. Unlike beta, the actual world is a *world*. Though unactualised possible worlds are not worlds, in my sense, Linsky and Zalta's actualism implies that each of them could have been.

Now, according to Linsky and Zalta, beta is a contingently non-concrete object that actually exists. For Linsky and Linsky, therefore, when we quantify over possible worlds, we need only quantify over actually existing objects like beta.

[23] Others have objected to Linsky and Zalta (1994), including Tomberlin (1996) and Bennett (2006). In my view, both Linsky and Zalta (1996) and Nelson and Zalta (2009) rebut these objections successfully. If my argument here fails, it does for different reasons.

[24] Linsky and Zalta (1994: 438) explicitly assume for the sake of argument that identifying possible worlds with abstract objects does not commit the actualist to quantifying over mere possibilia. So, to be fair, I'm not objecting to them but instead to this undefended assumption which they may not even accept. And I suspect they may not accept it because they self-identify as possibilists in the essay's first page.

[25] See Linsky and Zalta (1994: 445; 1996: 289) and, in a similar vein, Zalta (1993: 419–20). According to Schnieder (2007), Bolzano has a similar view.

But beta is contingently non-concrete precisely because it could have been a world and at least partially concrete. So, in general, for each actually existing, non-concrete object in Linsky and Zalta's theory that could have been a world, in my sense, there is a non-actual possibility, namely, a non-concrete object's being a world and at least partially concrete. There's an important difference between something's being non-concrete and possibly a world, on the one hand, and that thing's being a concrete world, on the other. Though beta actually exists, its being a world is possible and not actual. Were beta to have been a world, things would have gone differently than they've actually gone. Beta-as-a-world is merely possible. And at least when I think about and discuss possibilities, I have in mind beta-as-a-world type of possibilities.

Linsky and Zalta might remind us that the actualist need only quantify over actually existing non-concrete objects like beta. Yet beta is *contingently* non-concrete and *could have been* a world. Even though quantifying over the contingently non-concrete amounts to quantifying over nothing but actualia, objects like beta would have been ill-suited for the possible world role unless they each harbored the potential of being a world, in my sense. But beta cannot harbor such a potential unless it is possible for beta to be a world. And beta's being a world is one possibility that hasn't come to fruition. Beta's being a world is merely possible. So for each contingently non-concrete object that could have been a world, in my sense, there is a mere possibility that we can meaningfully discuss and quantify over, namely, the possibility of a possible world's being a real world. Linsky and Zalta's actualism cannot account for real contingency in the world unless it acknowledges the mere possibilities it was designed to avoid. Therefore, the move to quantify over actually existing non-concrete objects doesn't save the actualist from quantifying over mere possibilia, whether those objects are necessarily or contingently non-concrete.

7. Aftermath

Insofar as actualists endorse both Worldhood and Contingency, they must quantify over mere possibilia. Does this mean that actualists are unwitting possibilists, as Parfit claims? Or is actualism consistent with quantifying over mere possibilia? Although actualism is often characterized as the view that there are no mere possibilia, I believe that it is compatible with saying that there are mere possibilia.

To fix terms, let *actualism* be the conjunction of these three claims:

(i) **Anti-Meinongianism.** Everything (there is) exists.
(ii) **Anti-Possibilism 1.** Everything (there is) is actual.
(iii) **Anti-Possibilism 2.** Everything that exists is actual.

Each of (i) through (iii) expresses an ontological or existential claim and not a linguistic or semantic one. But the claims expressed partly depend on the sense of 'exists' expressed in (i) and (iii) and the sense of 'there is' expressed in (i) and (ii). So if an actualist has good reason to use different senses of 'exist' and 'there is' to express claims *different* from but *compatible* with (i) through (iii), then we have no reason to suppose that she's contradicted herself or rejected actualism.

Let the 'there is' in (i) and (ii) express what I'll call the *ontological* sense of 'there is'. In this sense, (i) amounts to saying that everything that *has being*, exists. So (i)'s negation says that something has being but doesn't exist. This expresses something like the Meinongian view that some objects do not exist but subsist, and therefore enjoy a form of being.[26] Our actualist does not endorse this Meinongian claim about non-existent beings. But she does endorse (ii), which, with the ontological sense of 'there is', says that everything that has being, is actual. She also endorses (iii) and understands it to imply that non-actual things don't exist.

In addition to the ontological sense of 'there is', our actualist also uses a *non-ontological* sense of 'there is'.[27] She uses this sense to express the sorts of claims discussed in the previous section about merely possible worlds. Yet, many since Quine (1948) have held that quantifying over anything commits one to its existence. And it would be foolish to try to undermine such a distinguished school of thought here. But Priest (2016: 339–42) has convinced me that Quine's arguments wither under scrutiny. Also, if we pay close enough attention to ordinary language, we will notice that we frequently quantify over things that presumably lack being and don't exist. If the argument in the previous section is successful, certain kinds of actualists must do the same. This doesn't mean that actualists are committed to the existence of non-existents but instead that the quantifiers are polysemous. So perhaps Quine was right that, in one sense of the quantifier, quantification is existentially loaded. But there also seem to be senses of the quantifiers that aren't.

Quineans have for the most part ignored research in linguistics that describes how we use 'there is' in ways that are consistent with quantifying over non-existents.[28] For example, the 'there is the...' locution is often used meaningfully to list things, and we can list things that don't exist. There are also uses of 'there is' which serve to recall something to mind, count, introduce something into conversation, and help us draw appropriate inferences. We can certainly both

[26] Meinong (1904).

[27] I borrow the term here from Parfit (2011: 481) but not his application of it. Parfit says that logical truths and numbers also exist in a non-ontological sense, and I disagree. For a view similar to Parfit's but much better developed, see Hofweber (2016: 55–101). Hofweber argues that quantifiers are polysemous, and I agree. His account of the additional inferential role sense of the quantifiers is one way to fill in the details of here.

[28] See Ward and Birner (1995) for an important paper neglected by philosophers, McNally (2019) for an overview, and Hofweber (2016: 55–101) for one way to situate research in linguistics on 'there is' within a broader philosophical project.

think and draw inferences about and discuss what doesn't exist. *How* we do that is fairly mysterious and ripe for philosophical speculation. But only philosophers could deny that we do it at all.

So, in addition to (i), our actualist endorses a claim that looks like the denial of (i). But the incompatibility is only superficial because the additional claim involves a different sense of 'there is', which we'll write as 'there is'. Hence, our actualist also endorses:

(i*) There is something that doesn't exist.

Then, in the same, non-ontological sense of 'there is', our actualist endorses,

(2*) There is something that isn't actual.

Finally, our actualist continues to endorse (3) in the same sense as before.[29] As a result, here we have someone who endorses actualism in the form of (i)–(iii). She also accepts (1*) and (2*). But since these are compatible with (i)–(iii), she hasn't reneged on her actualism or endorsed an inconsistency. So she remains an actualist even though she sometimes quantifies over things that have no being or existence. In other words, she is an actualist but also a *noneist*, someone who quantifies over things that lack both being and existence.[30]

This particular combination of actualism and noneism gives rise to *ostrich actualism*. Ostrich actualists share certain metaphysical commitments with other actualists. But they quantify over non-existents that lack all being whatsoever. They are actualists, then, but often sound like possibilists. As the view's name suggests, ostrich actualism shares certain structural features with ostrich nominalism.[31] Ostrich nominalists grant that objects have certain features in common but refuse to explain this commonality. So although they sometimes speak like realists about universals, they nonetheless endorse nominalism. Adherents of both ostrich views use language that seemingly commits them to entities that they otherwise deny to exist.

In certain circles, ostrich nominalism has a shady reputation. Despite some similarities, however, ostrich actualism deserves better. We can reasonably charge

[29] But our actualist may also grant an additional, *non-existential* sense to 'exist'. For example, when our actualist says that such-and-such a possibility exists, she doesn't necessarily mean that the possibility exists in the same sense that my chair exists, or that it exists as an abstract object. She means little more than that there is such a possibility, in the non-ontological sense of 'there is'. In this sense of 'exist', to say that 'there exists the possibility that *p*' basically means that 'it is possible that *p*.' Compare Hofweber (2016: 87). Schnieder (2007) attributes a similar view, one about the ambiguity of 'exists', to Bolzano. Thanks to Jacob Zimbelman for raising this point.

[30] Priest (2016). But our actualist needn't say that abstract objects don't exist, as Priest (2016: 106) does. Our actualist thinks that some abstract objects both exist and have being.

[31] For ostrich nominalism, see Armstrong (1980), Devitt (1980), and, more recently, Pickel and Mantegani (2012). For ostrich presentism, see Torrengo (2014).

ostrich nominalists with an explanatory deficit because they refuse to explain something that seems to require an explanation. But the ostrich actualist pays her explanatory debts. How could anyone quantify over mere possibilia without attributing to them some sort of being? Instead of refusing to answer, the ostrich actualist appeals to polysemy. She sticks her head back in the sand not to avoid her explanatory debt, but to retire in peace after successfully discharging it.

8. Conclusion

Parfit contends that actualism implies necessitarianism because actualism implies that there are no non-actual possibilia. However, Parfit's case leaves an opening for traditional actualists to argue that in some sense they, too, can say that there are possible but non-actual possibilia. When it comes to possible worlds, they can borrow a suggestion from Lewis (1986: 138–41) and say that "merely possible" worlds are actually existing but unactualised abstract objects.

However, this response doesn't succeed, not if the actualist thinks that things really could have gone differently. If the actualist thinks that reality really could have been different and that an abstract object is actualisable in the sense that it could have been actualised, then we can ask about what would have to happen for it to be actualised. Then, we're off to the races doing the very thing these abstract objects were supposed to help us avoid—discussing and quantifying over mere possibilia. Does this mean that actualism is inconsistent in that it both precludes and requires quantification over mere possibilia? And if actualists must quantify over merely possible worlds anyway, why should they bother with these actually existing abstracta at all?

I've argued that actualism allows for a certain kind of noneist quantification over mere possibilia. As long as the actualist uses an additional, non-ontological sense of 'there is', she's in the clear. But if the actualist can quantify over merely possible worlds in this way, why should she believe in unactualised abstract objects in the first place? Unactualised abstract objects may continue to play a crucial role for the actualist in an account of intentionality. When I think about a possible world in which, say, the Reds soon win the World Series, I'm not thinking *about* a set of propositions. I'm thinking about the Reds, not propositions about the Reds. So even if unactualised abstract objects often aren't the targets of our thoughts of mere possibilia, these abstract objects may be the lenses through which we think about them. Propositions may serve as intermediaries between us and what they represent and the windows into possible reality. If some represent what's merely possible, we may think about what's merely possible by grasping them. If I'm right, certain actualists may not have to do anything to their metaphysical systems except repurpose them.

References

Adams, Robert Merrihew (1974). "Theories of Actuality", *Noûs* **8** (3): 211–31.

Adams, Robert Merrihew (1981). "Actualism and Thisness", *Synthese* **49** (1): 3–41.

Armstrong, D. M. (1980). "Against 'Ostrich' Nominalism: A Reply to Michael Devitt", *Pacific Philosophical Quarterly* **61** (4): 440–9.

Bennett, Karen (2005). "Two Axes of Actualism", *Philosophical Review* **114** (3): 297–326.

Bennett, Karen (2006). "Proxy 'Actualism'", *Philosophical Studies* **129** (2): 263–94.

Cameron, Ross P. (2016). "On Characterizing the Presentism/Eternalism and Actualism/Possibilism Debates", *Analytic Philosophy* 57(2): 110–40.

Devitt, Michael (1980). "'Ostrich Nominalism' or 'Mirage Realism'?", *Pacific Philosophical Quarterly* **61** (4): 433–9.

Fine, Kit (1977). "Postscript", in *Worlds, Times, and Selves*, ed. A. N. Prior and Kit Fine (Amherst. MA: University of Massachusetts Press).

Hofweber, Thomas (2016). *Ontology and the Ambitions of Metaphysics* (Oxford: Oxford University Press).

Jubien, Michael (1996). "Actualism and Iterated Modalities", *Philosophical Studies* **84** (2–3): 109–25.

Kripke, Saul A. (1980). *Naming and Necessity* (Cambridge, MA: Harvard University Press).

Lewis, David (1970). "Anselm and Actuality", *Noûs* 4 (2): 175–88.

Lewis, David (1986). *On the Plurality of Worlds* (Oxford: Basil Blackwell).

Lewis, David (1990). "Noneism or Allism?", *Mind* **99** (393): 23–31.

Linsky, Bernard and Edward N. Zalta (1994). "In Defense of the Simplest Quantified Modal Logic", *Philosophical Perspectives* **8**: 431–58.

Linsky, Bernard and Edward N. Zalta (1996). "In Defense of the Contingently Nonconcrete", *Philosophical Studies* **84** (2–3): 283–94.

McDaniel, Kris (2017). *The Fragmentation of Being* (Oxford: Oxford University Press).

McMichael, Alan (1983). "A Problem for Actualism about Possible Worlds", *Philosophical Review* **92** (1): 49–66.

McNally, Louise (2019). 'Existential Sentences'. In *Semantics: Sentence and Information Structure*, ed. Paul Portner, Claudia Maienborn, and Klaus von Heusinger (Berlin: De Gruyter), pp. 281–305.

Meinong, A. (1904). "Über Gegenstandstheorie", in *Untersuchungen zur Gegenstandstheorie und Psychologie*, ed. A. Meinong (Leipzig: J. A. Barth), pp. 1–51; translated as "The Theory of Objects" in *Realism and the Background of Phenomenology*, ed. Roderick M. Chisholm (Glencoe, IL: Free Press, 1960), pp. 76–117.

Menzel, Christopher (2014). "Classical Possibilism and Lewisian Possibilism": Supplement to "Actualism", *Stanford Encyclopedia of Philosophy*, ed. Edward N. Zalta (Summer 2018 edn).

Menzel, Christopher (2020). "In Defense of the Possibilism–Actualism Distinction", *Philosophical Studies* **177** (7): 1971–97.

Murray, Adam and Jessica Wilson (2012). "Relativized Metaphysical Modality", in *Oxford Studies in Metaphysics*, vol. 7, ed. Karen Bennett and Dean W. Zimmerman (Oxford: Oxford University Press), pp. 189–226.

Nelson, Michael and Edward N. Zalta (2009). "Bennett and 'Proxy Actualism'", *Philosophical Studies* **142** (2): 277–92.

Parfit, Derek (2011). *On What Matters*, vol. 2 (Oxford: Oxford University Press).

Pickel, Bryan and Nicholas Mantegani (2012). "A Quinean Critique of Ostrich Nominalism", *Philosophers' Imprint* **12** (6): 1–21.

Plantinga, Alvin (1974). *The Nature of Necessity* (Oxford: Clarendon Press).

Plantinga, Alvin (1976). "Actualism and Possible Worlds", *Theoria* **42**: 139–60.

Plantinga, Alvin (1987). "Two Concepts of Modality: Modal Realism and Modal Reductionism", *Philosophical Perspectives* **1**: 189–231.

Priest, Graham (2016). *Towards Non-Being: The Logic and Metaphysics of Intentionality*, 2nd edn (Oxford: Oxford University Press).

Quine, Willard V. (1948). "On What There Is", *Review of Metaphysics* 2 (5): 21–38.

Schnieder, Benjamin (2007). "Mere Possibilities: a Bolzanian Approach to Non-Actual Objects", *Journal of the History of Philosophy* **45** (4): 525–50.

Stalnaker, Robert (1976). "Possible Worlds", *Noûs* **10** (1): 65–75.

Stalnaker, Robert (1986). "Counterparts and Identity", *Midwest Studies in Philosophy* **11**: 121–40.

Tomberlin, James E. (1996). "Actualism or Possibilism?", *Philosophical Studies* **84**: 263–81.

Tomberlin, James E. (2001). "How Not to be an Actualist", *Philosophical Perspectives* **15**: 421–25.

Tomberlin, James E. and Frank McGuinness (1994). "Troubles with Actualism", *Philosophical Perspectives* **8**: 459–66.

Torrengo, Giuliano (2014). "Ostrich Presentism", *Philosophical Studies* **170** (2): 255–76.

van Inwagen, Peter (1980). "Indexicality and Actuality", *Philosophical Review* **89** (3): 403–26.

van Inwagen, Peter (1986). "Two Concepts of Possible Worlds", *Midwest Studies in Philosophy* **11**: 185–213.

Ward, Gregory and Betty Birner (1995). "Definiteness and the English Existential", *Language* **71** (4): 722–42.

Williamson, Timothy (1998). "Bare Possibilia", *Erkenntnis* **48**: 257–73.

Williamson, Timothy (2013). *Modal Logic as Metaphysics* (Oxford: Oxford University Press).

Zalta, Edward N. (1993). "Twenty-Five Basic Theorems in Situation and World Theory", *Journal of Philosophical Logic* **22**: 385–428.

13
Saying Nothing and Thinking Nothing

John A. Keller and Lorraine Juliano Keller

1. Introduction

Lapsing into nonsense is an occupational hazard of philosophy. But, unless they've been drinking, the sort of nonsense philosophers are liable to lapse into is (usually) not pure gibberish. Rather, it's nonsense that has the illusion of making sense: "deceptive nonsense". Deceptive nonsense is sometimes accompanied by what Gareth Evans (1982) called "illusions of thought": cognitive events that *seem* to have content, but don't. But if nonsense sentences, assertions, and thoughts don't mean *anything*, it's hard to see how such illusions could arise. As Carnap famously asked,

> And how could one account for the fact that metaphysical books have exerted such a strong influence on readers up to the present day, if they contained not even errors, but nothing at all? (Carnap 1959: 78)

In this paper we defend the existence of deceptive nonsense and illusions of thought by (i) sketching a general framework for thinking about them, (ii) clarifying the sense in which they lack meaning, (iii) providing arguments for their existence, and (iv) responding to some arguments against them.

2. Overview

In her classic *Encyclopedia of Philosophy* article, Annette Baier distinguishes six species of nonsense, claiming that an utterance is nonsense if it has one or more of the following features:[1]

1. It is obviously false.
2. It is wildly inapposite.
3. It involves a category error.
4. It is syntactically ill-formed.

[1] See Baier (1967: 521). She calls these the "main" ways of departing from sense, leaving open whether there are other ways. We avoid the word 'sentence' here because it's controversial that all of the relevant examples count as sentences.

John A. Keller and Lorraine Juliano Keller, *Saying Nothing and Thinking Nothing* In: *Non-Being: New Essays on the Metaphysics of Non-Existence.* Edited by: Sara Bernstein and Tyron Goldschmidt, Oxford University Press (2021).
 DOI: 10.1093/oso/9780198846222.003.0013

5. It is an otherwise meaningful sentence containing nonsense words (meaningless words).
6. It is a string of nonsense words.

Examples of each species include:

1. The dog is a mathematician.
2. The dog chased the cat [spoken by a defendant in court, in answer to the question of where they were at the time of the crime].
3. The piano has been drinking.
4. cats Blargs chase.
5. Blargs chase cats.
6. Slithy toves brillig.

One way of grouping these species into genera would be to contrast what Keller (2017) calls **silly nonsense**—utterances whose contents are bizarrely false, like 'I have 2.3 children', or perhaps 'Caesar is a prime number'—with what he calls **semantic nonsense**: utterances that fail to have a meaning and hence cannot be evaluated for truth or falsity at all. (It is a matter of some dispute whether category mistakes—nonsense belonging to species 3 in Baier's taxonomy—belong to the genus of silly or semantic nonsense.[2]) Examples (4)–(6) are semantic nonsense, since (5) and (6) contain meaningless expressions and (4) combines words in a meaningless way. Examples (1) and (3), on the other hand, are silly nonsense. It would be nice if we could fit (2) into the category of silly nonsense, but it probably belongs to a genus of its own. Our focus is on semantic nonsense: specifically, examples like (4)–(6).[3] Or rather, we're interested in examples that are like (4)–(6) in lacking meaning, but unlike (4)–(6) in not *obviously* lacking meaning. Utterances like (4)–(6) are what might be called **gibberish**: semantic nonsense that is *obviously* nonsense, nonsense that no vaguely competent speaker could think was anything other than nonsense. Consider, however,

7. Das Nichts nichtet.[4]
8. Vulcan is a planet.[5]
9. Witches cast spells.[6]

[2] See Magidor (2013) for discussion.

[3] Cora Diamond (1981: 10) has argued that according to (certain) forms of Fregeanism, category (5) collapses into category (6). Her thought is that if the Context Principle is true—if words only have meanings when they are embedded in (meaningful) sentences, since the meanings of words are partially *determined* by the meanings of the sentences in which they appear—then *none* of the words in 'Blargs chase cats' means anything, since the sentence itself doesn't mean anything.

[4] See Carnap (1959), discussing Heidegger (1929). Often translated as 'the nothing noths', this is widely viewed as a neologism that makes no grammatical or conceptual sense.

[5] See Braun (1993). Since 'Vulcan' is an empty name, it has no content on Millian views where the contents of names are their referents. This plausibly entails that the sentence as a whole lacks content. (But see n. 11 below.)

[6] See Braun (2015). Plausibly, 'witch', 'cast', and 'spell' lack determinate meanings: since there are no witches, acts of casting, or spells, we cannot rely on the world to settle ambiguities and contradictions in how these terms are used and in the concepts we associate with them.

These are sentences that some people have sincerely uttered but that others think lack meaning. If these sentences are nonsense, they are (at least potentially) **deceptive nonsense**. While controversial assumptions are required to deliver the verdict that (7), (8), and (9) are nonsense, examples of deceptive nonsense can be generated from less contested principles. Consider Maeve (age 4), and Ava and Jake (her parents), in a context where nobody named 'John' is salient.

Case 1: Suppose Ava and Jake are talking about names for their new baby, and Ava suggests 'John'. In response, Jake says, "'John' is abominable." Maeve, passing by, hears Jake but doesn't realize that the name 'John' is being mentioned rather than used. Excited to learn this bit of apparently salacious news, she rushes off and tells her friend Izzy (who believes her), uttering:

10. John is abominable.

Case 2: Discussing potential names, Jake suggests 'Donald', and Ava says in reply, "Your suggestion is abominable." Maeve, passing by halfway through Ava's utterance, thinks she hears Ava say (10) (/CH(ə)n/ is abominable), and rushes off to tell her friend Izzy, who believes her.

Case 3: Maeve (age 8 now) comes across a piece of paper with some derogatory statements printed on it, including (10), and takes those sentences to be assertions by some speaker about somebody named 'John'. Maeve then repeats (10) to Izzy, intending to defer to that speaker about the referent of 'John'. Sadly, however, those inscriptions were randomly generated by a computer, and there is nobody that 'John' refers to.

The recipe for generating cases of deceptive nonsense like (1)–(3) is simple: a speaker *S* (thinks she) learns a name *N* from some other speaker *S** (by deferring to *S**'s referential intentions), but unbeknownst to *S*, *S**

a) was mentioning *N* (the word, without an intended referent)
b) was using another expression *N** (in another grammatical category) that *S* mistakes for *N*
c) does not exist (*N* isn't actually produced by a speaker), or
d) was engaging in deliberate nonsense (had no intended referent).

In such cases, by attempting to defer to *S**, *S* ensures that her use of the name is meaningless, and that the sentences uttered by *S* using that name will fail to express propositions.[7]

[7] *Mutatis mutandis* for nonsense predicates. Consider **Case 4:** Ava says to Jake, "What's an example of nonsense?", and Jake says in reply

11. Geeshjohn is minabobable.

Maeve, passing by, hears Jake's utterance of (11), which she then repeats to Izzy, who believes her.

One interesting thing about deceptive nonsense is that it can give rise to:

Illusions of Meaning: sentences that seem to have meanings, but that really don't.

Illusions of Assertion: speech acts that seem to say or assert something, but that really don't.

Illusions of Thought: cognitive episodes that seem to have contents, but that really don't.

By definition, a sentence *s* that is deceptive nonsense has the illusion of being meaningful; and so an utterance of *s* would, in typical cases, create an illusion of assertion, and giving one's assent to such an utterance—or denying it, for that matter—would, it seems, typically result in an illusion of thought. (See section 2.2 for an explanation of the need for hedging these claims.) Furthermore, as Jim Pryor writes, "Rehearsing sentences to yourself is one way of having occurrent thoughts" (Pryor 2006: 329 n. 1) Since such silent talking to ourselves is a cognitive episode, it follows that if one silently rehearses *s* to oneself (while taking *s* to be meaningful), one is suffering from an illusion of thought.

2.1 Kinds of Meaning

We've said that sentences (and assertions and thoughts) that fail, in whole or in part, to have meanings are semantic nonsense. But 'meaning' is said in many ways. The relevant kind of meaning that nonsense expressions lack is *content*.

There is wide disagreement about the *nature* of content, but wide agreement about the roles it plays. Contents that are *truth-apt*—susceptible of truth and falsity—are commonly referred to as **propositions.** Propositions are the fundamental bearers of truth value, the semantic contents of univocal, declarative sentences (in contexts of utterance), and the contents of thoughts (beliefs, hopes, etc.) and assertions. Sentences are true (or false) by virtue of expressing true (or false) propositions. Sub-sentential expressions—lexical items, such as 'girl', and phrases, such as 'down the street'—do not have propositions as contents. The specific nature of sub-sentential content is contentious (see sections 3.1–3.3), but the contents of sub-sentential expressions are generally taken to determine, in conjunction with syntax, the propositions expressed by sentences in which those expressions occur. Typically, if a constituent of a sentence lacks content, the sentence itself will fail to express a proposition, as is the case in (5) above: since 'blarg' does not have a content, (5) does not express a proposition. (If an argument for this claim is needed, note that (5) cannot be evaluated for truth or falsity.)

Because natural languages like English contain indexicals ('I', 'now') and demonstratives ('that', 'those'), the semantic contents of which are partially determined by the context in which they are used, content must be distinguished from **standing meaning.**[8] Standing meaning is the context-invariant linguistic information associated with expression-*types*—the conventional, linguistic rules mastery of which characterizes language acquisition. But the standing meaning of expression-types is not always sufficient for determining content.

For example, the sentence type 'I am hungry' does not express a proposition (a truth-evaluable content), since it might be true when Maeve utters it and false when Izzy does (or true when Maeve utters it before lunch and false when she utters it after): that is, some tokens of it are true and others are false. Whether it is true or false depends on features of the context, such as the speaker and time of the utterance. Content, then, is the sort of meaning that is associated with sentence *tokens* in contexts of utterance. In general, the semantic content of an expression is determined by its standing meaning, plus the relevant features of context (e.g. speaker, time, and place). In what follows, we use double brackets as a device for referring to the content of the expression within: just as putting quotation marks around an expression *e* creates a name for *e*, putting double brackets around a denoting expression *e* creates a name for the content of *e*'s denotation. Hence for example [['Grass is green']] = the proposition that grass is green (that is, '[['Grass is green']]' *refers* to the proposition that grass is green).

One hallmark of semantic nonsense is that it is not truth-evaluable, even in context. That's why the most important sense in which semantic nonsense lacks meaning is that it lacks content. Of course, any sentence that lacks a standing meaning will also lack content, since content is determined by standing meaning (in context). Certain paradigm instances of nonsense—what we have called "gibberish"—lack a standing meaning, and hence lack content. Because gibberish lacks a standing meaning, it is generally not deceptive nonsense: since gibberish lacks *any* sort of linguistic meaning, speakers (and listeners) will generally *know* that gibberish lacks content. But it is not always the case that (even relatively competent) speakers know whether expressions lack a standing meaning (according to many views about standing meaning): 'John' (in (10)) is not gibberish to Maeve, for example, even though it lacks both a standing meaning and a content. So we cannot simply say that the difference between gibberish and deceptive nonsense is that gibberish lacks a standing meaning in addition to lacking content.

2.2 Kinds of Semantic Nonsense

Propositions, as fundamental bearers of truth-conditions, are the (potentially) shared contents of sentences, assertions, and thoughts. For example, the sentence, 'Ava is human', Izzy's assertion of that sentence, and Izzy's belief that Ava is

[8] See Heck (2002).

human all have the proposition [['Ava is human']] as their content. This partly explains why we use that sentence to report Izzy's belief. Since the bearers of truth-evaluable content (things that express propositions) fall into three main categories, there are three main categories of nonsense:[9]

Sentential Nonsense: a *sentence* that lacks content.

Assertoric Nonsense: an *assertion* that lacks content.

Cognitive Nonsense: a *thought* (belief, hope, desire, etc.) that lacks content.

Note that these three types of semantic nonsense can give rise to the three types of illusion outlined in section 2: sentential nonsense can give rise to illusions of meaning, assertoric nonsense to illusions of assertion, and cognitive nonsense to illusions of thought. It is worth noting, however, that these different kinds of nonsense do not necessarily go hand-in-hand. Depending on one's views about about assertion, it might be possible to have sentential nonsense without assertoric or cognitive nonsense, since we don't put our thoughts into words perfectly: one could, for example, misspeak in a way that results in sentential nonsense, perhaps by speaking ungrammatically. Still, there would be a thought one was trying to express, and one's interlocutors might readily grasp one's meaning (and so, on at least some theories of assertoric content, one would have performed a contentful speech act). For example, the second sentence before this one is ungrammatical, and hence (arguably) does not express a proposition, even in context. But there was a thought we were trying to communicate when we wrote it, and we plausibly managed to successfully assert or express that thought. After all, you probably took our meaning: indeed, you probably didn't even notice the error.[10] If small grammatical errors prevented us from saying or asserting anything, we could gain the benefits of lying by theft over honest deceit (asserting something we believe to be false) by inserting slight grammatical infelicities into our speech.

Conversely, if (mental) content externalism is false but semantic externalism is true, there will be cases where people sincerely utter meaningful sentences but only have the illusion of thought, since they don't really know or understand the meaning of the sentences they're uttering. When students first learn about Einstein's theory of relativity, they learn to say things like, 'Simultaneity is relative to reference frame'. That sentence is meaningful and true, as are assertions of it. But it's unlikely that beginning students actually know what the sentence means: there's no thought or belief of theirs that that sentence expresses. They are merely parroting their teachers.

Such students may suffer from what Keller (2017: 2) calls "illusions of nonsense": cases where something meaningful or true seems to be semantic or silly

[9] For related discussion, see Cappelen (2013: 26).

[10] If you don't see it, 'about' is repeated.

nonsense (perhaps a category mistake). As an example of an illusion of *silly* nonsense, consider, 'There are as many even numbers as numbers', a sentence that a precocious seventh grader might dismiss out of hand. Examples of illusion of *semantic* nonsense can be brought to mind by thinking of cases where someone conflates the fact that *she* doesn't know what an expression means with the expression's lacking a meaning. This is... not uncommon in philosophy.

3. In Favor of Deceptive Nonsense and Illusions of Thought

It's trivial to provide examples of gibberish: we already have. But those examples weren't assertions, they were *deliberate* nonsense. The interesting questions are (i) whether there is *deceptive* nonsense: utterances that are intended to be meaningful, and that one takes to be meaningful, but which nonetheless fail to be meaningful; and (ii) whether such deceptive nonsense is accompanied by illusions of thought. The cases we gave in section 2 suggest so, as does the testimony of philosophers. Consider:

- Natural rights is simple nonsense: natural and imprescriptible rights, rhetorical nonsense—nonsense upon stilts. (Bentham 1843)
- The book will... draw a limit to... the expression of thoughts.... The limit can... only be drawn in language and what lies on the other side of the limit will be simply nonsense (Wittgenstein 1961: Preface)
- The alleged statements of metaphysics which contain [words like 'God' and 'essence'] have no sense, assert nothing, are mere pseudo-statements (Carnap 1959: 67)
- Philosophical problems arise when language goes on holiday. (Wittgenstein 1953: §38)
- It is obviously a perfectly significant statement, whether true or false, to say that Romulus existed. If Romulus himself entered into our statement, it would be plain that the statement that he did not exist would be nonsense, because you cannot have a constituent of a proposition which is nothing at all. (Russell 1956: 242)
- If a sentence makes no statement at all, there is obviously no sense in asking whether what it says is true or false. And we have seen that sentences which simply express moral judgments do not say anything. (Ayer 1952: 108).
- Reflective persons unswayed by wishful thinking can themselves now and again have cause to wonder what, if anything, they are talking about. (Quine 1960: 242)

- I do not understand what philosophers say... and I think the reason I do not understand them is that they have failed to explain what they mean... And I think the reason they have failed to explain what they mean is that there is nothing, or nothing coherent, that they do mean. (van Inwagen 1980: 285)
- Lacanians believe that the unconscious is structured like a language. They are not sure what this means, but they trust Lacan, who said so. (Recanati 1997: 84)
- [I have made] the *prima facie* case that uses of 'intuitive' and cognate terms are so defective that they should be classified as nonsensical. (Cappelen 2013: 40)

This is but a small sample of the accusations of nonsense philosophers have leveled against each other. While we don't agree with all of them, we do think that deceptive nonsense is not only possible, but actual, and probably common. That is, we think that philosophers (and others) have unintentionally spoken nonsense and suffered from the illusions that typically accompany it. Even if the philosophically interesting accusations of nonsense listed above are all incorrect, the mundane cases given in section 2 seem to illustrate how easy it is for deceptive nonsense to arise. In the remainder of this section we support this intuitive verdict by showing how it can be vindicated from a number of different theoretical perspectives.

3.1 Millianism

According to Millianism, the semantic content of a name is its bearer. Thus, Millianism is a version of Direct Reference Theory, according to which the sole semantic function of a certain class of expressions (proper names, and perhaps indexicals and demonstratives) is to refer to an individual. Millianism entails that empty names—names without bearers, such as 'John' in (10), or 'Vulcan'—lack semantic content. Since the semantic content of a sentence is determined by the semantic contents of its constituents (plus syntax), sentences containing empty names will plausibly lack content as well. As David Braun says,

> According to Direct Reference, if 'Vulcan' does not refer, it has no semantic value. Even worse, it seems that sentences containing 'Vulcan' cannot express propositions, since there is no semantic value to "fit into the subject position" of the proposition (Braun 1993: 451).

To get from this view of empty names to deceptive nonsense and illusions of thought, all that is needed is the anodyne observation that sometimes we don't know that a name is empty. Thus, Millianism leads, fairly straightforwardly, to deceptive nonsense and illusions of thought.[11]

3.2 Neo-Fregeanism

The existence of deceptive nonsense isn't tied to Millian semantics, however. Any view about the contents of singular terms according to which they are *object-dependent* will entail that content is lacking in these cases. According to Neo-Fregean theories, for example, the contents of singular terms are the senses of those terms rather than their referents, but those senses are object-dependent.[12] On such views, 'John' (in (10)) and 'Vulcan' will lack senses as well as referents. Thus, deceptive nonsense and illusions of thought arise on Neo-Fregean theories.

3.3 Fregeanism

On traditional Fregean views, senses are not object-dependent, and so empty names will generally have contents, since they will have senses. But only generally: on Fregean views, it's still *possible* for a speaker to unknowingly use a name without a sense and thus fail to express a proposition. Cases (1)–(3) illustrate this. All that is required for the existence of deceptive nonsense on Fregean views is for *semantic deference* to be possible: for it to be possible to use an expression to mean whatever other speakers use it to mean (if anything). Failed attempts at deference will thus lead to a lack of content. In Cases (1)–(3), nobody was actually using 'John', and no sense of 'John' was expressed or made contextually salient; hence there was no content for Maeve's use of 'John' to inherit, her deferential intentions notwithstanding. Thus, deceptive nonsense and illusions of thought arise on Fregean theories.[13]

[11] It is worth noting, however, that some Millians (including Braun 1993) defend a view according to which sentences like 'Vulcan is a planet' express "gappy propositions". On such views, 'Vulcan is a planet' isn't semantic nonsense as we've defined it, since it has content. There's still a cognitive and semantic illusion though, since 'Vulcan is a planet' doesn't have the kind of content that it seems to have. Its content is a gappy proposition, rather than a fully-saturated one, and it's the *same* gappy proposition as the one expressed by 'Phlogiston is a planet'. On this view, then, someone might still suffer from the illusion of thinking that [['Vulcan is a planet']] and [['Phlogiston is a planet']] were different.

[12] This view is defended by e.g. Gareth Evans (1982) and John McDowell (1986), perhaps the most (in)famous defenders of illusions of thought.

[13] We do not claim that this assessment is true to the historical Frege—we are not engaged in exegesis. Most likely, our analysis conflicts with Frege's more stringent requirements on semantic competence (see Frege 1956), which are not widely held today. But even if Maeve isn't competent with 'John', that wouldn't affect the plausibility of the claim that Maeve thinks she is thinking and talking about (somebody named) John. And that's all that's needed for her to suffer from illusions of meaning, assertion, and thought.

3.4 The Big Picture

We've been considering cases involving expressions having too few, i.e. zero, meanings. So long as our attempts to introduce meaning can go wrong—and what in human affairs can't go wrong?—there will be failed introductions, resulting in expressions with too few (i.e. zero) meanings. And so long as things can go wrong in this way without our *immediately* realizing it—and what can't go wrong without it being immediately realized?—we will get cases of deceptive nonsense. As Cappelen (2013: 32) puts it, no plausible semantic or meta-semantic theory has a "built-in *guarantee of infallibility*" with respect to the introduction of new meaningful expressions, or our beliefs about whether such introductions have been successful.

But deceptive nonsense can arise from the other direction as well: not from an expression having too few meanings, but too many. If Maeve defers to Ava and Jake about the referent of 'John', but Ava and Jake are (perhaps unwittingly) using 'John' to refer to different people, Maeve's deference will result in her use of 'John' lacking (a determinate) content. More generally, if Maeve intends to use an expression to mean whatever others are using it to mean, but there are multiple things others have used it to mean and neither Maeve nor her circumstances have done anything to differentiate between those possible meanings, her use will lack a determinate content. This is, roughly, Cappelen's view of the word 'intuition' as used by many philosophers.[14]

4. Against Deceptive Nonsense and Illusions of Thought

We think that Cases (1)–(3) are compelling examples of deceptive nonsense and illusions of thought. Some philosophers have argued, however, that illusions of thought are *impossible*. Even though we are ultimately unpersuaded by these arguments, we think extant discussions do not do them justice. We focus here on a recent discussion by Herman Cappelen: in this section, we argue that Cappelen's responses to these arguments are unsatisfying, and in section 5 we'll give what we take to be a more convincing response to them.

4.1 The Original Argument

Cappelen presents the first argument as follows:[15]

[14] See Cappelen (2013).

[15] According to Cappelen, this argument was inspired by an argument Paul Boghossian presented (but didn't endorse) in conversation.

The Original Argument

O1. Illusion requires falsity. The thinker must have a false belief (or other attitude) about her propositional attitudes.

O2. A natural development of (1) is that for a subject to have the illusion of thought, she must have a false belief of the form *I was thinking that p*.

O3. Suppose *p* is nonsense.

O4. Then: *I was thinking that p* can't be nonsense, since it has to be false.

O5. Step 4 requires an account of (or semantics for) the second-order thought that allows the complement *p* to be nonsense and the entire self-attribution to be not-nonsense and false.

O6. The correct account or semantics for second-order thoughts requires that the complement in *A thinks that p* be propositional.

O7. So: illusion of thought is impossible. (Cappelen 2013: 29)

4.2 Cappelen's Objections

Cappelen's discussion of The Original Argument is brief and clear, so we reproduce it here with minimal editorializing. He says that (O1)

> is dubious or at least in need of further argument: we could think of an illusion of thought along the lines of an illusion of a dagger, where that is not to be construed as having a false belief about the presence of a dagger, but simply as what it looks like grammatically: the illusion of a dagger. Similarly, we can have the illusion of a thought and not construe that as the having of a false belief about a thought-like event. (Cappelen 2013: 30)

He says that (O2)

> is dubious because even if you think illusions of thought require false beliefs about thoughts, the false thoughts need not be of the form, *I was thinking that p*. It could be a demonstrative thought of the form, *That was a thought* (accompanied by a demonstration of the cognitive event that was not a thought). A demonstrative thought of that form would be false if the demonstrated event wasn't a thought (i.e. we have no reason to think the demonstrative thought is nonsense just because it demonstrates nonsense). (Cappelen 2013: 29)

Finally, Cappelen says that (O6) is dubious given that

> there's no consensus on what the correct semantics for belief reports is (and no consensus on the correct account of second-order thought), and so any claim

> about what the correct semantics *allows* will be controversial and require very substantive theoretical commitments... Putting that lack of consensus aside, I know of no view that rules out an account of second-order thought according to which such a thought *presupposes* that the complement is propositional, and if that presupposition fails, the thought is false. (Cappelen 2013: 29–30)

4.3 The Revised Argument

We think that Cappelen's objections are fairly decisive against The Original Argument. But we also think that The Original Argument is unnecessarily weak, and that there's a version of the argument that is simpler, more intuitively compelling, and less susceptible to Cappelen's objections. It runs as follows (where '*p*' stands for a sentence or sentence-like linguistic string):

The Revised Argument

R1. If it's possible for a subject *S* to have the illusion of thinking that *p* in context *c*, then it's jointly possible that, in *c*, *p* lacks content and *S* falsely (or truly) says or thinks something of the form *I was thinking that p*. $(\Diamond I \supset \Diamond(L \bullet (S \vee B))$

R2. It's not jointly possible, in a single context *c*, for *p* to lack content and for *S* to falsely (or truly) say or think something of the form *I was thinking that p*. $\sim\Diamond(L \bullet (S \vee B))$

R3. So, it's not possible for a subject *S* to have the illusion of thinking that *p*. $\sim\Diamond I$

The Revised Argument seems to straightforwardly avoid Cappelen's first two criticisms of The Original Argument. In short, it does this by changing the modality of the claims from necessity to possibility. Contra (O1) (and in line with Cappelen's critique of it), The Revised Argument grants that illusions of thought are not *necessarily* accompanied by false second-order thoughts about them: one *might* have an illusion of thought without having a false (or true!) second-order belief about it. Instead, (R1) merely insists that, given that there are illusions of thought, it's *possible* to have false (or true) second-order beliefs about them, and in particular second-order beliefs of the form *I was thinking that p* (and likewise for one to falsely or truly say something of the form *I was thinking that p*.)

Similarly, contra (O2) (and in line with Cappelen's critique of it), The Revised Argument grants that false (or true) second-order beliefs (and their verbal expressions) about the cognitive episodes that are illusions of thought *needn't* be of the form *I was thinking that p*. Rather, (R1) merely insists that, given that there are illusions of thought, it's *possible* to have false (or true) second-order beliefs about those episodes that *are* of the form *I was thinking that p* (and to report those beliefs with sentences of the form *I was thinking that p*).

The Revised Argument does *not* avoid Cappelen's critique of (O6), since that critique applies equally well to (R2). But applying equally well does not entail applying well, and it's not clear that Cappelen's critique applies well to either (O6) or (R2). While it's true that there's no *consensus* about the semantics of belief reports (including second-order thoughts), and that not *every* going theory requires that-clauses embedded in belief reports to be meaningful for the belief-reports themselves to be meaningful, it's also true that many of the most popular theories do. And that seems intuitively correct. But then, as David Bell writes,

> The difficulty is, crudely speaking, that either the non-existence of the embedded, merely apparent thought will contaminate the second-order thought of which it is a part, or, conversely, the intelligibility of the second order thought will bestow respectability on its first order component. (Bell 1988: 51)

These considerations seem sufficient for making The Revised Argument interesting and important, even if not compelling. We don't claim that The Revised Argument is knockdown; after all, we reject its conclusion. But if resisting The Revised Argument forces us to reject many, or even *any*, going theories about second-order thought and the semantics of belief reports, that's enough to give it some force.

4.4 The Action-Explanation Argument

Cappelen also considers a line of argument against illusions of thought presented in Segal (2000), Wikforss (2007), and O'Brien (2009). In a nutshell, the argument is that illusions of thought would undermine our ability to explain certain aspects of agents' behavior. As Segal puts it:

> The main argument for attributing empty concepts [as the contents of expressions that defenders of illusions of thought would say have no content] in all these cases... is simply that by so doing, and only by so doing, can we make psychological sense of a very wide variety of human activity and cognition. (Segal 2000: 37)

Along similar lines, Wikforss says:

> From the point of view of the individual, after all, it is as if there was a thought available, one that she reasons with and acts on... How can this be explained if one endorses [the claim that these "thoughts" are mere illusions]?
> (Wikforss 2007: 173)

And O'Brien writes,

> Even when A fails to suppose that *P*, due to content failure, it seems to her that she supposes that *P*, and she can act and infer as she would, were she...

> supposing that *P*. On the gap view we have no explanation of why it seems to A that she is supposing that *P*, or of her actions consequent on its seeming to her she is supposing 'that glass is heavy', or of her inferring that 'if my supposition is true, there is at least one heavy glass'. We seem to need *something* to do the normative and epistemic work—some associated act, or some remnant or degraded version of the act one gets in the good case. (O'Brien 2009: 219)

In response to what we'll call **The Action-Explanation Argument**, Cappelen says that

> when there's illusion of thought, there is illusion of reasoning, and so illusion of practical reasoning. It should come as no surprise that we are fallible with respect to the nature of the cognitive mechanisms that trigger action. (Cappelen 2013: 31)

He goes on to point out that this doesn't rule out explaining our behavior by appeal to our beliefs, desires, and reasoning:

> Such an explanation could, at least in principle, appeal to the illusion of thought, i.e. that there was an illusion can be part of the explanation. It can also . . . appeal to second-order thoughts about the nonsensical cognitive event. Here is [a story about why someone would (sincerely) write a nonsensical sentence]: *Martin thought that* that *(demonstrating some cognitive event) was a thought, he wanted to communicate that thought to others, he thought that the sentence S expressed that thought and that by writing down S his desire could be fulfilled.* So, he had a bunch of false second-order beliefs, and those, combined with his desires, explain his actions. (Cappelen 2013: 31)

4.5 Contra Cappelen on The Action-Explanation Argument

Our concern with this response to The Action-Explanation Argument isn't that the argument can be reformulated so as to avoid it. Rather, our concern is that it's implausible to explain actions ostensibly arising from illusions of thought by appeal to second-order thoughts about those illusions. First, we don't think that *every* illusion of thought is accompanied by a second-order thought about it: we don't think the second-order thoughts required by this explanation are always *there*. Second, it's hard to see how the sort of demonstrative second-order thoughts countenanced by Cappelen (e.g. *that was a thought*) could be sufficient to explain one's behavior: to explain why Martin said 'das Nichts nichtet' (rather than, say, 'Witches cast spells'), it's not enough to point out that he had the second order thought *that was a thought*. For a second-order thought to do the job, it seems like it needs to be a thought of the form *I was thinking that das Nichts nichtet.*[16] But it's unclear how such second-order thoughts could be meaningful if 'das Nichts nichtet' isn't. (Recall our discussion of (R2) above.) Finally, Cappelen's response involves a kind of disjunctivism, treating the springs of action as categorically different when they are illusions of thought: in "good" cases, first-order thoughts bring about actions (in conjunction with our desires), but in illusion-of-thought cases, second-order thoughts do all the work. While we are happy to grant that there are illusions of practical reasoning and that we don't always know the reasons why we do the things we do, it seems implausible that the role of illusions of thought in our mental economy is so different than that of proper thoughts: if so, why are such illusions so hard to detect? At the very least, it seems, this hypothesis would predict some sort of experimentally measurable difference in cognitive functioning when our thoughts are nonsensical, since we would have to abandon our normal *modus operandi* and fall back on our second-order thoughts (even assuming they're there). While we are hesitant to endorse empirical claims from the armchair, we don't find the existence of such a difference plausible.[17]

5. Resolving the Antinomy

We think that the considerations presented in section 3 (and the cases in section 2) fairly conclusively establish the existence of deceptive nonsense and illusions of thought. But we also think that the arguments in section 4 for the *impossibility* of illusions of thought are intuitively compelling, and that Cappelen's objections to them are unsatisfying. In this section, we explain how to resolve this apparent antinomy. We think the most satisfying resolution appeals to the Language of Thought Hypothesis (LOT), according to which thought occurs in a (non-conventional) mental language ("Mentalese") with its own syntax and semantics.[18] But we think a similar, if slightly less satisfying, resolution strategy is available to those who reject LOT.

[16] Cappelen says that Martin also believes that 'das Nichts nichtet' expresses the thought he just had. But *why* does Martin think this? How do you get from the belief that *that was a thought* to the belief that 'das Nichts nichtet' expresses that thought? Presumably that's not just a brute fact. But the demonstrative second-order thought could be referring to anything, and so it's hard to see what leads Martin to choose the words he does.

[17] See O'Brien (2009: sect. 3.3) for further discussion of problems with disjunctivist responses to The Action-Explanation Argument.

[18] Fodor (1975) is the *locus classicus* of contemporary discussion of LOT, although our proximate inspiration is Braun (1993). Similar views were prominent in medieval philosophy: see e.g. Karger (1996) and Bulthuis (2020).

5.1 Thinking Nothing

To set the stage for what follows, consider a final mystery confronting defenders of illusions of thought: explaining how it could be that "nothing at all" is going on when we talk, think, and write nonsense. Carnap, for example, says,

> how could it be explained that so many men in all ages and nations, among them eminent minds, spent so much energy, nay veritable fervor, on metaphysics if the latter consisted of nothing but mere words, nonsensically juxtaposed? And how could one account for the fact that metaphysical books have exerted such a strong influence on readers up to the present day, if they contained not even errors, but nothing at all? (Carnap 1959: 78)

And Cappelen writes,

> [Wittgenstein and Carnap] thought that many of those we consider great thinkers were not thinking at all. According to Carnap, what some considered the high points of human intellectual achievement are no more than a bunch of people making noises and marks on paper. Those who read, commented on, and developed their work suffered from the same illusion. They had what appear to be discussions; they wrote books and papers apparently responding to each other. But it was all the most fundamental kind of failure: it was neither true nor false, no thoughts were expressed, and there was no agreement or disagreement. It was all just a complete waste of time, energy, ink, and paper. (Cappelen 2013: 23)

5.2 Thinking Empty Thoughts

We think that these passages are at least potentially misleading, and the mystery they raise about illusions of thought is itself an illusion. There is no need to say that those who "think nonsense" are not really thinking: all we're committed to is saying that their thinking lacks content. That is, we can simply deny what O'Brien (2009: 215) calls The Dependence of Thought on Content Thesis, according to which there is no thinking without content.[19]

Words like 'belief' and 'thought' can be used to refer both to the *act* of believing or thinking something (a concrete mental state or event), and to the *contents* of such states (abstract propositions). When we talk about *sharing beliefs*, we are (usually) talking about numerically identical belief *contents* (propositions) shared by people in numerically distinct belief *states*. So 'beliefs' sometimes refers to

[19] O'Brien claims that this principle is widely endorsed by defenders of illusions of thought.

propositions believed—the *contents* of our belief states—but it can also refer to belief states themselves, as when we say things like "My belief that John is abominable was influenced by yours". To keep things clear in what follows, we'll use 'beliefs$_S$' to refer to states of believing and 'beliefs$_C$' to the contents (if any) of such states.

With this distinction in hand, we can say that while illusions of thought do not, by definition, involve thought *contents*, they may still involve beliefs$_S$ (and hopes$_S$ and desires$_S$). It's just that those beliefs$_S$, hopes$_S$, and desires$_S$—those thoughts$_S$—are *empty*.[20] We can thus give a unified account of thinking, speaking, and writing nonsense: thinking nonsense involves *actually* thinking empty thoughts; speaking nonsense involves *actually* speaking empty words, and writing nonsense involves *actually* writing sentences that don't (actually) express contents. Such writers aren't writing *nothing*—they might be very productive—it's just that the writing they produce lacks content. On this picture, thinking, speaking, and writing nonsense is clearly possible. And contrary to what the above quotes suggest, it isn't even always a waste of time: thinking, writing, and speaking nonsense on the way to formulating some important truth (or falsehood) wouldn't be. But even when it *is* a waste of time, it isn't doing *nothing*.

5.3 The Language of Thought

According to one way of thinking about beliefs$_S$ and their relation to beliefs$_C$, to have a belief$_S$ is to have a Mentalese sentence in one's "belief box",[21] and to have a belief$_C$ is to have a Mentalese sentence in one's belief box that has that belief$_C$ as its content, a situation we will sometimes describe as the belief$_S$ having the belief$_C$ as its content. It is straightforward to consistently describe illusions of thought on this view: in Cases (1)–(3), 'John is abominable' is semantic nonsense, and Maeve has a Mentalese translation of 'John is abominable' ('JOHN IS ABOMINABLE') in her belief box, a translation that is also nonsense. So 'John is abominable' is (deceptive) sentential nonsense, Maeve's utterance of 'John is abominable' is (deceptive) assertoric nonsense, Maeve's belief$_S$ is (deceptive) cognitive nonsense. Maeve is thus suffering from an illusion of thought.

[20] O'Brien herself seems to endorse something like distinction. She writes, "Given the possibility of a similar such ambiguity in the case of thought, [one] can very reasonably suggest that the thinking involved in rehearsing a contentless syntactic string is not the kind of thinking at issue in the Dependence of Thought on Content Thesis." (O'Brien 2009: 228)

[21] See Schiffer (1981). This is obviously a metaphor: having a sentence in one's "belief box" is supposed to be something like the mental analogue of assertion: something like assent. Merely having a Mentalese sentence "in one's head" is insufficient for belief, since there are other propositional attitudes one might have towards its content: one might hope that *p*, fear that *p*, wonder whether *p*, and so on.

Nonetheless, Maeve's utterance of 'John is abominable' was (partially) caused by one of her (first-order) beliefs$_S$. Indeed, there's a sense in which it was (partially) caused by her belief$_s$ that John is abominable. After all, it was caused by 'JOHN IS ABOMINABLE' being in her belief box. That's true even though Maeve doesn't have the belief$_C$ that John is abominable, since there *is* no such belief$_C$, and *a fortiori* no such belief$_c$ expressed by 'JOHN IS ABOMINABLE'. We can say, then, that illusions of thought involve real beliefs$_S$: real sentences in one's belief box. It's just that those beliefs/sentences don't have contents. That's the illusion: you are deceived about your belief$_s$ having content. But you are not deceived about the existence of the belief$_s$ itself. From this perspective, illusions of thought mislead us about the existence of beliefs$_C$ but not the existence of beliefs$_S$.

5.4 Unmediated Belief

According to an alternative conception of beliefs$_s$ and their relation to beliefs$_C$, a belief$_s$ is simply a relation between a subject and proposition (the relevant belief$_C$), unmediated by any sort of mental sentence. On this view, if there's no proposition (no belief$_C$), there's no belief$_S$. Hence, when one suffers from an illusion of thought, there's no belief there *at all*—not just no belief$_C$, but no belief$_s$ either: nothing analogous to having a Mentalese sentence in one's belief box. Of course, one is presumably in a mental state similar to the state one is in when one believes something: that's what generates the illusion. But that mental state isn't a belief, and so, strictly speaking, illusions of thought don't involve beliefs in either the act or content sense, but merely illusions of belief. On this view, illusions of thought mislead us about the existence of both beliefs$_s$ and beliefs$_C$. Perhaps those drawn to O'Brien's Dependency of Thought on Content Thesis are presupposing a view like this. Note, however, that this unmediated picture of belief doesn't entail that there's *nothing* going on when one is having an illusion of thought: there's still a belief-*like* state that one is in, even if that state isn't actually a belief, due to its lack of content.

5.5 Belief Reports

To see how the above considerations allow us to respond to the arguments from section 4, note that there are three ways of thinking about **nonsense reports** (statements of the form *S believes that p*[22] in cases where *p* lacks content, such

[22] Including second-order thoughts of the form *I believe that p*.

as 'Maeve believes that John is abominable'): they might be considered true, false, or nonsense (neither). Here are sets of truth-conditions illustrating these possibilities for both LOT and Unmediated views of belief:

	LOT	Unmediated
False	(LF) 'Maeve believes that John is abominable' is true iff the proposition expressed by 'John is abominable' is expressed by some sentence in Maeve's belief box.	(UF) 'Maeve believes that John is abominable' is true iff Maeve stands in the relation of believing to the proposition expressed by 'John is abominable'.
Nonsense	(LN) 'Maeve believes that John is abominable' is true iff Maeve has a Mentalese sentence *m* in her belief box such that [[*m*]] = [['John is abominable']].	(UN) 'Maeve believes that John is abominable' is true iff Maeve stands in the relation of believing to [['John is abominable']].
True	(LT) 'Maeve believes that John is abominable' is true iff Maeve has a Mentalese sentence *m* in her belief box that is an adequate translation of 'John is abominable'.	(UT) 'Maeve believes that John is abominable' is true iff Maeve is disposed to accept or assent to 'John is abominable'.

Given relevant plausible background assumptions (including a Russellian treatment of definite descriptions), LF and UF would make 'Maeve believes that John is abominable' false, since there is no proposition satisfying the description on their right-hand sides. LN and UN, on the other hand, would make the nonsense report itself nonsense, since if 'John is abominable' has no content, then [['John is abominable']] doesn't exist, which makes the right-hand-sides of the biconditionals in LN and UN nonsense (neither true nor false), which makes the left-hand-sides nonsense (neither true nor false) as well.[23] For obvious reasons, LT and UT would make the nonsense report true.

5.6 The Revised Argument Revisited

What, then, about The Revised Argument? Consider this instance of it:

R1. If it's possible for Maeve to have the illusion of thinking that John is abominable (in context *c*), then it's jointly possible that, in *c*, 'John is abominable' lacks content and Maeve falsely (or truly) thinks or says something of the form *I was thinking that John is abominable.*

[23] At least if the expression '[['John is abominable']]' has a Millian semantics. The example would need to be modified in a Fregean context.

R2. It's not jointly possible, in any context (including *c*), for 'John is abominable' to lack content and for Maeve to falsely (or truly) think or say something of the form *I was thinking that John is abominable.*

R3. So, it's not possible for Maeve to have the illusion of thinking that John is abominable.

If nonsense reports are *false*, then (R1) is true but (R2) is false. If, however, nonsense reports are *nonsense*, then (R1) is false but (R2) is true. Finally, if nonsense reports are *true*, then (R1) is true but (R2) is false. There is, then, no way of assigning truth-conditions to nonsense reports that makes both premises of The Revised Argument true.[24] This is happy news, since defending some particular set of truth conditions for nonsense reports would require a paper unto itself, and we're past our word limit. But if there is *no* way of assigning truth-conditions to nonsense reports that makes both premises of The Revised Argument true, we can conclude without further ado that it is unsound. It is worth noting, however, that each of The Revised Argument's premises is vindicated by *some* (vaguely plausible) way of assigning truth conditions to nonsense reports. This might explain the intuitive appeal of the argument.

5.7 The Action-Explanation Argument Revisited

Now let's reconsider the problem of explaining actions based on illusions of thought. From the perspective of LOT, the solution is straightforward: Maeve utters 'John is abominable' because of the presence of 'JOHN IS ABOMINABLE' in her belief box, its lack of content notwithstanding. There is no need to appeal to second-order thoughts at all, and hence no need to appeal to them too extensively. And the presence of 'JOHN IS ABOMINABLE' in Maeve's belief box seems like a significantly *better* explanation of her utterance of 'John is abominable' than the second order thought that *that was a thought*, even if she has that second-order thought. In general, the presence of nonsense beliefs$_S$ (nonsense Mentalese sentences in one's belief box) allows us to explain how nonsense functions in our mental economy without constant appeal to second-order thought: nonsense thoughts$_S$ (beliefs$_S$, desires$_S$, etc.) exist, and so can play their normal role as springs of action, even though they have no content.

Things are more complicated on the unmediated picture of belief, according to which illusions of thought don't involve *any* (relevant) beliefs$_S$ *or* beliefs$_C$. On that view, we can't say that Maeve's beliefs$_S$ are playing their normal role in

[24] Hence, resisting the argument doesn't require us to reject *any* view about the truth-conditions of belief reports.

bringing about her behavior. The obvious fallback position is to appeal to the belief_S-like cognitive states—*failed* beliefs_S—that give rise to Maeve's illusions of thought in our explanation of Maeve's speech and deeds. But since these belief_S-like cognitive states will not be beliefs_S, this story will be less elegant and straightforward than the one that adherents of LOT are able to give. When it comes to the Action-Explanation Argument, then, LOT theories seem to have an advantage over unmediated conceptions of belief.

5.8 O'Brien's Objection

Are there objections to LOT theories that blunt this advantage? Lucy O'Brien (2009) gives an argument against a LOT-like account,[25] contending that it doesn't provide a satisfactory account of the cognitive significance of illusions of thought. She begins by considering a case involving the veridical perception of two similar glasses, where one engages in distinct episodes of successful demonstrative thought about each, which the LOT theorist would gloss as 'THAT GLASS IS HEAVY' appearing twice in one's belief box. (Call these two token Mentalese sentences T1 and T2.) Despite their lexical similarity, the reason T1 and T2 count as two beliefs, rather than one, is that their *contents* are different: those Mentalese sentences have different truth-conditions, since the tokens of 'THAT' that appear in them denote different glasses.

O'Brien goes on to consider a phenomenologically indistinguishable case where one merely *hallucinates* the two glasses, and hence where there is no demonstrated content to distinguish T1 and T2. Nonetheless, "it seems to [one] that she is thinking two thoughts and it seems to her that they are distinct thoughts." (O'Brien 2009: 227) Here, T1 and T2 have no content, and so cannot be distinguished by their contents. Hence, it seems that LOT cannot explain the fact that the subject is suffering from two illusions of thought, rather than one.

It's worth noting that a similar problem might arise for names: if Maeve knows *two* horrible people named 'John', she might have two tokens of the LOT sentence 'JOHN IS ABOMINABLE' in her belief box, which "count as" two beliefs since those instances of 'JOHN' refer to different people (and so have different contents). But Maeve could also suffer from two illusions of thought she would express with the sentence 'John is abominable'. In that case, we could not distinguish the LOT sentences in her belief box by their contents, since *ex hypothesi*, they don't have any. Thus, LOT may not seem to be able to explain the fact that Maeve is suffering from two illusions of thought, rather than one.

[25] Essentially a "silently rehearsing sentences to oneself" view. In what follows, we have adapted her argument to apply directly to LOT.

5.9 The Indexing Response

The response to both the demonstrative and nominal versions of this objection is, we think, the same: indexing. If names are *singular* terms, they cannot refer to more than one thing. Hence, it is common to think that there is not one English name, 'John', but indefinitely many homonymous names spelled J-O-H-N.[26] If that's right, we should presumably say the same thing about LOT names: we don't have two instances of 'JOHN', but one instance of '$JOHN_1$' and another of '$JOHN_2$'. There is obviously no problem distinguishing '$JOHN_1$ IS ABOMINABLE' from '$JOHN_2$ IS ABOMINABLE'.

A similar "indexing approach" to natural-language demonstratives is endorsed by Kaplan, and by Fodor when it comes to Mentalese demonstratives.[27] If this is correct, the response to O'Brien's demonstrative case is the same as it was in the case of names: '$THAT_1$ GLASS IS FULL' and '$THAT_2$ GLASS IS FULL' are distinct LOT sentences (composed of distinct LOT words), and so there's no need to appeal to their (missing) contents to distinguish them.[28] And if that's right, LOT theories maintain their advantage over unmediated conceptions of belief with regard to the Action-Explanation Argument.

6. Conclusion

We have argued that while deceptive nonsense and illusions of thought exist, they are not as bad as they might seem. Some authors suggest that illusions of thought involve ignorance of what is going on in our heads, of what mental states we are in and whether we are even thinking at all. While we're as pessimistic as anyone about the extent of human ignorance, we have argued that *this* sort of ignorance is not, in fact, entailed by illusions of thought: such objections to illusions of thought are much ado about nothing. We *do* know we're thinking. We might even know *what* we're thinking. We just don't always know what, if anything, we are thinking about.[29]

[26] For example, Kripke says that "uses of phonetically the same sounds to name distinct objects count as distinct names', at least 'for theoretical purposes'." (Kripke 1980: 8) This view is endorsed in Kaplan (1990), and elsewhere.

[27] See e.g. Kaplan (1989), and Stojnić and Lepore (2020) for discussion.

[28] Stojnić and Lepore (2020) raise significant objections to the Kaplan-Fodor strategy, but they ultimately offer a response to those objections on Fodor's behalf that is friendly to our appeal to LOT in explaining the cognitive episodes underlying illusions of thought.

[29] To avoid any confusion, we intend the previous three sentences to communicate the following claims: we (reliably) know *whether* there are sentences in our belief, desire, etc. boxes; we might even (reliably) know *what* sentences are in our belief, desire, etc. boxes; but we less reliably know what, if any, *contents* those sentences have.

Acknowledgments

Thanks to Audre Brokes, Nathaniel Bulthuis, Herman Cappelen, Joseph Corabi, Todd Moody, Andrew Payne, and Brice Wachterhauser for helpful comments and discussion.

References

Ayer, A. J. (1952). *Language, Truth, and Logic*, 2nd edn (New York: Dover).

Baier, Annette (1967). "Nonsense", in *The Encyclopedia of Philosophy*, ed. P. Edwards, 8 vols (New York: Macmillan), vol. 5, pp. 520–2.

Bell, David (1988). "Phenomenology, Solipsism, and Egocentric Thought", *Proceedings of the Aristotelian Society Supplementary Volume* **62** (1): 45–60.

Bentham, Jeremy (1843). "Anarchical Fallacies", in *The Works of Jeremy Bentham*, vol. 2, ed. John Bowring (Edinburgh: William Tait), pp. 489–534.

Braun, David (1993). "Empty Names", *Noûs* **27** (4): 449–69.

Braun, David (2015). "Wondering About Witches", in *Fictional Objects*, ed. Stuart Brock and Anthony Everett (Oxford: Oxford University Press), pp. 71–113.

Bulthuis, Nathan (2020). "Walter Burley on Co-Signification in Opaque Contexts", in *Oxford Studies in Medieval Philosophy*, vol. 8, ed. Robert Pasnau (Oxford: Oxford University Press), pp. 221–47.

Cappelen, Herman (2013). "Nonsense and Illusions of Thought", *Philosophical Perspectives* **27**: 22–50.

Cappelen Herman and Ernie Lepore (2005). *Insensitive Semantics: A Defense of Semantic Minimalism and Speech Act Pluralism* (Oxford: Blackwell).

Carnap, Rudolf (1959). "The Elimination of Metaphysics through Logical Analysis of Language", in *Logical Positivism*, ed. A. J. Ayer, trans. Arthur Pap (New York: The Free Press), pp. 60–81.

Diamond, Cora (1981). "What Nonsense Might Be", *Philosophy* **56** (215): 5–22.

Evans, Gareth (1982). *The Varieties of Reference* (Oxford: Oxford University Press).

Fodor, Jerry (1975). *The Language of Thought* (Cambridge, MA: Harvard University Press).

Frege, Gottlob (1956). "The Thought: A Logical Inquiry", *Mind* **65** (259): 289–311.

Heck, Richard (2002). "Do Demonstratives Have Senses?", *Philosopher's Imprint* **2** (2): 1–33.

Heidegger, Martin (1929). "Was Ist Metaphysik?" [What Is Metaphysics?], inaugural addeess to the combined faculties of the University of Freiburg, July 24.

Kaplan, David (1989). "Demonstratives: An Essay on the Semantics, Logic, Metaphysics, and Epistemology of Demonstratives and Other Indexicals", in *Themes from*

Kaplan, ed. Joseph Almog, John Perry, and Howard Wettstein (New York: Oxford University Press), pp. 481–563.

Kaplan, David (1990). "Words", *Proceedings of the Aristotelian Society, Supplementary Volumes* **64**: 93–119.

Karger, Elizabeth (1996). "Mental Sentences According to Burley and to the Early Ockham", *Vivarium* **34** (2): 192–230.

Keller, John A. (2017). "Introduction", in *Being, Freedom, and Method: Themes from the Philosophy of Peter van Inwagen*, ed. John A. Keller (Oxford: Oxford University Press), pp. 1–8.

Kripke, Saul A. (1980). *Naming and Necessity* (Cambridge, MA: Harvard University Press).

Magidor, Ofra (2013). *Category Mistakes* (Oxford: Oxford University Press).

McDowell, John (1986). "Singular Thought and the Extent of Inner Space", in *Subject, Thought, and Context*, ed. Phillip Pettit and John McDowell (New York: Oxford University Press), pp. 137–68.

O'Brien, Lucy (2009). "Mental Actions and the No-Content Problem", in *Mental Actions*, ed. Lucy O'Brien and Matthew Soteriou (Oxford: Oxford University Press), pp. 215–30.

Pryor, James (2006). "Hyper-Reliability and Apriority", *Proceedings of the Aristotelian Society* **106** (1): 329–46.

Quine, W.V. (1960). *Word and Object* (Cambridge, MA: Harvard University Press).

Recanati, François (1997). "Can We Believe What We Do Not Understand?", *Mind and Language* **12** (1): 84–100.

Russell, Bertrand (1956). "The Philosophy of Logical Atomism", in *Logic and Knowledge*, ed. Robert Marsh (London: George Allen & Unwin), pp. 175–282.

Schiffer, Stephen (1981). "Truth and the Theory of Content", in *Meaning and Understanding*, ed. Herman Parret and Jacques Bouveresse (New York: De Gruyter), pp. 204–22.

Segal, Gabriel (2000). *A Slim Book About Narrow Content* (Cambridge, MA: MIT Press).

Soames, Scott (2005). "Naming and Asserting", in *Semantics vs. Pragmatics*, ed. Zoltan Szabó (Oxford: Oxford University Press), pp. 356–82.

Stojnić, Una and Ernie Lepore (2020). "Fodor and demonstratives in LOT", *Theoria* **35** (1): 75–92.

Van Inwagen, Peter (1980). "Philosophers and the Words 'Human Body'", in *Time and Cause: Essays Presented to Richard Taylor*, ed. Peter van Inwagen (Dordrecht: D. Reidel Publishing Company), pp. 283–99.

Van Inwagen, Peter (2017). "Concluding Meditation", In *Being, Freedom, and Method: Themes from the Philosophy of Peter van Inwagen*, ed. John A. Keller (Oxford: Oxford University Press), pp. 343–94.

Wikforss, Åsa (2007). "Semantic Externalism and Psychological Externalism", *Philosophy Compass* 3 (1): 158–81.

Williamson, Timothy (2000). *Knowledge and Its Limits* (Oxford: Oxford University Press).

Wittgenstein, Ludwig (1953). *Philosophical Investigations*, trans. G. E. M. Anscombe (Oxford: Basil Blackwell).

Wittgenstein, Ludwig (1961). *Tractatus Logico-Philosophicus*, trans. D. F. Pears and B. F. McGuinness (London: Routledge and Kegan Paul).

14

Why It Matters What Might Have Been

Arif Ahmed

What there actually is exhausts all things; there are not in addition things that do not but might have existed. What is actually the case exhausts all facts: there are no further matters of fact that do not obtain but might have. And yet we do often think, and care, about what might have existed or obtained even if it doesn't, in fact even when we know that it doesn't. This essay asks why it matters that we entertain these *counterfactual* thoughts. Sections 1 and 2 give the background. Section 3 denies that thinking about counter-facts helps us learn *actual* facts. Sections 4 and 5 argue that counterfactual thought makes us more cautious in the face of risk.

1. Counterfactuals and the State of Nature

Much philosophical work on counterfactuals concerns their semantics, i.e. the conditions in which they are true or false (Stalnaker 1987: ch. 8; Lewis 1973), acceptable or unacceptable (Mackie 1980: 52–4) or have this or that *chance* (Edgington 2004). One payoff of this work is that we now have a better idea of their truth- (/acceptability-) conditions than when Goodman's classic paper first raised these questions (Goodman 1947). Another is that we can pose sharply a question that Goodman didn't raise: why do they exist?

This question connects to the semantic issue because the latter says *what it is* that wants explanation. We want to explain a conditional with just *these* truth-conditions/acceptability-conditions/chances: . . . (fill the dots with your favourite semantics). The need for some explanation is especially acute if the semantic theory makes counterfactuals about circumstances that at best do not actually exist and at worst do not exist at all.

For instance, Lewis's best-known semantics (1979: 44) makes the counterfactual a variably strict conditional reflecting a lexicographic partial ordering of possible worlds that certainly have no *actual* existence and may have no existence at all. Why should there be generally available, even in languages with no single marker of it,[1] a thought that is sensitive to this ordering of such things?

[1] For instance in Chinese: see Au 1984.

Arif Ahmed, *Why It Matters What Might Have Been* In: *Non-Being: New Essays on the Metaphysics of Non-Existence.* Edited by: Sara Bernstein and Tyron Goldschmidt, Oxford University Press (2021). © Arif Ahmed.
DOI: 10.1093/oso/9780198846222.003.0014

What might help is a state-of-nature explanation: one that vindicates a concept, practice, or institution on the grounds that only *it* meets certain general human needs or desires, so we should not be surprised at its existence or ubiquity. Philosophers have defended such explanations of the institutions of justice (Hume 1739–40: III.ii.2) and the state (Nozick 1974) and of the concepts of knowledge (Craig 1990) and truth (Williams 2002). Here I discuss a state-of-nature explanation of this particular kind of 'thought about the non-existent'.

Explanations of this sort vary over whether they take the practice in question (a) to have been adopted *deliberately* by practitioners who were both aware of and motivated by the need to which it answered,[2] (b) to have arisen quite independently of anyone's intentions,[3] (c) to lie somewhere in between—for instance, having been deliberately adopted for one end, a practice might survive because it serves another.[4] But when the practice is a precondition of thought, or certain kinds of thought, (a) might seem impossible (Williams 2002; Quine 1936); and anyone proposing (b) or (c) would have to say what evolutionary pressures would fix in a population those concepts, institutions, or practices that no member of it is actively seeking to preserve. I won't address those problems here.

The narrower purpose of this paper is to meet an important *necessary* condition on any such explanation. If the explanation of counterfactual thought is that it meets some general need then it must be true that it meets some general need. If it's true and explanatory that it meets some general need then it must also be true that without them that need would have gone unsupplied. So it must be true that we'd have been worse off without counterfactuals—worse off in some way for not having the capacity to think about certainly non-actual circumstance. So the capacity to think about the non-existent makes *some* difference for good or ill. What difference?[5]

2. Causal Counterfactuals and the Function of Modal Thought

To explain counterfactualizing is to explain the existence of judgments with the content that the preferred semantic theory determines. Here I'll specify a semantic theory and underline the prima facie oddity of a practice of judgments about the corresponding contents.

[2] See e.g. Hobbes (1640: V.1) on the origin of language.

[3] See e.g. Nozick (1974: 18–22) on 'invisible hand' explanations.

[4] See e.g. Weber (1992: 122–3) on the ascetic character of modern labour.

[5] For discussion of this whole problem area, and for a different answer, see Divers (2010); for a third answer see Price (2013).

2.1 Edgington's Semantics for Counterfactuals

Let A and C be propositions describing possible events at dates T_A and T_C respectively. If A is false, consider the last pre-T_A time at which some otherwise inconspicuous—and possibly miraculous—deviation from the actual course of events would, if it had occurred, have brought about A. Call this the time of the fork, T_F. If A is true let T_F be just before T_A. Then consider the counterfactual A > C ('If it had been the case that A then it would have been the case that C'). According to Edgington:

> The objectively correct [probability] to assign to such a counterfactual is... the conditional chance, at [T_F], of C given A ∧ S, where S is a conjunction of those facts concerning the time between antecedent and consequent which are (a) causally independent of the antecedent and (b) affect the chance of the consequent. (Edgington 2004: 21)

For instance, let an indeterministic coin be tossed at T_A, at which time I call heads, A being the false proposition that I call tails. Let the coin land tails at T_C, at which time also I lose the bet, C being the false proposition that I win it. T_F is some time shortly before my bet (perhaps it is the last time at which the miraculous firing of a neuron in my brain would have caused me to call tails instead of heads). The correct probability for A > C is the conditional chance, at T_F, of my winning the bet given (A) that I call tails *and* (S) that the coin lands tails, the latter being an actual event that is causally independent of how I bet. So the semantics recommends high confidence in A > C. This agrees with intuition.

Two asides. First, note that this rule is independent of the truth-value of A itself. So the semantics diverges from the more standard Lewis semantics, which equates A > C with A ∧ C when A is true. Our rule allows A > C to get a low value even though A ∧ C is true. For instance, if C describes some actual but chancy event whose occurrence was causally dependent on that of another actual event A. But we can set that aside—nothing in the main argument depends upon it.[6]

Second, the rule hardly covers all English uses of 'If it had been the case that ___ it would have been the case that ___' and related constructions. For one thing, we often slip into contexts where the assessment of A > C depends on how things are in possible situations that differ significantly from actuality *before* T_A or T_F. One

[6] Lewis (1973: 26–31). Edgington's view thus agrees with Bennett's Irrelevance Principle (2003: 240). Edgington nowhere explicitly signs up to this point. But it is implicit in her discussion: for instance, she frequently uses the semantic rule as stated to evaluate H > E in cases where we don't know H's truth-value; presumably it is meant to apply whether H is false *or* true (2004: 23–4). I agree with Edgington but for an argument against the Irrelevance Principle based on other principles that Bennett himself accepts, see Walters (2009) and the discussion in Ahmed (2011) and Walters (2011).

might create such a context by using or even mentioning some other counterfactual B > D, where T_B is significantly earlier than T_A.

For instance, suppose I start wondering aloud *whether Hoover would have become a communist if he had been born in Russia*. I thereby create a wholly new context for assessing whether, *if Hoover had become a communist he would have betrayed his native country*. I create a context for assessing the second conditional in which the time of the relevant conditional chances is that of Hoover's actual birth, not that of his non-actual conversion (Wright 1983).

But here I'll assume with Lewis that notwithstanding this possible interpretation, there *is* a standard interpretation of counterfactuals to which we revert absent cues to the contrary, and whose semantics roughly conforms to Edgington's theory just outlined (Lewis 1979: 34).

Whether or not that theory agrees everywhere with intuition, it does so in most central cases and therefore suffices for present purposes. For it's close enough to being right to make it plausible that those basic needs (if any) that our *actual* counterfactualizing practice supplies are those that Edgington-style counterfactualizing would also supply. From now on I'll reserve the term *causal counterfactual* for the conditional that the Edgington semantics governs. And I'll use *subjunctive conditional* for the type of construction with which English speakers most usually express it.

2.2 Zero-Tolerance and Modal Thought

The most obvious puzzle about causal counterfactuals from the present perspective is that we routinely apply serious standards of rightness and wrongness to ones whose antecedents describe certainly non-existent situations.[7]

What manifests this is that causal counterfactuals are sometimes *zero-tolerant*: sometimes A > C is true (or not obviously false) even though A itself is known false. A (possibly controversial) type of example would be where A describes a priori contingencies (Kripke 1980: 54–7). The mass of the standard kilogram cylinder in Paris (supposing there is one) is 1 kg. That cylinder might have had a slightly greater mass. And it cannot be settled trivially or a priori whether, if it had now weighed (say) 1.01 kg, we should still have used it as the standard kilogram. How things would have been in that case depends on the (in this instance uncertain) application of Edgington's criteria; for all *I* know things might have gone either way. But we know for certain that we are in fact *not* in that case—that the standard cylinder (if there is one) weighs exactly 1 kg.

[7] This may even apply to causal counterfactuals with causally or metaphysically impossible antecedents (see e.g. Bernstein 2016), although my argument doesn't turn on that.

If you doubt that the antecedent there is both certainly and contingently false, it's easy to imagine other cases. It is a real question whether Napoleon would have abdicated in 1814 if he hadn't invaded Russia in 1812, though we know for a fact that he did both. It is a real question whether Kennedy would have been assassinated in November 1963 had the Cuban missile crisis not occurred in October of the previous year, though again we know that it did and he was. These causal counterfactuals illustrate zero-tolerance.

The fact that some causal counterfactuals are zero-tolerant implies that some say nothing *at all* about the actual world, in the sense that their semantic value depends on chances conditional on circumstances that are certainly not actual. But this makes especially pressing the question of why we should care. Why care about what certainly does not exist?

Blackburn once raised a similar question:

> The possibilities we allow or which we rule out, determine how we conduct our inferences, and eventually our practices in the actual world. It is quite inexplicable how they should do that, if we relied upon the image of different spheres of real facts or states of affairs. Why should it interest me if in such spheres something holds, or in some it does not, or if in ones quite similar to the actual world things do or do not hold, if I want to discover and use truths about the way things actually are? (Blackburn 1984: 214–15)

That question isn't quite mine, but two points of comparison may put mine into focus.

(i) In a way Blackburn's problem looks more general: he is expressing puzzlement about *modal talk in general*, not only its (causal) counterfactualizing species. But maybe it is only through causal counterfactual talk that modality entered into the thoughts of non-philosophers, and the more general and abstract notion of possibility and necessity is a philosophical abstraction. After all, we can analyze 'bare' modality in counterfactual terms (Williamson 2007: Appendix 1), and maybe we can analyze modal belief in terms of counterfactual beliefs (McFetridge 1990: 153–4). So the present enquiry may have the *same* scope as Blackburn's.

(ii) In another way Blackburn's problem seems narrower; for what concerns him is apparently not the point of modalizing as such, but the point of modalizing *given that* the semantically relevant sets of possible worlds are 'different spheres of real facts' as in Lewis's analysis (1973: ch. 4). Whereas the problem about counterfactualizing arises from any semantics that entails zero-tolerance, including the Edgingtonian one in terms of past conditional chances, which doesn't mention possible worlds at all. Again the difference is less than it seems: what troubles Blackburn is still troubling if you reject 'possible worlds'. Whether or not *they* exist, modal and counterfactual talk are frequently about possibilities that everyone *knows* not to obtain (e.g. about what would have happened had Napoleon not invaded Russia in 1812 or had Kennedy not been assassinated in 1963) and so is in *that* sense not about anything real. *This* fact, on which Edgington's and Lewis's semantics agree, is what causes the problem. That 'what might have been is an abstraction, remaining a perpetual possibility, only in a world of speculation' is a problem for *any* philosopher of modality, not only a Lewisian realist.

3. The Epistemic Approach

3.1 Updating on Counterfactual Knowledge

I'll start with Edgington's 2004 solution to the problem. She writes (2004: 23) that "we use counterfactuals in empirical inferences to conclusions about what is [actually] the case". Consider two examples.

(1) You are driving one evening in the dark close to the house of some friends and have considered paying a visit. You turn the corner. "They are not at home", you say, "for the lights are off. And if they had been at home the lights would have been on." (Edgington 2004: 23)

(2) It is suspected that a patient, Jones, was poisoned. 'If he had been poisoned he would have shown exactly the symptoms that he actually does show', says the doctor. 'So he probably was poisoned.' (Cf. Anderson 1951: 37)

In both cases—in (1) by modus tollens, in (2) by abduction—the subjunctive functions as a premise in an argument whose point is to extend our knowledge of the actual world.

Edgington writes that these two forms of inference are in fact one. This general description covers both: there is some hypothesis of interest H, e.g. that they are at home. You get some new evidence E. And you update your beliefs on that evidence by answering two questions: (a) How likely is it that E would have been true if H had been true? (b) How likely is it that E would have been true if H had been false? More particularly: if Cr is a probability function representing your prior degrees of belief then the relation between your prior odds O (H) and your rational posterior odds O_{new} (H) in H is as follows:

(3) $O_{new}\ (H) = O\ (H)\ Cr\ (H > E)\ /\ Cr\ (\neg H > E)$[8]

[8] Edgington 2004: 24. The *odds* that a credence function Cr gives a proposition H are defined as Cr (H) / Cr (•H).

Thus in (1) the explicit premise "If they had been in the lights would have been on" expresses your low Cr (H > E). Assuming that your Cr (¬H > E) is not also low (as it might be if e.g. the lights are on a timer), you are justified in becoming more confident that they are not in. And in case (2) the explicit premise "If he been poisoned he would have shown exactly the symptoms that he actually does show" expresses the doctor's high Cr (H > E). Assuming that Cr (¬H > E) is not also high (as it would be if e.g. the victim had developed these symptoms before the supposed poisoning), the doctor is justified in becoming more confident that it was arsenic poisoning than he formerly was. This is Edgington's account of the point in asserting and thinking about subjunctive conditionals with these semantics.

3.2 Causal Counterfactuals Unsuitable for This

There isn't any doubt that we do use subjunctive conditionals in this way. What *is* in doubt is whether the ones that we use that way express *causal* counterfactuals. There are two reasons to doubt it.

First, their application to instances of (3) cannot explain the zero-tolerance of causal counterfactuals. Observe first that in (1) and (2), in fact in all of the applications by which Edgington illustrates (3), H itself is never *known* false or *known* true. That is the point: H is a *hypothesis* and you evaluate H > E and ¬H > E so as to improve your opinion on whether H *is* true. Next, observe that in (1) and (2), and again in all of Edgington's other examples, E *is* known true. Again that is the point: E constitutes your *evidence* and you evaluate H > E and ¬H > E in order to update your opinion on the basis of E.

But as we saw, the causal counterfactual *is* zero-tolerant: its instances often have a truth-value or an objective chance in which one takes an interest even *after* one has come to know that the antecedent is false, in which case one's prior O (H) is zero, so that no amount of reasoning from any amount of empirical evidence could change one's opinion about O via the mechanism that Edgington describes. So its role in (3) cannot *explain* why the causal counterfactual is zero-tolerant.

The second reason for doubt is that the semantics of the causal counterfactual is *incompatible* with the use of it that (3) sets out. Roughly, this is because the background evidence that supports rational updating on a hypothesis H essentially includes propositions that *would not have been true if H had been true*, not if that subjunctive expresses a causal counterfactual.

For instance: Edward and Harry are completely unrelated but happen to share a gene that determines eye colour. But I don't know what colour it is. So I don't know whether Harry has blue eyes, and he is not around for me to check. But Edward *is* around, and I see that he *does* have blue eyes. 'They have the same gene, so if Harry had had blue eyes then so would Edward; and if Harry had not had blue eyes then neither would Edward. So Harry must have blue eyes.' That looks perfectly rational.

But according to Edgington's semantics there is something wrong with it: we should *not* take Edward's eye-colour to confirm Harry's. To see this, let H be the proposition that Harry has blue eyes and let E be the proposition that Edward has blue eyes.[9] According to Edgington our prior Cr (H > E) should aim at the objective chance, at around the time when Harry's eye colour was settled (e.g. shortly after his conception), that Edward has blue eyes conditional on Harry's having blue eyes, given whatever is causally independent of the latter. Similarly our prior Cr (¬H > E) should aim at the objective chance, at that same time, that Edward has blue eyes conditional on Harry's *not* having blue eyes, again holding fixed everything that is causally independent of the latter.

But Edward's eye colour is *itself* causally independent of Harry's. They have (we can imagine) no common ancestor and there's no other common cause of their having the same gene, let alone any direct causal connection between them. (There was not, for instance, any single source of radiation that promotes the gene and to which both were prenatally exposed.) So Edgington semantics equates *both* one's prior Cr (H > E) *and* one's prior Cr (¬H > E) with one's prior Cr (E). Hence by (3) one's new odds on H should be identical to one's old odds on H. So on Edgington's own theory the evidence that Edward has blue eyes does *nothing* to confirm the hypothesis that Harry does.

This consequence is absurd and invalidates Edgington's approach. In particular it shows that if there *is* a conditional that conforms to her semantics then *its* use cannot be to rationalize updating in the face of new evidence.[10] But since there *is* such a causal counterfactual, the question remains what *difference* its existence

[9] Plainly H and E don't describe *events* but states of affairs. Still, if Edgington's semantics doesn't allow for this then it is plainly inadequate. Obviously we do (and should) use subjunctive reasoning to update our credences in propositions describing states of affairs as much as we use it to update our credences in those describing events.

[10] Although the argument based upon this counterexample has been applied to the pure Edgingtonian approach, on which the chance-based semantics apply to H > E whether or not H is true, it works equally well on the alternative Lewis-style approach to true-antecedent cases. On that approach $H \wedge (H > E)$ is equivalent to $H \wedge E$ for any H and E (so Bennett's Irrelevance Principle fails—see n. 6 above).

Applying this semantics to the example requires that we decompose each of Cr (H > E) and Cr (¬H > E) as follows:

(i) $Cr\ (H > E) = Cr\ (H > E \mid H)\ Cr\ (H) + Cr\ (H > E \mid \neg H)\ Cr\ (\neg H)$
(ii) $Cr\ (\neg H > E) = Cr\ (\neg H > E \mid H)\ Cr\ (H) + Cr\ (\neg H > E \mid \neg H)\ Cr\ (\neg H)$

In (i) the Lewis semantics applies to the first summand on the RHS and the Edgington semantics to the second; conversely in (ii). And since the conditional chance of E on H is just the unconditional chance of E, this simplifies those clauses to which the Edgington semantics *does* apply. Combining these points:

(iii) $Cr\ (H > E) = Cr\ (H \wedge E) + Cr\ (E)\ Cr\ (\neg H)$
(iv) $Cr\ (\neg H > E) = Cr\ (\neg H \wedge E) + Cr\ (E)\ Cr\ (H)$

According to (3) the ratio of (iii) and (iv) is the factor for updating one's odds on H on learning that E.

But this cannot be right: (3), (iii), and (iv) jointly set an upper limit on one's posterior odds on H given one's priors in E and H that makes us too slow to learn from experience: one's new odds on H cannot exceed one's posterior odds by a factor that itself exceeds $(1 + Cr\ (E)\ Cr\ (\neg H)) / Cr\ (E)\ Cr\ (H)$.

makes to us. That is, why does it matter that we follow the discipline of the Edgingtonian semantics when thinking about non-existent events and states of affairs? what follow are my own speculations on this.

4. Counterfactuals and Regret

The first connection that I want to highlight is between retrospective feelings about decisions and the counterfactuals describing how things would have gone had we acted otherwise. Regret is the most obvious such feeling: you'd regret that you called heads if (and to the extent that) you now think that you'd have done better by calling tails. There is also the opposite of regret, for which 'relief' is perhaps the best word: I mean the feeling that you'd get about doing such-and-such when reflecting how much *worse* things would have been (for you now) had you acted otherwise.

Clearly regret is negative and (what I'm calling) relief is positive. Furthermore, the strength of these feelings is an increasing function of the *distance* in value terms between actual and counterfactual scenarios. Suppose I am sitting an exam and am torn between two answers, A and B. I go for A, but it should be B, so I fail. How much I then regret writing A depends on what turns on the exam. If it has no external consequences, then I'll feel mild regret. I am comparing the actual situation, in which I write A and fail, to a counterfactual one, in which I write B and pass but am no better off. But if it is, say, my final chance at an important qualification then I might greatly regret it—because now the relevant comparison is (for example) between a counterfactual situation in which I am a qualified engineer and the actuality that I never will be.

One thing that counterfactuals tell us is *what* possible situation to compare with actuality for purposes of regret and relief. I regret writing A in the second version of the story more than in the first because only in the second version do I think that, had I written B instead of A, I'd now be qualified. In the first version things would not have been much different.

Far from making them inexplicable, this function of counterfactual thought practically requires that counterfactuals concern non-existent situations. After making one choice I normally know that I have made it and not another. So of

On the other hand, as we approach prior certainty that E and H are both true if either is, then our posterior odds on H (that is, after we learn E) should increase without limit.

For instance, suppose that you have prior odds of 4:1 that Harry has blue eyes, and also that Edward has blue eyes. Then as we approach certainty that they have the same gene (and hence the same eye colour), learning that Edward has blue eyes ought to make us practically certain that Harry does too—the posterior odds on H should approach infinity. But as is easily checked, (3), (iii), and (iv) imply that one's posterior odds should never exceed 6:1.

course to compare the actual situation with one in which I chose differently is to compare the former with one that I *know* is not actual.

The function is also compatible with the Edgingtonian semantics. According to it, A > B depends (roughly) on the chance of B given A, back when A had a chance of occurring. This means that when A describes some action or performance, what *counterfactually* depends on A depends (roughly) on what *causally* depends on it, regardless of our background knowledge of what actually happened. The function of counterfactuals that I am highlighting entails that you'll regret an action A if, at the time of doing it, the *chance* of some preferable-to-the-actual outcome, conditional on some alternative A* and on everything causally independent of A*, was high. And this more or less fits the facts. We regret our actions for their causal *effects* and not because of other causally independent events that co-vary with them (such as their own causes).

The story so far is therefore that the subjunctive expresses *two* types of counterfactual proposition. One, which we might call *epistemic*, plays the role in updating on new evidence that (3) describes. That it plays this role is (I argued) incompatible with its having the Edgingtonian semantics.

A second type of counterfactual, what I am calling the *causal* counterfactual, *does* fit the Edgington semantic. Its role is to guide not beliefs but feelings. It determines when and how much you regret doing something. Only creatures that have causal counterfactual thoughts can share our reasons for feeling regret and relief.

But this story is superficial because it raises without answering two further questions. (i) Why is it the causal counterfactual, rather than anything else, that governs the apportioning of regret? One could imagine worlds of any of the following types:

(a) Worlds at which nobody regretted anything that they did.
(b) Worlds at which everyone regretted everything that they did.
(c) Worlds at which everyone regretted just those actions that were *evidence* of undesirable states. For instance, you might not remember where you spent last night, but you regret smoking a cigar because it is evidence that you spent it somewhere disreputable.
(d) Worlds at which everyone regretted just those actions that were themselves (causally) counterfactually dependent upon undesirable states or events e.g. upon unbecoming motives.

Here, (a)–(d) describe ways in which regret might have lacked its actual connection to the causal counterfactual. In (a) and (b) it's unconnected to any counterfactual; in (c) it stands to (what I called) the epistemic counterfactual in the same relation as it actually stands to the causal counterfactual; in (d) it stands in a non-actual relation to the causal counterfactual. So why doesn't the allocation of regret accord with these, or with any of many other, possibilities?

(ii) What is the point of regret? How does it contribute to human flourishing that people do often have this feeling? It seems futile because retrospective. *Regret* that one has done A can have no effect on its object because what A describes cannot be undone; whereas (e.g.) the *hope* that B and the *fear* that C might help to bring about B or to prevent C.[11]

What connects (i) and (ii) is that our rules for assigning regret have something to do with the point of the feeling. In conclusion then, I want briefly to address this—i.e. the question about its point.

5. Regret and Risk-Aversion

My answer turns on the fact that a prevailing practice can make a difference not only directly but also indirectly by virtue of general belief in it. These two kinds of effect sometimes work in opposite directions. For instance, raising the prison sentences for some crimes has a direct effect of increasing the prison population. But general belief that the sentence is longer may have the opposite effect: it may deter some people who would risk a shorter sentence. Of course there is no way of saying a priori which effect is dominant.

Regret is retrospective. But its *indirect* effect is prospective: before making a choice one might expect to regret one or another option. If so this is a disincentive to perform it. How strong a disincentive should it be, for rational agents? Some philosophers have argued that it is *always* irrational to do something that one expects later to regret (see e.g. Arntzenius 2008: 277).

This seems too strong, for two reasons. First, it wrongly counts *both* options irrational if knows that one will regret whichever one chooses. To take Kierkegaard's example: suppose that you are deciding whether or not to marry and know that whatever you choose you will regret not choosing otherwise. Does this mean that you will choose irrationally?

Second, it wrongly counts as irrational any decision taken in the knowledge that one's future self will accept values that would rule it out. In Nagel's example (1970: 74 n. 1), a young man who now values spontaneity and creativity above status and security might know that in twenty years' time his preferences will reverse. Still, there is nothing irrational about his now choosing to be a painter rather than (say)

[11] You might say that to regret doing something might motivate you not to do it again. But that only holds if you already know that doing X again will again produce the undesired outcome Y or make it more likely. But if you already know that then *prospective* feelings—your desire to avoid Y—will already motivate you not to do X again, and then the regret is redundant in this connection. Creatures that feel no regret (and grasp no counterfactuals) can still learn from experience.

a doctor. If somebody objects that he will almost certainly regret it in twenty years, he might retort that that is because in twenty years he will adhere to values that he now holds in contempt; and which values should concern him *now* if not his *present* ones?

So the anticipated cost of regret need not be prohibitive. But anticipation of regret still makes a difference to decision under *risk*.

You can toss a coin at a fixed stake: if you toss it and it lands heads you win your stake; if you toss it and it lands tails you lose your stake. Or you can walk away. For an agent who is incapable of regret or relief—or who doesn't *expect* to feel either—the foreseeable reward of winning is the stake and the foreseeable penalty for losing is also the stake. But for an agent that expects one of these feelings, the foreseeable reward of winning is the stake *plus* the relief that one gambled. And the foreseeable penalty of losing is the stake *plus* the regret that one gambled. These additional rewards and penalties add to the risk involved in the gamble.

Now for the more general story. People are in general risk averse when it comes to money. This means that for any fixed amount $\$x$ they would prefer not gambling at all to a 50/50 bet that wins or loses $\$x$. (That is why insurance companies exist and make profits.) It also means that this preference becomes stronger as x increases:

(4) As x increases a typical person gets *more* averse to a 50/50 gamble on which they stand to win or lose $\$x$.

Now suppose a person expects to feel regret if they bet and lose, and relief if they bet and win. In particular, suppose that if one has a 50/50 gamble that wins or loses $\$n$, then the regret associated with losing is equivalent to a loss of a further $\$rn$—this represents the additional disutility of that feeling—and the relief associated with winning constitutes is equivalent to winning a further $\$rn$, for some constant $r > 0$. Then for someone who expects to feel regret or relief, a 50/50 gamble of $\$x$ will look to them like a 50/50 gamble of $\$x(1 + r)$ to someone who does not expect to feel regret or relief. The expectation of these feelings increases the apparent size of a gamble. It follows from (4) that the anticipation of regret and relief will make such a person even *more* risk-averse than they would have been had they not expected to have these feelings. (For formal details see the appendix.)

All this applies to gambles other than coin tosses. The capacity to feel regret will make agents less willing than they otherwise would be to stake large amounts on any gamble (e.g. investing in shares). And it applies to gambles in which the stakes and rewards are not monetary. The capacity to feel regret will in general make risk-averse agents even *less* willing to invest large amounts of time and effort in adventures. In short: the effect of regret is to make conservative agents even more conservative.

So it makes a difference that we can think about what would happen in circumstances that we *know* are non-existent. In particular it makes a difference that we care about chances conditional on certainly non-actual circumstances, or about what goes on at certainly non-existent possible worlds. The difference is that creatures that can think these modal thoughts can feel regret; and creatures that can feel regret will expect to feel regret; and risk-averse creatures that expect to feel regret are more risk-averse. In short, caring about these things makes us *more* cautious than we would otherwise have been in our *actual* dealings with the *real* world.

6. Conclusion

Probably the most I'd want to claim for this proposal is that it represents an improvement on the epistemic solution. On the other hand (i) both proposals raise questions that I have not discussed in this paper, and (ii) my own proposal has problems of its own.

(i) The proposal is empirically disconfirmable. Disconfirmation might take the form of psychological studies of the ontogenesis of counterfactual judgment, of regret and of risk aversion. If it turned out, for instance, that children both felt and anticipated regret long before they were able to make counterfactual judgments, or if it turned out that feelings of regret had no tendency to make them more risk-averse, then these would be strong grounds against my view. So this is an obvious candidate for the next line of enquiry.[12]
(ii) I have spoken only of the *difference* that causal counterfactuals make. I haven't said anything about why making this difference might benefit anyone. Why (or when) *is* it a good idea to take a more cautious attitude toward gambles, and what (reproductive or other) benefits exactly does this confer?[13]

There are also features of the causal counterfactual that my proposal leaves unexplained. For instance, it seems to make it mysterious that our standards for assessing counterfactuals seem independent of whether the antecedent describes what was ever under anyone's control: 'If such-and-such asteroid had not struck

[12] Much of the (extensive) psychological literature on *ex post* counterfactual judgments focuses on their role in helping agents learn from mistakes in similar future cases—e.g. Roese (1994); see Byrne (2002: 427) for general discussion. There is some evidence that anticipated regret influences decisions: see e.g. McConnell et al. (2000). There is evidence (Guttentag and Ferrell 2008) that children do not anticipate regret until around age 10, at around which time they are generally able to reason counterfactually (Perner and Rafetseder 2011).

[13] For one possible answer see Galor and Michalopoulos (2012).

the Earth then the dinosaurs would have survived for millions of years'. Still, this is not *only* a problem for my story about counterfactuals: the well-known manipulationist account of causation faces analogous concerns. And it may be that a similar development, in terms of either (a) a historical hypothesis about the projection of agency onto inanimate entities, or (b) some form of analogical reasoning by which we extend the sphere of our judgments in either case, is promising in both cases.[14] But if the proposal deserves elaboration at all then these are two areas where the need for it is most pressing.

Appendix: Regret and the Attitude towards Risk

Assume first a regret-*free* agent A with twice-differentiable increasing utility function u for dollar amounts and gambles over dollars satisfying the standard axioms[15] and hence also:

(1) If a gamble G gives the agent net wealth $\$x$ with probability $p_G(x)$, then $u(G) = \sum_x u(x)p_G(x)$

Assume, as is fairly realistic over a reasonably wide range of numbers (Joyce 1999: 35 n. 19), that A is risk-averse, i.e. that A's marginal utility is everywhere decreasing, so for any x within the range of possible dollar prizes:

(2) $u''(x) < 0$

Now consider gambles G_p^x, $x \geq 0$, where G_p^x wins a stake $\$x$ with fixed probability p and loses it with fixed probability $1 - p$. For simplicity we assume (without loss of generality) that A's initial wealth is \$0. Let M_A^p be the *maximum* (or supremum) of the positive stakes that A prefers to not betting at all given probability p of winning.
By (1) the utility of G_p^x, call it $v(x)$, is:

(3) $v(x) =_{def.} u\left(G_p^x\right) = u(x)p + u(-x)(1 - p)$

By definition A is indifferent between M_A^p and not gambling; so:

(4) $M_A^p > 0$ by definition

(5) $v\left(M_A^p\right) = v(0)$

It follows from (3) and (4) respectively that for any x in the range of dollar prizes:

(6) $v''(x) = pu''(x) + (1 - p)u''(-x)$

(7) $v'(0) > 0$

It follows from (4), (5), and (7) that some $y \in \left(0, M_A^p\right)$ satisfies $v'(y) = 0$ and hence by (2) and (6) that:

(8) $v'\left(M_A^p\right) < 0$

[14] On the manipulationist theory see Woodward (2003), esp. ch. 3. On the explanatory roles of projection see Russell (1912); on the role of that and/or analogical reasoning see Menzies and Price (1993).

[15] i.e. the von Neumann–Morgenstern axioms: see Kreps (1988: 43–4).

Now consider an agent B like A in all respects except that B expects to feel regret and relief. For B, winning \$$x$ on a gamble will have the same utility (up to affine transformations) as A's utility of winning \$$(x + rx)$ for some (we assume for simplicity) constant $r > 0$, where rx represents *relief* if $x > 0$ and *regret* if $x < 0$. It follows that A's maximal stake M_A^p has utility *for B* of $v((1 + r)M_A^p)$; so by (5) and (8) that B's utility of not gambling, i.e. $v((1 + r).0)$ exceeds this. So B will *not* accept a gamble $G_p^{M_A^p}$ that A *will* accept, and in that sense can be said to take a more conservative attitude than A towards risk.

References

Ahmed, Arif (2011). "Walters on Conjunction Conditionalization", *Proceedings of the Aristotelian Society* **111** (1 pt 1): 115–22.

Anderson, Alan Ross (1951). "A Note on Subjunctive and Counterfactual Conditionals", *Analysis* **12** (2): 35–8.

Arntzenius, Frank (2008). "No Regrets, or: Edith Piaf Revamps Decision Theory", *Erkenntnis* **68** (2): 277–97.

Au, Terry Kit-Fong (1983). "Chinese and English Counterfactuals: The Sapir–Whorf Hypothesis Revisited", *Cognition* **15** (1–3): 155–87.

Bennett, Jonathan (2003). *A Philosophical Guide to Conditionals* (Oxford: Clarendon Press).

Bernstein, Sara (2016). "Omission Impossible", *Philosophical Studies* **173** (10): 2575–89.

Blackburn, Simon (1984). *Spreading the Word: Groundings in the Philosophy of Language* (Oxford: Clarendon Press).

Byrne, Ruth M. J. (2002). "Mental Models and Counterfactual Thoughts about What Might Have Been", *Trends in Cognitive Sciences* **6** (10): 426–31.

Craig, Edward (1990). *Knowledge and the State of Nature: An Essay in Conceptual Synthesis* (Oxford: Clarendon Press).

Divers, John (2010). "Modal Commitments", in *Modality: Metaphysics, Logic and Epistemology*, ed. Bob Hale and Aviv Hoffman (Oxford: Oxford University Press), pp. 189–219.

Edgington, Dorothy (2004). "Counterfactuals and the Benefit of Hindsight", in *Cause and Chance: Causation in an Indeterministic World*, ed. Phil Dowe and Paul Noordhof (London: Routledge), pp. 12–27.

Galor, Oded and Stelios Michalopoulos (2012). "Evolution and the Growth Process: Natural Selection of Entrepreneurial Traits", *Journal of Economic Theory* **147** (2): 759–80.

Goodman, Nelson (1947). "The Problem of Counterfactual Conditionals", *Journal of Philosophy* **44** (5): 113–28.

Guttentag, Robert and Jennifer Ferrell (2008). "Children's Understanding of Anticipatory Regret and Disappointment", *Cognition and Emotion* **22** (5): 815–32.

Hobbes, Thomas (1640). *Human Nature* [repr. in *Human Nature and De Corpore Politico*, ed. J. C. A. Gaskin (Oxford: Oxford University Press, 1994)].

Hume, David (1739–40). *A Treatise of Human Nature.*

Joyce, James M. (1999). *Foundations of Causal Decision Theory* (Cambridge: Cambridge University Press).

Kreps, David M. (1988). *Notes on the Theory of Choice* (Boulder, CO: Westview).

Kripke, Saul A. (1980). *Naming and Necessity* (Oxford: Blackwell).

Lewis, David (1973). *Counterfactuals* (Oxford: Blackwell).

Lewis, David (1979). "Counterfactual Dependence and Time's Arrow", *Noûs* **13** (4): 455–76.

Mackie, J. L. (1980). *The Cement of the Universe: A Study of Causation* (Oxford: Clarendon Press).

McConnell, Allen R., Keith E. Niedermeier, Jill M. Leibold, Amani G. El-Alayli, Peggy P. Chin, and Nicole M. Kuiper (2000). "What if I Find it Cheaper Someplace Else? Role of Prefactual Thinking and Anticipated Regret in Consumer Behavior", *Psychology and Marketing* **17** (4): 281–98.

McFetridge, I. G. (1990). "Logical Necessity: Some Issues", in *Logical Necessity and other Essays*, ed. John Haldane and Roger Scruton (London: Aristotelian Society), pp. 135–44.

Menzies, Peter and Huw Price (1993). "Causation as a Secondary Quality", *British Journal for the Philosophy of Science* **44** (2): 187–203.

Nagel, Thomas (1970). *The Possibility of Altruism* (Oxford: Clarendon Press).

Nozick, Robert (1974). *Anarchy, State and Utopia* (New York: Basic Books).

Perner, Josef and Eva Rafetsader. (2011). "Counterfactual and Other Forms of Conditional Reasoning", in *Understanding Counterfactuals, Understanding Causation: Issues in Philosophy and Psychology*, ed. Christoph Hoerl, Teresa McCormack, and Sarah R. Beck (Oxford: Oxford University Press), pp. 90–109.

Price, Huw (2013). "Where Would We Be Without Counterfactuals?", in *New Directions in the Philosophy of Science*, ed. Maria Carla Galavotti, Dennis Dieks, Wenceslau J. Gonzalez, Stephan Hartmann, Thomas Uebel and Marcel Weber (London: Springer), pp. 589–607.

Quine, W. V. 1936. "Truth by Convention", in *Philosophical Essays for A. N. Whitehead*, ed. O. H. Lee (New York: Longmans), pp. 90–124.

Roese, Neal J. (1994). "The Functional Basis of Counterfactual Thinking", *Journal of Personality and Social Psychology* **66** (5): 805–18.

Russell, Bertrand (1912). "On the Notion of Cause", *Proceedings of the Aristotelian Society* **13**: 1–26.

Stalnaker, Robert C. (1987). *Inquiry* (Cambridge, MA: MIT Press).

Walters, Lee (2009). "Morgenbesser's Coin and Counterfactuals with True Components", *Proceedings of the Aristotelian Society* **109** (1 pt 3): 365–79.

Walters, Lee (2011). "Reply to Ahmed", *Proceedings of the Aristotelian Society* **111** (1 pt 1): 123–33.

Weber, Max (1992). *The Protestant Ethic and the Spirit of Capitalism*, trans. Talcott Parson (London: Routledge; first published London: George Allen & Unwin, 1930).

Williams, Bernard (2002). *Truth and Truthfulness: An Essay in Genealogy* (Princeton, NJ: Princeton University Press).

Williamson, Timothy (2007). *The Philosophy of Philosophy* (Oxford: Blackwell).

Woodward, James (2003). *Making Things Happen: A Theory of Causal Explanation* (New York: Oxford University Press).

Wright, Crispin (1983). "Keeping Track of Nozick", *Analysis* **43** (3): 134–40.

15

Explanatory Relevance and the Doing/Allowing Distinction

Jacob Ross

One kind of negativity that plays an important role in our ordinary moral thinking is *non-doing*. If something bad happens, the simplest way for us to avoid blame is to maintain that *we didn't do it*. And it seems we can maintain this defense even in cases where we could have prevented the harm. For it is generally thought that there is an important distinction between doing a harm and merely allowing the harm to occur, and that the former is worse than the latter, or requires stronger reasons to justify. This commonsense view is known as the Doctrine of Doing and Allowing.[1] In this paper, I will focus on the distinction between doing and allowing that figures in this commonsense view. That is, I will ask how the distinction between doing and allowing should be understood insofar as we want to maintain that doing harm is worse than merely allowing it.

It seems that, when we do harm, rather than merely allowing it, the harm *depends* on us, or on our actions or behavior, in some distinctive way. But what kind of dependence is at issue here? In the first section of this paper, I will consider three possible answers to this question: I will consider the view that the dependence in question is *counterfactual*, the view that it is *probabilistic*, and the view that it is *causal*. And I will argue that none of these views is satisfactory. In the second section, I will argue that our judgments about doing and allowing are complex, and that they are sensitive to a number of factors that are hard to explain on standard conceptions of the distinction. Finally, in the third section, I will propose an alternative view of the doing/allowing distinction, which I will call the *explanatory relevance* view. I will argue that whether an agent counts as doing a given harm depends on the *explanatory connection* between this harm and the agent's *motivations*. I will argue that this view solves the problems facing the alternative views, and that it makes sense of the complex set of factors to which our judgments of doing and allowing are sensitive.

[1] See Rickless (1997) and Woollard (2012). Note that the doing/allowing distinction is not the same as the act/omission distinction. For a recent discussion of the latter, see Sartorio (this volume Chapter 16).

Jacob Ross, *Explanatory Relevance and the Doing/Allowing Distinction* In: *Non-Being: New Essays on the Metaphysics of Non-Existence.* Edited by: Sara Bernstein and Tyron Goldschmidt, Oxford University Press (2021). © Jacob Ross.
DOI: 10.1093/oso/9780198846222.003.0015

1. Three Kinds of Dependence

In this section, I will consider three attempts to understand the doing/allowing distinction in terms of some kind of dependence between the harms and our actions. On the first view I will consider, the harms we do are the harms that depend *counterfactually* on our actions. On the second view, the harms we do are the harms that depend *probabilistically* on our actions. And, on the third view, the harms we do are the harms that depend *causally* on our actions.

1.1 The Counterfactual Dependence View

In order to get a grip on the distinction between doing and allowing, let us begin by considering paradigmatic cases of each. Consider the following two cases:

> **Falling:** A and B are standing precariously at the very edge of a 100-foot cliff. A begins to fall and, to prevent this, A grabs hold of B's arm and pulls himself to safety, pulling B off the cliff and to her death in the process.
>
> **Stable:** Again, A and B are standing precariously at the very edge of a 100-foot cliff. This time it is B who begins to fall. A has the option of grabbing hold of B's arm and pulling her to safety, but in so doing A would be pulled off the cliff. And so A does nothing and B falls to her death.

In both cases, A faces a choice between one way of acting that would result in A's living and B's dying, and another way of acting that would result in A's dying and B's living. And, in both cases, A chooses to act in such a way that A lives and B dies. And yet it seems that, in Falling, A kills B, whereas, in Stable, A merely allows B to die. Consequently, it seems that A acts wrongly in the first case but permissibly in the second. And so this pair of cases illustrates the Doctrine of Doing and Allowing.

So what accounts for the difference between these cases? It seems that, in the first case, B's death depends counterfactually on A's action. That is, there is an action of A's, namely pulling B's arm, such that, had A not performed this action, B would not have died. By contrast, in the second case, there is no such action. This suggests the following view:

> **But-For View:** S counts as doing harm H just in case harm H occurs and there is some action ϕ such that, had S not done ϕ, harm H would not have occurred.

Unfortunately, this view suffers from a problem of overdetermination. Suppose there is more than one action that would result in a given harm, and if the agent hadn't performed one of these actions, he would have performed another. As an illustration, consider:

> **Resourceful Terrorist:** a terrorist detonates a bomb, and destroys a hospital, by pulling a trip line. But if he hadn't pulled the trip line, he would have activated the remote, which would likewise have detonated the bomb and destroyed the hospital.

In this case, at least on a natural understanding of how to individuate actions, there may be no action performed by the terrorist such that, if he hadn't performed it, the hospital would not have been destroyed. And yet surely he counts as doing rather than allowing the harm.

We can solve this problem by moving to the following, revised view:

> **Inaction-Default View:** S counts as doing harm H just in case harm H occurs and there is some interval t such that, had S done nothing during t, H would not have occurred.[2]

This view, however, suffers from a similar overdetermination problem. Recall that the problem with the But-For View was that the harm under consideration might be overdetermined by the *agent's* action—for it may be that, if the agent hadn't done the action in question, she would have done something else resulting in the same harm. Similarly, the problem with the Inaction-Default View is that the harm under consideration might be overdetermined by the actions of *other agents*—for it may be that, if the agent under consideration hadn't done anything to bring about the harm, someone else would have. Consider, for example:

> **Terror Duo:** There are two terrorists, Terrence and Theresa, each of whom can detonate the bomb and destroy the hospital by activating a remote. Terrence's remote has a 10-second delay whereas Theresa's remote has a 5-second delay. Terrence activates his remote at noon, thereby detonating the bomb at 10 seconds after noon. However, if Terrence hadn't done so, Theresa would have activated her remote 5 seconds after noon, thereby detonating the bomb at 10 seconds after noon.[3]

In this case, it seems that Terrence destroys the hospital, in spite of the fact that the destruction would have occurred even if Terrence had done nothing.

In this example, the harm done by Terrence doesn't depend counterfactually on his action because of preemption. There are other cases where the harm we do

[2] A similar view can be found in Donagan (1977).
[3] Similar cases are discussed in Woollard (2012).

does not depend counterfactually on our action because it depends on our *inaction*. Consider:

> **Bad Uncle:** An earthquake knocks giant Gian onto his one-year-old niece Nancy who was crawling on the floor next to him. Gian so enjoys the sound of his niece gasping for breath that he continues to lie on her, not moving a muscle, until she suffocates to death.[4]

In this case, it seems clear that Gian kills Nancy. However, it is not the case that, had he done nothing, she would have lived. To the contrary, it is precisely *because* he does nothing at the time in question that she dies.

Thus, it seems that the Inaction-Default View fails to provide necessary conditions for doing harm. For we can do harm even if the harm would have occurred had we done nothing. Morever, this counterfactual view also fails to provide sufficient conditions for doing harm. For in some cases in which a harm depends counterfactually on our actions, we count as merely allowing it rather than as doing it. As an illustration:

> **Raging Bull:** A bull is charging toward Charlie. Standing between Charlie and the bull is Wayland, who runs out of the way, moments before the bull gores Charlie. If, however, Wayland had stayed put, the bull would have gored Wayland en route to Charlie, and Charlie would have had time to escape.[5]

In this case, if Wayland had done nothing, the harm to Charlie would have been avoided. And yet it seems clear that Wayland does not *do* this harm to Charlie; he merely allows it to occur.

Thus, it seems that if the harms we do depend on our actions in a distinctive way, this way cannot be counterfactual dependence. Perhaps the relevant kind of dependence is instead probabilistic. We will briefly consider this view in the next subsection, before turning to a more plausible alternative.

1.2 The Probabilistic Dependence View

Perhaps what distinguishes the harms we do from the harms we merely allow is that the harms we do depend *probabilistically* on our actions in a distinctive way. Thus, one might hold that we count as doing harms when our actions raise the probabilities of these harms, or raise these probabilities to a sufficient degree.

[4] A somewhat similar case is discussed in Quinn (1989).
[5] Similar cases can be found in Woollard (2015).

Unfortunately, this view runs into many of the same problems as the counterfactual view. Like the counterfactual view, the probabilistic view gets the wrong results in preemption cases, such as Terror Duo. For, in Terror Duo, Terrence clearly destroys the hospital, and hence does harm to the hospital, in spite of the fact that the hospital's destruction would have been just as likely to occur if he had done nothing. And so it seems that, in pressing the button, Terrence does harm to the hospital without raising the probability of this harm.

Second, like the counterfactual dependence view, the probabilistic view gets the wrong results in cases like Raging Bull, in which an agent *actively allows* a harm to occur which would not have occurred if the agent had done nothing. In moving out of the way of the bull, Wayland significantly raises the probability of harm to Charlie, and yet it seems that Wayland merely allows this harm to occur.

In addition, the probabilistic view faces further problems that are not shared with the counterfactual dependence view. One problem arises from what we may call *inert probability raisers*. Consider:

> **Fuse Malfunction:** There is a bomb set up to destroy a hospital, and this bomb has two fuses: one reliable fuse that has a 1% chance of malfunctioning, and one unreliable fuse that has a 99% chance of malfunctioning. At the very same moment, Theresa lights the reliable fuse and Terrence lights the unreliable one. Strangely, the reliable fuse lit by Theresa malfunctions whereas the unreliable fuse lit by Terrence works and activates the bomb.

In this case, Theresa does no harm to the hospital, and yet her action significantly raises the probability of its destruction.

A final problem facing the probabilistic dependence view is what we may call the *directionality problem*. For any two events A and B, A raises the probability of B whenever $Prob(B|A) > Prob(B)$, which is true whenever $Prob(A|B) > Prob(A)$. In other words, the probability-raising relation is symmetric: if A raises the probability of B, then B raises the probability of A. And this creates problems for the probabilistic view of the doing/allowing distinction, as can be illustrated by the following case:

> **Ironing Incident:** Suppose Ira likes ironing his clothes. He is very skillful at doing so, and almost never burns his finger. However, he keeps a bowl of water on the ironing board, and, on the very rare occasions when he does burn his finger, he dunks his finger into the water to cool it off. Under no other circumstances does he dunk his finger in the water.

In this case, the unconditional probability of Ira's burning his finger is very low, whereas the probability of his doing so conditional on his dunking his finger into the water is 100%. Thus, Ira's dunking his finger into the water significantly raises the probability of his burning his finger. And yet the harm to his finger was not done by dunking his finger into the water.

Thus, the probabilistic dependence view, like the counterfactual dependence view, faces numerous difficulties. Perhaps we can solve these problems by moving to the causal dependence view.

1.3 The Causal Dependence View

What do all the counter-examples we have considered have in common? It appears that they are all cases in which counterfactual or probabilistic dependence comes apart from *causal* dependence. Consider, first, the counter-examples to the counterfactual view. In Terror Duo, Terrance's behavior *causes* the destruction of the hospital, even though this destruction does not depend counterfactually on his action. Similarly, in Bad Uncle, Gian's behavior *causes* his niece's death, despite the fact that there is no action of his on which her death counterfactually depends. And, in Raging Bull, Charlie's death depends counterfactually on Wayland's action, and yet it seems that Wayland *does not cause* Charlie's death. Rather, the bull causes Charlie's death, and Wayland simply avoids using his causal influence to prevent Charlie's death.

Next, consider the additional counter-examples facing the probabilistic dependence view. In Fuse Malfunction, Theresa's lighting her fuse raises the probability of the hospital's destruction, but it *does not cause* this destruction. Similarly, in Ironing Incident, Ira's dunking his finger in the water raises the probability of his burning his finger, but *it does not cause* his finger to burn.

And so it seems that, in all the cases we've considered in which the counterfactual and probabilistic dependence views seem to get things wrong, they do so because these kinds of dependence come apart from causal dependence. And this suggests that the doing/allowing distinction should instead be understood in terms of causal dependence, as follows:

> **Causal Dependence View:** S counts as doing harm H just in case S causes H through S's voluntary behavior.[6]

Note that it's important to add the qualification "through S's voluntary behavior" in order to rule out cases like the following:

> **Hulkamania:** Hulk Hogan is fighting Andre the Giant and Sergeant Slaughter. Slaughter is already on the ground, and Hogan picks up Andre and body-slams him onto Slaughter, breaking Slaughter's ribs.

[6] Views of this kind are defended in Callahan (1989) and in Åqvist and Mullock (1989).

In this case, there is a clear sense in which Andre causes the breaking of Slaughter's ribs. Even so, since Andre's voluntary behavior plays no role in Slaughter's harm, Andre does not *do* the harm to Slaughter.

Prima facie, the causal view of the doing/allowing distinction seems very plausible. But I will argue that it must nonetheless be rejected. For a major problem with the causal view is that it cannot make sense of cases in which an agent *does* a harm by causing another agent to *merely allow* this harm. Consider, for example, the following case:

> **Heat Ray**: Hanna is hanging from a chain over the edge of a cliff. Helda is holding the chain and attempting to pull Hanna to safety. Heather, however, wants Hanna to fall from the cliff, and so she fires her heat ray at the chain near to where Helda is holding it. As the pain becomes very intense, Helda lets go of the chain, and Hanna falls to her death. (Assume that, while the pain is very intense, it doesn't make it *impossible* for Helda to continue holding the chain. Thus, assume that Hanna's release of the chain is under her voluntary control.)

It seems clear that, in this case, Heather kills Hanna by causing Helda to let Hanna die. But the causal theorist cannot allow for this. For, according to the causal dependence view, to kill Hanna is simply to cause Hanna's death. Hence, on this view, if Heather kills Hanna by causing Helda to let her die, then Heather must *cause Hanna's death* by causing Helda to let her die. But for this to be true, there would have to be a chain of causation proceeding from Heather's firing the heat ray to Helda's letting go of the chain to Hanna's death. Hence, Helda's letting go of the chain would need to cause Hanna's death. And this would imply, according to the causal view, that Helda killed Hanna. Thus, the causal dependence theorist cannot consistently maintain that Heather killed Hanna by causing Helda to let her die.

The Heat Ray case creates problems not only for the simple causal dependence view of the doing/allowing distinction that we have been discussing, but also for more sophisticated views with a similar structure. For example, it creates problems for the view proposed by Fiona Woollard that "an agent counts as doing harm if and only if a fact about the agent's behavior is part of the sequence leading to the harm" (Woollard 2015: 35). Woollard provides a very sophisticated account of what it is for a given fact to count as "part of the sequence leading to the harm." But the details of this account don't matter for our purposes. For, regardless of how this sequence is defined, the Heat Ray case presents Woollard's view with a dilemma.

Here's the dilemma. Either Helda's releasing the chain is part of the sequence leading to Hanna's death, or it isn't. If it *is* part of the sequence, then a fact about Helda's behavior is part of the sequence leading to Hanna's death, and so Woollard's view implies that Helda killed Hanna, contrary to intuition.

Suppose, on the other hand, that Helda's releasing the chain *is not* part of the sequence leading to Hanna's death. Now there is no plausible candidate for "the sequence leading to Hanna's death" that includes Heather's firing the heat ray but that does not include Helda's releasing the chain. Consequently, if Helda's releasing the chain is not part of the sequence leading to Hanna's death, then Heather's firing the heat ray is likewise not part of this sequence. And so Woollard's view has the counter-intuitive implication that, in behaving as she does in this scenario, Heather does not kill Hanna.

I have argued that we should reject the causal dependence view of the doing/allowing distinction because it does not allow for the possibility of doing a harm by causing someone to merely allow this harm. In the next section, I will argue that the causal dependence view faces another problem as well. The problem is that, if the causal dependence view were true, then whether an agent does a harm would depend solely on whether the agent's voluntary behavior caused the harm. Consequently, only factors bearing on the causal relation between the agent's behavior and the harm could affect the doing/allowing distinction. However, as I will argue presently, whether an agent counts as doing harm depends on number of other factors.

2. Anomalous Factors Influencing our Doing/Allowing Judgments

In what follows, I will argue that our judgments of doing and allowing are affected by a number of factors that don't seem to bear on the causal relationship between the agent's behavior and the harm—or, as I will put it, these judgments are affected by a number of *anomalous factors*. Of course, from the fact that our intuitive judgments about doing and allowing are affected by these factors, it doesn't follow that the doing/allowing distinction *itself* is affected by these factors. It could be, after all, that some of our intuitive judgments are mistaken. Still, other things being equal, a theory of the doing/allowing distinction that is consistent with more of our judgments is preferable to a theory that is consistent with fewer of our judgments. Consequently, outlining the factors that affect our intuitive judgments about doing versus allowing will be helpful in formulating and evaluating theories of this distinction.

In order to argue that a given factor influences our intuitive judgments about doing and allowing, I will consider pairs of cases, C1 and C2, that differ with respect to the factor in question, and that are as similar as possible in other respects. In each case, I will argue that the agent seems more like she is *doing* the harm in C1 than in C2 and, conversely, that the agent seems more like she is *merely allowing* the harm in C2 than in C1. In saying this, I do not mean to be making the non-comparative claim that the action in C1 *seems like a doing*, whereas the action in C2 *seems like an allowing*. For different readers may evaluate the cases differently. Thus, some readers may be inclined to classify both as cases of doing, while others may be inclined to evaluate both as cases of allowing. Even so, in saying that the action in C1 seems *more like a doing*, and that C2 seems *more like an allowing*, I mean to imply that even readers who classify both as doings will be less confident in C2's being a doing, and that even readers who classify both cases as mere allowings will be less confident in C1's being a mere allowing. If this is right, then it will follow that most readers' judgments of doing and allowing are indeed influenced by the factor in question, even if this influence is not decisive in the case in question.

I will now consider five factors affecting our judgments of the doing and allowing that are anomalous, in the sense that they don't seem to bear on the question of whether the agent's behavior caused the harm. Here is one such factor:

2.1 Independently Prohibited Behavior

One factor that seems to influence our judgments of doing versus allowing concerns whether the action in question is independently prohibited. When I say that an action is *independently prohibited*, I mean that it violates a moral requirement, and that it does so independently of its relation to the harm question. In general, other things being equal, we are more inclined to judge that an agent *does* a given harm in behaving in a given way if the behavior in question violates a moral requirement independently of its relation to the harm. As an illustration, consider the following pair of cases:

> **Belongs to Victim:** A, B, and C each have a deadly illness. There is only one available dose of the medicine needed to survive the illness, and it belongs to C. It's too late for C to save herself, and so, with her dying words, she says "I bequeath this medicine to B." In defiance of C's wishes, A takes the medicine himself, and B dies.
>
> **Belongs to Agent:** As in the first case, except, this time C bequeaths the medicine to A. In accordance with C's wishes, A takes the medicine, and B dies.

In Belongs to Victim, since A steals from B the medicine B needs to survive, it seems more like A kills B. By contrast, in Belongs to Agent, since A simply takes her own medicine, it seems more like A lets B die. In Belongs to Victim, A's action is independently prohibited, since, independently of any harm that may result from it, it violates a requirement against stealing. And this violation of an independent moral requirement may explain why, in Belongs to Victim, we are more inclined to regard A as doing harm to B.

Of course, we could explain this difference by appealing to a much more specific difference between the two cases, namely, that in Belongs to Agent, A is using his own property whereas, in Belongs to Victim, he is using the Victim's property.[7] But there is reason to doubt that the property relations are what fundamentally matters. For consider a variant of Belongs to Victim in which, while the victim owns the medicine, she gives the agent permission to use it. In this case, we will be less inclined to judge that the agent kills the victim in using the medicine, since the agent doesn't violate any independent moral requirements. Conversely, consider a variant of Belongs to Agent in which, while the agent owns the medicine, he has promised not to use it himself but to allow the agent to use it. In this case, we will be more inclined to judge that the agent kills the victim in using the medicine, since the agent violates an independent moral requirement against promise-breaking. Hence, it appears to be the independent moral requirements, and not the property relations *per se*, that are doing the work.

Note, further, that whether an action is independently impermissible doesn't seem to have any bearing on whether this action causes a given harm. And so it appears to be an anomalous factor influencing our judgments of doing versus allowing.

2.2 Prior Obligation to Protect

A second factor that seems to influence our judgments of doing versus allowing harm concerns the agent's prior obligations to the victim. In particular, other things being equal, we are more inclined to judge that an agent harms a victim if the agent has *a prior obligation* to protect the victim from the harm in question. As an illustration, consider the following pair of cases:

Promise: A tells B that A is about to travel by boat to his vacation home in Antarctica. B wants to come along, but fears that she will die of starvation and exposure. And so B says "If I come with you, do you promise to provide food and shelter?" A agrees, and so B travels with A to his home in Antarctica. But when they arrive, A refuses to provide B with food and shelter. And so B dies.

No Promise: As before, A tells B that he is heading to his vacation home in Antartica, and B wants to come along. This time, however, B has the false and unwarranted belief that A will do precisely the opposite of whatever he says he will do. And so B says to A "If I come with you, do you promise to provide food and shelter?" Then A replies "I will *absolutely not* provide food and shelter." And, because of B's peculiar belief about A, B infers that A *will* provide food and

[7] For a sophisticated discussion of the relevance of ownership to the doing/allowing distinction, see McMahan (1993).

shelter. And so B travels with A to his home in Antarctica. However, As A warned B he would do, A he refuses to provide B with food and shelter. And so B dies.

In Promise, it seems more like A kills B, whereas, in No Promise, it seems more like A merely allows B to die. Note that, in both cases, A's behavior appears to stand in the same causal relation to B's death. In both cases, A makes an utterance, and this utterance motivates B to come with A to Antartica. And in both cases, upon their arrival, A refuses to provide B with food and shelter, and so B dies. The difference between the two cases is not causal, but moral. In Promise, A's promise to B generates a moral obligation to protect B from certain kinds of harm, and when this obligation is violated, it seems that A harms B. And so it seems that our judgments of doing and allowing are influenced by prior obligations to protect.

2.3 Intentionality of Harm

A third factor that seems to influence our judgments of doing versus allowing harm concerns the motivation of the agent. In particular, other things being equal, we are more inclined to judge that an agent does a given harm to a victim if the agent behaves in the way they do *in order* that the victim should undergo this kind of harm. As an illustration, consider the following pair of cases:

Unintended Harm: A is the grandson of B. One day, they are watching the news together, and they see a story about a grandmother and grandson who attempt to cross the English Channel together, with the grandson swimming and the grandmother holding onto him for support. Sadly, they were unsuccessful, and the grandmother perished. Fearing that his own grandmother might be tempted to try the same feat with him, A says to B: "That is definitely *not* something we should try!" Unfortunately, since B is highly countersuggestible, A's warning has the opposite of the intended effect, and B begs A to swim across the English Channel with her in a similar manner. A fears that he won't have the strength to pull his grandmother all the way across the Channel and that, without his support, she may die. However, since her heart is set on the journey, he reluctantly agrees. Just as he feared, as they cross the Channel, he eventually reaches the point where it seems extremely unlikely that he could make it the rest of the way supporting her, and where it seems like attempting to do so would result in both drowning. At that point, he abandons her and swims the rest of the way alone. Unable to swim on her own, B drowns.

Intended Harm: There is nothing A desires more than the death of his grandmother, B. But he doesn't want to be charged with murder, and so he needs to find a way to avoid being blamed for her death. One day, he finds the perfect solution. He and his grandmother are watching the news together, and they see a story about a grandmother and grandson who unsuccessfully attempt to cross the English Channel together. Hoping to implant the idea in her mind, he says "That is definitely *not* something we should try!" As he hoped, she takes the bait. Being highly countersuggestible, she begs him to swim across the English Channel with her. Feigning reluctance, he agrees. His plan is to abandon her during the journey and leave her to drown. However, he fears he may be observed, and so he decides he will abandon her only if and when he reaches the point where it seems extremely unlikely that he could make it the rest of the way supporting her, and where it seems like attempting to do so would result in both drowning. To his delight, he does indeed reach such a point. Hence, in accordance with his plan, he abandons B and swims the rest of the way alone. Unable to swim on her own, B drowns.

The first case seems more like a case of letting die, while the second seems more like a case of killing. Note that these two cases differ only with respect to A's motivations. In Intended Harm, A did what he did *in order* that B should die, whereas, in Unintended Harm, A did what he did for other reasons. Hence, our judgments of doing versus allowing harm seem to be affected by whether the agent intended the harm in question.

2.4 Deliberateness of Behavior

A fourth factor that seems to influence our judgments of doing versus allowing harm concerns the deliberateness of the behavior resulting in the harm. In particular, other things being equal, we are more inclined to judge that an agent does a given harm to a victim if the harm resulted from behavior that the agent engaged in deliberately rather than carelessly or accidentally. As an illustration, consider the following pair of cases:

Deliberate: A is told that, if he hunches his shoulders at any time in the next day, a bomb will go off killing B. And he knows that, if he set his mind to it, he could avoid hunching his shoulders by continuously focusing on keeping them back. However, he also knows that he will soon be auditioning for the role of the hunchback of Notre Dame. And so, in order to prepare for the audition, he hunches his shoulders, fully aware that doing so will result in the detonation of the bomb. Consequently, the bomb goes off, and B dies.

Careless: As in the previous case, A knows that, if he hunches his shoulders at any time in the next day, a bomb will go off killing B. And he also knows that he

> can avoid this by continuously focusing on keeping his shoulders back. As the day goes on, however, he begins thinking about a philosophy paper he is working on, and he becomes more and more absorbed by the dialectic. Eventually, while pondering a particularly thorny issue, he begins to hunch his shoulders. Consequently, the bomb goes off and B dies.

In Deliberate Action, it seems more like A kills B, whereas in Careless Action it seems more like A merely lets B die. And the relevant difference appears to be that, in Deliberate Action, A acts deliberately, whereas, in Careless Action, A acts carelessly or accidentally. And so this factor appears to influence our judgments about doing versus allowing harm.

2.5 Latitude

One further factor that seems to influence our judgments of doing versus allowing harm concerns the specificity of the type of action required to avoid the harm. In particular, other things being equal, we are less inclined to judge that an agent does harm to a victim if, in order to avert the harm, the agent would need to act in some highly specific manner. As an illustration, consider the following pair of cases:

> **Anything But "My Way":** A is a jazz singer with a repertoire of 100 songs. As A is walking to the stage to perform, a Mafioso warns him that a bomb will go off killing B unless A sings some song other than "My Way." Nonetheless, when A appears on stage, he sings "My Way." Consequently, the bomb goes off and B dies.
>
> **"My Way" or the High Way:** Like the previous case, except this time the Mafioso warns A that a bomb will go off killing B unless A sings "My Way." Nonetheless, when A appears on stage, he sings "New York, New York." Consequently, the bomb goes off and B dies.

The first case, where A insists on playing the one song he knows will result in B's death, seems more like a case of killing than the second case, where A simply refuses to play the one song demanded by the Mafioso. Here the relevant difference seems to be that there are far more ways in which the agent can avoid the harm in the first case than in the second. And so our judgments of doing and allowing appear to be sensitive to this kind of difference.

I have mentioned five anomalous factors that appear to affect our judgments of doing versus allowing harm. I expect, however, that there are many more anomalous factors affecting our judgments besides these. In particular, I believe that, when an agent acts in such a way that a harm results, other things being equal, we are more inclined to judge that the agent did the harm when any of the following obtain:

(i) The action required significant effort;
(ii) The action involved a change in the agent's behavior;
(iii) The action was unusual or non-standard;
(iv) At the time of the action, the agent was aware of the harm that would result from it;
(v) At the time of the action, the agent was aware of an alternative that would avoid this harm.

Furthermore, I believe that, other things being equal, we are *less* inclined to judge that the agent did the harm when:

(i) Preventing the harm would have required significant effort;
(ii) Preventing the harm would have required a change in the agent's behavior;
(iii) Preventing the harm would have required behaving in an unusual or non-standard way;
(iv) Preventing the harm would have been costly for the agent.

I expect the list of relevant factors could be extended still further. I will not, however, discuss any of these additional factors here. The five factors I have already discussed and illustrated suffice to show that our judgments of doing and allowing have a complexity that the causal view cannot account for. In the next section, I will propose an alternative view that does a better job accounting for them.

3. The Explanatory Relevance View

I will conclude by presenting and defending what I will call the *explanatory relevance* view of the doing/allowing distinction. In section 3.1, I will present an outline of this view. In section 3.2, I will discuss the relationship of explanatory relevance that figures in the formulation of this view. And in sections 3.3 and 3.4, I will argue that this view compares favorably to the other accounts of the doing/allowing distinction we have considered.

3.1 Outline of the View

I will argue that the doing/allowing distinction should be understood, not in terms of causal or counterfactual dependence, but rather in terms of *explanatory relevance*. Moreover, I will argue that, in determining whether an agent did a harm or merely allowed it to occur, while the relationship between the harm and the agent's action matters derivatively, what matters fundamentally is the relationship between the harm and the agent's *motivation*. More specifically, I will argue that, when an agent A performs an action or voluntary behavior B, then A counts as doing a harm H just in case:

(i) Harm H can be explained, in part, on the basis of A's motivational state M, through the mediation of behavior B.
(ii) Motivational state M plays a sufficiently major role in this explanation.

Or, to put this more simply, an agent counts as doing a given harm by voluntarily behaving in a given way just in case, through the mediation of this behavior, *the agent's motivational state is sufficiently explanatorily relevant* to the harm.

This proposal will require some unpacking. I understand *explanatory relevance* as a relationship that obtains between the values of two parameters. Thus, we might talk about someone's education level being explanatorily relevant to their income, or to the income of their children. By a *parameter* I mean any respect in which two objects can be similar or different. Some parameters, such as a height and weight, are one-dimensional, so that the value of such a parameter can be represented by a real number. Other parameters, such taste in music, are multi-dimensional, so that their values cannot be represented by a single number. Still, musical taste qualifies as a parameter, since it makes sense to talk about two individuals being more or less similar with respect to their musical taste.

To say that the value of one parameter (for one individual) is *explanatorily relevant* to the value of another parameter (for the same or for a different individual) is to say that the value of the first parameter plays a role in explaining the value of the second, or, in other words, that knowing the value of the first parameter contributes to the intelligibility or explicability of the value of the second parameter. Further, we can think of the *degree* of explanatory relevance of one parameter to another as the size or magnitude of the role that the first parameter plays in explaining the second, or, in other words, as the degree to which the knowledge of the first parameter contributes to the intelligibility or explicability of the second.

(For brevity, I will sometimes speak of *one parameter* as being explanatorily relevant to *another parameter*. But this should be understood to mean that *the value of this parameter for an individual* is relevant to *the value of the second parameter for an individual.*)

On the view of the doing/allowing distinction I am proposing, an agent counts as doing a given harm by behaving in a given way just in case, through the mediation of this behavior, the agent's motivational state is sufficiently explanatorily relevant to the harm. Since I am understanding explanatory relevance as a relationship that obtains between the values of parameters for individuals, it follows that, on the proposed account of the doing/allowing distinction, the motivational state of the agent, the behavior of the agent, and the harm undergone by the victim must each be thought of as the value of a parameter for an individual. Each of these parameters will be complex and multidimensional. The first parameter, *motivation*, will comprise all the desires, fears, and other motivational attitudes and dispositions of the agent. This can be thought of as a single parameter, since we can intelligibly speak of two agents as being more or less similar with respect to how they are motivated. The second factor, *behavior*, comprises all the movements and other bodily comportments that are under the voluntary control of the agent. Again, this can be thought of as a single parameter, since we can think of two agents as more or less similar in their behavior. And the third factor, *harm*, concerns the manner and degree to which a victim suffers a loss of welfare. Once again, this can be thought of as a single parameter, since two people can be more or less similar with respect to the loss of welfare they undergo.

In order to apply this account, we will need some principles for determining degrees of explanatory relevance.

3.2 Three Principles of Explanatory Relevance

In this subsection, I will propose three principles concerning the degree to which the value of one parameter is explanatorily relevant to the value of another. To motivate these principles, it will be helpful to consider some cases. Let's begin with the following pair:

> **Case 1:** Talia is six feet tall. Throughout her childhood, she was on an all-dairy diet. If she had been on any other diet, she would not have been nearly so tall. And, provided that she had been on an all-dairy diet, she could not have been much taller or shorter than 6 feet tall.
>
> **Case 2:** Talia is six feet tall. Throughout her childhood, she was on an all-dairy diet. However, no matter what kind of diet she had been on, she would have been very close to 6 feet tall.

It seems clear that Talia's diet plays a bigger role in explaining her height in Case 1 than in Case 2. For in Case 1, her diet seems to be making all the difference to her height, whereas in Case 2, it seems to be making very little difference. Thus, in Case 1, without knowing Talia's diet, you would have very little understanding of why she is six feet tall and not some very different height. By contrast, in Case 2, you could, in principle, fully understand why her height must be very close to six feet without knowing her diet.

To generalize from this example, for any two parameters P1 and P2 such that P2 is dependent on P1, let us say that P2 is *sensitive* to P1 to the extent that changes to the value of P1 would result in changes to the value of P2. If P2 is *highly sensitive* to P1, then *even small* changes in the value of P1 would result in large changes in P2. If P2 is *highly insensitive* to P1, then *no* change in P1 would result in large changes to P2. And if P2 is *moderately sensitive* to P1, then *only large* changes in P1 would result in large changes to P2.

In Case 1, Talia's height is highly sensitive to her diet, whereas in Case 2, Talia's height is highly insensitive to her diet. Hence, we can explain why Talia's diet is more relevant to her height in Case 1 than in Case 2 by appealing to the following principle:

> **Sensitivity Principle:** For any two parameters P1 and P2 such that P2 depends on P1, other things being equal, the more sensitive P2 is to P1, the more explanatorily relevant P1 is to P2.

Now consider a third case:

> **Case 3:** Talia is six feet tall. Throughout her childhood, she was on an all-dairy diet. If she had been on any other diet, she would not have been nearly as tall. However, she could easily have been on an all-dairy diet without being anywhere near six feet tall. For, if her diet were held fixed but the values of other parameters (e.g. rest, exercise, etc.) were altered, then her height would have been very different.

Once again, it seems that Talia's diet plays a bigger role in explaining her height in Case 1 than in Case 3. For, in Case 1, her diet on its own determines her height, whereas in Case 3 her diet is only one among a multitude of factors that combine to determine her height.

To generalize from this example, for any two parameters, P1 and P2, such that P2 is dependent on P1, let us say that P1 *secures* P2 to the extent that, holding fixed the value of P1, varying the values of other, independent parameters would have little effect on the value of P2. If P1 secures P2 to a *high degree*, then, holding P1 fixed, *even large* changes in the values of other parameters would not produce large changes to the value of P2. If P1 secures P2 to a *low degree*, then, holding P1 fixed, *even small changes* in the values of other parameters would result in large changes in the value of P2. And if P1 secures P2 to a *moderate degree*, then, holding P1 fixed, *only large* changes in the values of other parameters would result in large changes to the value of P2.

In Case 1, Talia's diet secures her height to a high degree, whereas in Case 3, Talia's diet secures her height to a low degree. Hence, we can explain why Talia's diet plays a bigger role in explaining her height in Case 1 than in Case 3 by appealing to the following principle:

> **Security Principle:** For any two parameters P1 and P2 such that P2 depends on P1, other things being equal, the greater the degree to which P1 secures P2, the more explanatorily relevant P1 is to P2.

A third important principle of explanatory relevance is the following:

> **Transitivity:** For any three parameters P1, P2, and P3, if P1 is highly explanatorily relevant to P2, and P2 is highly explanatorily relevant to P3, then P1 is highly explanatorily relevant to P3.

This principle allows for indirect explanation. If the value of a first parameter plays an important role in explaining the value of a second, and if the value of the second parameter plays an important role in explaining the value of the third, then the value of the first parameter will play an important, though indirect, role in explaining the value of the third.

These three principles will suffice for our purposes. I should, however, give three caveats before proceeding. The first is that these principles are not meant to *define* the explanatory relevance relation. To the contrary, in order to apply these principles, we must already know that these relations obtain. The second caveat is that none of these principles is meant to provide a necessary condition for explanatory relevance. I want to allow for the possibility that a first parameter could be explanatorily relevant to a second without the first parameter securing the second, and without the second parameter being sensitive to the first. The final caveat is that these principles are not meant to be exhaustive. There may well be other important principles governing the relation of explanatory relevance.

I will now put these principles into use in defending the explanatory relevance account of the doing/allowing distinction. I will do so in two steps. In section 3.3, I will argue that the explanatory relevance view shares the advantages of the causal view over the counterfactual and probabilistic views. And in section 3.4, I will argue that the explanatory relevance view solves the problems facing the causal view.

3.3 Why the Explanatory Relevance Account Shares the Advantages of the Causal View

We saw, in sections 1.1 and 1.2, that the counterfactual and probabilistic views face a number of counter-examples. And we saw, in section 1.3 that the causal view avoids these counter-examples. I will now argue that the explanatory relevance view can avoid these counter-examples in a similar manner, since, in the cases in question, we should expect the causal and explanatory relevance views to have similar implications.

The counter-examples to the counterfactual and probabilistic views are of two kinds. First, there are cases where these views falsely predict that the agent does harm, and, second, there are cases where these views falsely predict that the agent does not do harm. Counter-examples of the first kind, in which the theory falsely predicts that the agent does harm, include Raging Bull, Fuse Malfunction, and Ironing Incident. In each of these cases, it is clear that the agent does not do the harm either because the agent's behavior plays *no role whatsoever* in bringing about the harm (as in Fuse Malfunction and Ironing Incident) or because the agent's behavior plays *very little* role in bringing about the harm (as in Raging Bull). But if the agent's behavior plays little or no role in bringing about the harm, then the explanatory relevance of the agent's behavior to the harm will be very low. And since the agent's motives are explanatorily relevant to the harm only through the mediation of the agent's behavior, it follows that if the agent's behavior has very little relevance to the harm, then the agent's motivations cannot have a high degree of explanatory relevance to the harm. And so, in all these case, the explanatory relevance view will imply that the agent does not do the harm.

Next, consider the counter-examples in which the counterfactual and probabilistic dependence views falsely imply that the agent *does not* do the harm. These counter-examples include Terror Duo and Bad Uncle. In each of these cases, it is clear that the agent does the harm because, first, it is clear that the agent's behavior is a major cause of the harm, and, second, it is clear that the agent's behavior is under the agent's voluntary control. But in any case in which it is clear that the agent's behavior is a major cause of the harm, the agent's behavior will be highly explanatorily relevant to the harm. And in any cases in which it is clear that the agent's behavior is under the agent's voluntary control, the agent's motivations will be highly explanatorily relevant to the behavior. And if the agent's motivation is highly explanatorily relevant to the agent's behavior, and if the latter is highly explanatorily relevant to the harm, then it follows from the Transitivity Principle that the agent's motivation is highly explanatorily relevant to the harm. And so the explanatory relevance view implies that the agent does the harm.

Thus, in all the counter-examples to the counterfactual and probabilistic views, the explanatory relevance view agrees with the causal view, and so it gets the right results.

What are the cases in which the causal dependence and explanatory relevance views come apart? If both the connection between the motivation and the behavior, and the connection between the behavior and the harm, are sufficiently strong, then both theories will imply that the agent does the harm. And if either the connection between the motivation and the behavior or the connection between the behavior and the harm is sufficiently weak, then both theories will imply that the agent does not do the harm. Hence, in order for the two theories to come apart, either the motivation must have an intermediate degree of influence over the behavior or else the behavior must have an intermediate degree of influence over the harm. And it is in cases of these kinds, as we will see in the next section, that the explanatory relevance theory does a better job than the causal theory.

3.4 How the Explanatory Relevance View Solves the Problems Facing the Causal View

In sections 1.3 and 2, we saw two problems facing the causal view of the doing/allowing distinction. First, this view does not allow for the possibility of doing a harm by causing another agent to merely allow this harm. Second, this view cannot account for the variety of factors to which our judgments of doing and allowing appear to be sensitive. In this final subsection, I will show how the explanatory relevance theory can solve both these problems, beginning with the second.

For any two cases that each involve an agent, a voluntary behavior, and a harm, if the agent does the harm in the first case but the agent merely allows the harm in the second case, then it follows from the explanatory relevance view that the agent's motivations are more relevant to the harm in the first case than in the second. The principles of Sensitivity and Security allow us to distinguish two reasons why this might be so:

(1) The harm is more sensitive to the agent's motivation in the first case.
(2) The agent's motivation secures the harm to a greater degree in the first case.

Moreover, given that the agent's motivations are relevant to the harm through the mediation of the agent's behavior, we can distinguish three reasons why (1) might be true:

1A. The agent's *behavior* is more sensitive to the agent's *motivation* in the first case.

1B. The *harm* is more sensitive to the agent's *behavior* in the first case.

1C. There is a greater sensitivity of the *harm* to the agent's *motivation* in the first case, in a way that cannot be explained simply as the result of 1A or 1B or both together.

Similarly, we can distinguish three reasons why (2) might be true:

2A. The agent's *motivation* secures the agent's *behavior* more in the first case.

2B. The agent's *behavior* secures the *harm* more in the first case.

2C. The agent's *motivation* secures the *harm* more in the first case, in a way that cannot be explained simply as the result of 2A or 2B or both together.

Thus, these six possibilities represent six ways in which we could explain why, in a given pair of cases, the agent counts as doing the harm in the first case and as merely allowing the harm in the second case.

Typically, if possibility 2B obtains, then the two cases will differ with respect to the causal connection between the agent's behavior and the harm. Consequently, both the explanatory relevance view and the causal view can explain why the two cases differ with respect to the doing/allowing distinction. However, as I will now argue, if any of the other five possibilities obtains apart from 2B, then the explanatory relevance view has the upper hand. For I will argue that these five ways in which cases can differ with respect to the explanatory relevance of the motivation to the harm correspond to the five anomalous factors discussed in section 2. Thus, the explanatory relevance theory is able to explain the relevance of these factors where the causal theory cannot.

To see this, let us consider, in turn, the five anomalous factors discussed in section 2. recall first that, in that section, I argued for the following claim:

> **Latitude Principle:** Other things being equal, we are less inclined to judge that an agent does a given harm to a victim if, in order for the agent to avert this harm, the agent would need to act in some highly specific manner.

And recall that we illustrated this principle by comparing *Anything but "My Way,"* where the agent could have prevented the bomb from going off by singing almost any song in his repertoire, with *"My Way" or the High Way*, where the agent could have prevented the bomb from going off only by singing "My Way." Note that this pair of cases instantiates possibility 1B: the harm is more sensitive to the agent's behavior in *Anything but "My Way."* For, in this case, if the agent were to change which song he sings in any way whatsoever, the harm would have been averted. In general, the more latitude an agent has in preventing a harm, the smaller is the range of possible behaviors that would result in the harm, and so the smaller is the shift in behavior that suffices to prevent the harm. Thus, in general, where a harm results from an agent's behavior, the greater the agent's latitude in preventing the harm, the more sensitive is the harm to the agent's behavior. And so, in general, pairs of cases to which the Latitude Principle applies will be pairs of cases illustrating possibility IB: the harm will be more sensitive to the agent's behavior in the case where the agent has more latitude in preventing the harm. Thus, the explanatory relevance view can make sense of why latitude matters to the doing/allowing distinction.

A second claim I argued for in section 2 is the following:

> **Independent Prohibitions Principle:** Other things being equal, we are more inclined to judge that an agent does a given harm in behaving in a given way if the behavior in question violates a moral requirement independently of its relation to the harm.

Recall that we illustrated this principle by comparing *Belongs to Victim*, where the agent takes the medicine belonging to the victim, with *Belongs to Agent*, where the agent takes his own medicine. Note that this pair of cases instantiates possibility 1A: the behavior is more sensitive to the agent's motivational state in Belongs to Victim than in Belongs to Agent. For, in Belongs to Victim, any agent who cares more about avoiding theft than about any competing consideration would avoid taking the medicine. By contrast, in Belongs to Agent, any level of concern for avoiding theft is compatible with the agent's taking the medicine. And so a greater range of possible motivational states is compatible with taking the medicine in Belongs to Agent than in Belongs to Victim. It follows that, in Belongs to Victim, the behavior of taking the medicine is more sensitive to the agent's motivational state.

In general, other things being equal, if an action violates an independent moral requirement in a first case, but not in a second case, then the action will be compatible with a smaller range of motivational states in the first case than in the second. For, in the first case, the action will be incompatible with motivational states that give more weight to satisfying the moral requirement than to any competing consideration. Consequently, other things being equal, if an action violates an independent moral requirement, it will be more sensitive to the agent's motivations. And so, in general, pairs of cases to which the Independent Prohibitions Principle applies will be pairs of cases illustrating possibility 1A: the agent's behavior will be more sensitive to the agent's motivational state in the case where the behavior violates an independent moral requirement. Thus, the explanatory relevance view can make sense of why independent prohibitions matter to the doing/allowing distinction.

A third claim I argued for in section 2 is the following:

> **Prior Obligation to Aid Principle:** Other things being equal, we are more inclined to judge that an agent does a given harm to a given victim when the agent has a prior obligation to protect the victim from this kind of harm.

Recall that we illustrated this principle by comparing *Promise*, where the agent promises to provide aid and then fails to provide it, with *No Promise*, where the agent fails to provide aid after making no such promise. Note that this pair of cases instantiates possibility 1C: the harm is more sensitive to the agent's motivation in *Promise* than in *No Promise*, and this difference cannot be explained purely in terms of the greater sensitivity of the harm to the behavior or of the behavior to the motivation. The harm is more sensitive to the agent's motivation in *Promise* than in *No Promise* because, in *Promise*, anyone who cares more about promissory obligation than about any competing considerations would provide the promised aid. By contrast, in *No Promise*, any level of concern for promissory obligation is compatible with failing to provide the aid. Moreover, the greater sensitivity of the harm to the agent's motivational state in Promise cannot be explained purely in terms of the greater sensitivity of the harm to the agent's actual behavior or of the agent's actual behavior to the agent's motivational state. For, in *Promise*, it isn't just the agent's *actual* behavior that is compatible with only a narrow range of possible motivational states. Rather, *any behavior resulting in the harm* would be compatible with only a narrow range of possible motivational states.

In general, in any two cases in which an agent fails to avoid a harm, other things being equal, if the agent has a prior obligation to protect the victim from this kind of harm in the first case but not in the second case, then the harm is more sensitive to the agent's motivational state in the first case than in the second. And this greater sensitivity will be irreducible. For, in the case where the agent has the prior obligation to avoid the harm, *any behavior resulting in the harm* will be compatible with a smaller range of possible motivational states. And so, in general, pairs of cases to which the Prior Obligation to Aid Principle applies will be pairs of cases instantiating possibility 1C. Thus, the explanatory relevance view can make sense of why prior obligations to protect matter to the doing/allowing distinction.

A fourth claim I argued for in section 2 is the following:

> **Deliberateness of Behavior Principle:** Other things being equal, we are more inclined to judge that an agent does a given harm to a victim if the harm resulted from behavior that the agent engaged in deliberately rather than carelessly or accidentally.

Recall that we illustrated this principle by comparing *Deliberate*, where the agent deliberately hunches his shoulders, with *Careless*, where the agent does so carelessly. Note that this pair of cases instantiates possibility 2A: the motivation secures the behavior to a greater degree in *Deliberate* than in *Careless*. For, if the agent in the example is deliberately hunching his shoulders, then he has an intention whose function is to guide his behavior in such a way as to ensure that he hunches his shoulders. Thus, he monitors his behavior, and he is disposed to correct this behavior whenever it begins to deviate from his intention. Hence, holding fixed the agent's motivational state, including his intention to hunch his shoulders, he could not easily have failed to hunch his shoulders. And so his motivation secures his behavior to a high degree. By contrast, when the agent hunches his shoulders carelessly, there is no such mechanism in place to ensure that he hunches his shoulders. And so, other things being equal, he could more easily have failed to hunch his shoulders. Thus, his motivational state secures his behavior to a lesser degree.

In general, other things being equal, if an agent's behavior is deliberate in a first case but careless or accidental in a second case, then the agent's motivational state will secure the agent's behavior to a greater degree in the first case than in the second. Thus, in general, pairs of cases to which the Deliberateness of Action

Principle applies will be pairs of cases illustrating possibility 2A. Thus, the explanatory relevance view can make sense of why deliberateness of action matters to the doing/allowing distinction.

A final claim I argued for in section 2 is the following:

> **Intentionality of Harm Principle:** Other things being equal, we are more inclined to judge that an agent does a given harm to a victim if the agent behaves in the way they do in order that the victim should undergo this kind of harm.

Recall that we illustrated this principle by comparing *Intended Harm*, where the grandson did what he did in order that his grandmother should die, with *Unintended Harm*, where the grandson did what he did for other reasons. Note that this pair of cases instantiates possibility 2C: the motivation secures the harm to a greater degree in Intended Harm, and this is not simply because the motivation secures the behavior more or because the behavior secures the harm more. The motivation secures the harm more in Intended Harm because, if the agent acts *in order* that his grandmother should die, then he has an *aim* in place which involves a *general tendency* to act in ways that promote his grandmother's death. By contrast, if he isn't aiming at his grandmother's death then, other things being equal, he will have less of a general tendency to act in ways that promote her death, and so he could more easily have failed to act in such ways. Thus, his motivational state will secure the death of his grandmother to a greater degree in the case where he aims at her death.

Moreover, the greater degree to which the agent's motivation secures the harm cannot be explained purely in terms of the greater degree to which his motivation secures his behavior or his behavior secures the harm. For, having the aim that his grandmother should die needn't secure *any particular course of action*. If there are several alternative courses of action that would promote his grandmother's death to an equal degree, and that are otherwise equally desirable, then what his aiming at his grandmother's death will secure is not that he engages in *any one* of these courses of action in particular, but rather that he engages in some course of action or other that promotes his grandmother's death. Hence, his aiming at his grandmother's death may secure the harm to a greater degree than it secures any particular behavior.

In general, in any two cases in which an agent's behavior results in a harm, other things being equal, if the agent aims at the harm in the first case but not in the second then the agent's motivations will secure the harm to a greater degree in the first case than in the second, and this increased securing will be irreducible. Hence, in general, pairs of cases to which the Intentionality of Harm Principle applies will instantiate possibility 2C. Thus, the explanatory relevance view can make sense of why the intentionality of harm matters to the doing/allowing distinction.

We have now explained all five anomalous factors discussed in section 2 by appealing different ways in which an agent's motivational state can be more or less explanatorily relevant to a harm. This is summarized in the following table:

Difference in Explanatory Relevance	Factor Affecting Doing/Allowing Judgments
1A Sensitivity of behavior to motivation	Independently Prohibited Behavior
1B Sensitivity of harm to behavior	Latitude
1C Irreducible sensitivity of harm to motivation	Prior Obligation to Protect
2A Securing of behavior by motivation	Deliberateness of Behavior
2C Irreducible securing of harm by motivation	Intentionality of Harm

And recall that the remaining possibility, 2B, corresponds to cases that can be explained equally by the explanatory relevance view and by the causal view.

Note that, while the table above includes only one anomalous factor corresponding to each of the possibilities in the left column, this was done only for the sake of brevity, and the table could be expanded to include more than one factor per possibility. As I mentioned at the end of section 2, I believe there may be numerous other anomalous factors, apart from the five listed above, that can affect our judgments of doing versus allowing harm. And I believe that the framework I have laid out has the potential to explain the relevance of these other factors as well. I will not, however, endeavor to show this here.

Instead, I will conclude by showing how the explanatory relevance view solves the structural problem facing the causal view. We saw that the causal view cannot allow for the possibility that an agent does a harm by causing someone else to merely allow this harm, as in the following case:

> **Heat Ray**: Hanna is hanging from a chain over the edge of a cliff. Helda is holding the chain and attempting to pull Hanna to safety. Heather, however, wants Hanna to fall from the cliff, and so she fires her heat ray at the chain near to where Helda is holding it. As the pain becomes very intense, Helda lets go of the chain, and Hanna falls to her death.

In this case, it seems that Heather kills Hanna by causing Helda to let Hanna die. The explanatory relevance theorist could offer the following explanation of how this could be so. If we consider the space of possible human motivational states, a wide range of them are compatible with acting in the manner in which Helda acts in this scenario—only the saint, the hero, or the masochist would act otherwise. By contrast, a much narrower range of motivational states is compatible with acting in the manner in which Heather acts. Consequently, Heather's behavior is much more sensitive than Helda's behavior to their respective motivational states. Consequently, the harm to Hanna is much more sensitive to Heather's motivational state than it is to Helda's motivational state. Accordingly, Heather's motivational state has greater explanatory relevance than Helda's motivational state to Hanna's death. And it is for this reason that Heather counts as killing Hanna whereas Helda counts as merely letting her die.

I have argued that what distinguishes the harms we do from the harms we merely allow is that the former are more tightly connected to our motivations. If this is right, then it would imply that the harms we do reflect more on our character than do the harms we merely allow. And this, in turn, could explain why the harms we do can reflect on us worse. And so the explanatory relevance account may not only help us to distinguish between doings and mere allowings, but it might also help us to understand why this distinction matters. This, however, would require further investigation.

Acknowledgments

In writing this paper, I benefited immensely from discussions with Jeremy Goodman, Jake Nebel, Matthew Osterberg, and Jonathan Quong. I am especially grateful to Ralph Wedgwood, conversations with whom have greatly influenced my thinking about these issues.

References

Åqvist, Lennart and Philip Mullock (1989). *Causing Harm: A Logico-Legal Study* (Berlin: De Gruyter).

Callahan, Daniel (1989). "Killing and Allowing to Die", *Hastings Center Report* **19** (1): 5–6.

Donagan, Alan (1977). *The Theory of Morality* (Chicago: University of Chicago Press).

McMahan, Jeff (1993). "Killing, Letting Die and Withdrawing Aid", *Ethics* **103** (2): 250–79.

Quinn, Warren S. (1989). "Actions, Intentions, and Consequences: The Doctrine of Doing and Allowing", *Philosophical Review* **98** (3): 287–312.

Rickless, Samuel (1997). "The Doctrine of Doing and Allowing", *Philosophical Review* **106** (4): 555–75.

Woollard, Fiona (2012). "The Doctrine of Doing and Allowing II: Relevance of the Doing/Allowing Distinction", *Philosophy Compass* 7 (7): 459–69.

Woollard, Fiona (2015). *Doing and Allowing Harm* (Oxford: Oxford University Press).

16
Responsibility and the Metaphysics of Omissions

Carolina Sartorio

1. Introduction

The metaphysics of omissions is a highly contested issue. If a selfish swimmer omits to save a child who is drowning right in front of him just because he doesn't want to be bothered, what does his omission to save the child amount to? Is it a special or negative kind of action (for example, an action involving a negative property—the non-saving of a child)? Or is it identical with an ordinary positive action, such as the swimmer's swimming away from the child at the time? Or is it, instead, not an action of any kind, but the absence of an action? These are difficult and highly controversial metaphysical questions. In this paper I will focus, not on those metaphysical questions themselves, but on the potential implications that questions of this kind might have for our moral responsibility (just "responsibility" hereafter). What kinds of consequences can we draw, if any, about our responsibility—and, in particular, about our responsibility for and in virtue of our non-doings—from the metaphysics of omissions?

One main way in which one might expect the metaphysics of omissions to be relevant concerns the role played by *causation* in grounding responsibility. Causation and responsibility seem to be importantly related. Imagine that Shooter aims his gun at Victim and shoots. A big storm had been building up, and as a result of the storm a sudden gust of wind deflects the bullet slightly and it never reaches Victim. Imagine that, at the same time, and also as a consequence of the storm, lightning strikes and kills Victim. In this case it seems clear that Shooter is not responsible for Victim's death because he didn't kill her (even if he tried); lightning did. And, surely, Shooter cannot be responsible for Victim's death if he didn't cause her death. In other words, responsibility for an event seems to require (among other things) having caused it.[1]

[1] Throughout this paper I will assume that agents can be responsible for outcomes in the world. Some people would reject this assumption on the grounds that it gives rise to unacceptable forms of moral luck. I will have to bypass this issue here.

Carolina Sartorio, *Responsibility and the Metaphysics of Omissions* In: *Non-Being: New Essays on the Metaphysics of Non-Existence.* Edited by: Sara Bernstein and Tyron Goldschmidt, Oxford University Press (2021). © Carolina Sartorio.
DOI: 10.1093/oso/9780198846222.003.0016

Now imagine that there was no storm, so Shooter (not lightning) causes Victim's death. Imagine, also, that there was another agent, Bystander, who could have easily stopped Shooter before he shot Victim and at no serious cost to himself, but failed to do so out of laziness. Bystander's contribution is an omission or a non-doing. As a result, in this case there is much disagreement about the underlying metaphysics, and, in particular, about how to understand Bystander's contribution. Some would say that Bystander is a cause of Victim's death, while others would say that Bystander's contribution is not causal.[2] Which of these views one is likely to embrace will depend on one's underlying metaphysics of omissions and on one's general views on causation. For example, if one thinks that causation is a relation between events, then whether one thinks that the relation between Bystander's omission and Victim's death is causal will depend on whether one thinks that omissions are events.

Now, regardless of which of these views one thinks is best, here is one argument that, I think, one *couldn't* plausibly make:

From lack of causal efficacy to lack of responsibility:

1. The true metaphysics of omissions and causation is one according to which omissions are never causes.
2. Responsibility for an outcome requires causing it.
3. Therefore, we can never be responsible for outcomes in virtue of our omissions.[3]

Why is this a bad argument? Because, even though both premises have some initial plausibility, it is still much clearer that 3 is false than that the conjunction of 1 and 2 is true. Surely, the selfish swimmer is responsible for the child's drowning and the lazy bystander is responsible for Victim's death (if we are ever responsible for anything that happens), and they are responsible in virtue of their omissions to act. In particular, compare Bystander's responsibility for Victim's death with his responsibility for the shooting of a second victim in a remote location (someone he didn't have the means to save). Clearly, there is a significant difference in that Bystander bears some responsibility in the former case but not in the latter. But, in order for this to be true, 3 must be false. Again, Bystander's responsibility for Victim's death doesn't seem negotiable.

[2] Views according to which omissions are causes include Mellor (1995), Thomson (2003), Lewis (2004), Schaffer (2004), and McGrath (2005). Views according to which omissions are not causes include Dowe (2001), Beebee (2004), Varzi (2007), and Bernstein (2014). For a survey of the main issues and views surrounding this topic, see Bernstein (2015).

[3] An argument of this kind was offered by Elazar Weinryb: "[O]missions have no consequences, since they lack the required causal efficacy. My argument sheds grave doubts on the appropriateness of holding someone responsible for the harm he omits to prevent." (Weinryb 1980: 3; here Weinryb seems to have in mind both moral and legal responsibility). Weinryb then concludes that any moral criticism towards agents' omissions or the apparent consequences of omissions should be directed instead towards the thoughts, and in particular the wants, of those agents who omitted to act (Weinryb 1980: 18).

In other words, we can be reasonably sure that the argument fails, even if we may not know which premise is false. For we know that Bystander is responsible (again, if anybody is ever responsible for anything that happens), even if we may not be sure about how to account for his responsibility: if in terms of causation or in other terms. This suggests that our responsibility involving omissions doesn't directly *hinge* on the metaphysics of omissions and causation by omissions. A certain metaphysics being true or false is not the kind of thing that could render us incapable of being responsible by omission. Our responsibility is, instead, "resilient" in that it would survive the discovery that a given metaphysics is true or false.

Note that this is in stark contrast with other metaphysical truths which could, at least potentially, have significant implications for our responsibility. Consider, for example, the truth of causal determinism (the thesis that our acts are determined by the state of the world in the remote past and the laws of nature). On some popular views of free will, whether determinism is true is something that can have significant implications for our free will and thus our moral responsibility. On libertarian views of free will, for example, we could only have free will and be responsible for our behavior and the consequences of our behavior if our acts were not causally determined. Of course, there are other views of free will according to which the truth of determinism is irrelevant to our free will and responsibility; in fact, many compatibilist views have this implication. But the point is that it is at least an *open question* (and one that has been intensely debated in the free will literature) whether the truth of determinism would undermine our responsibility. The situation is different with the metaphysics of omissions: it is less clear what implications, if any, the metaphysics of omissions could even potentially have for our responsibility.

The goal of this paper is to make some progress in addressing this difficult issue. I will focus on issues of the following kind: When faced with questions about our responsibility concerning omissions, is it at all appropriate to look into the metaphysics of omissions and omission-involving causation for an answer? When is it appropriate and when is it not? And why? Aside for a few exceptions, these questions have largely been ignored in the literature. But they are important questions with potentially significant implications for our theorizing about responsibility.

2. The Question about Causal Powers

Let us start by looking more closely into why our responsibility appears to be immune to the discovery that omissions cannot be causes. Here it is relevant to note that those who have argued that omissions cannot be causes have suggested other ways in which omissions can make us responsible. These other ways appeal to concepts that are broader than causation and that as a result can accommodate omissions. Interestingly, however, those alternative proposals typically end up appealing to causation in some way or other—just not to causation *by omission*.

For example, Phil Dowe's view is that omissions cannot be causes because only positive events enter in causal relations and omissions are not positive events. However, Dowe suggests that omissions can be "quasi-causes" (Dowe 2001). Quasi-causation is a concept that appeals partly to causal processes that obtain, not in the actual world, but in merely possible or counterfactual worlds. On Dowe's view, for example, the selfish swimmer's omission doesn't cause the child's drowning in the actual world but it quasi-causes it, and this is partly in virtue of the fact that, in some relevant possible world where the omission *doesn't* obtain (some world in which the swimmer acts selflessly and attempts a rescue), this starts a causal process that ends in the child's being rescued. So, on this view, one can say that the responsibility that the swimmer bears to the child's death is grounded, at least partly, in that quasi-causal connection and thus on the relevant counterfactual causal process. The same goes for the lazy bystander and his responsibility for Victim's death.

Helen Beebee developed a view that is similar in some important respects (Beebee 2004). Beebee suggests that omissions can never be causes but they can still be explanatorily relevant, for they can play a role in causal explanations of events. On Beebee's view, for example, Victim died partly *because* of the lazy bystander's omission, even if the omission wasn't a cause of Victim's death. According to Beebee, when we learn of this explanatory connection, we learn at least something minimal about the causal history of Victim's death—namely, that it did not include an event of Bystander's intervening. But we also learn something about the causal structure of some nearby possible worlds: we learn that in some worlds where Bystander does intervene, his intervention does not cause Victim's death; in fact, it causes Victim's survival. Note that, again, just like in Dowe's view, the full account of Bystander's contribution ends up appealing to the existence of causal processes in counterfactual worlds (Beebee 2004: 305–6).

As these two examples suggest, causation is likely to play *some* important role in an account of the responsibility of agents in these cases, regardless of which metaphysics of omissions is true. If omissions are causes, the responsibility of agents will be simply grounded in actual causal processes. But, if omissions are not causes, the responsibility of agents can still be grounded in a combination of actual causal facts and counterfactual causal facts. In other words, the connection between responsibility and causation for omissions could be broadly characterized as follows:

Causation as grounds for responsibility:

When agents are responsible by omission, their responsibility is partly grounded in causal facts that obtain in the relevant possible world(s) (the actual world and/or some relevantly similar counterfactual world(s)).

Given this, it is not surprising that responsibility is immune to discoveries about the metaphysical nature of omissions. Even if the thought that responsibility requires causation (of the actual kind) is initially plausible, there is a readily available "fallback" option in case it ends up being misguided. And it is one that is not too far removed from the initial thought to count as a significant departure from it.

Now, one might think that the introduction of merely possible causal connections carries with it a commitment to certain views of responsibility as opposed to others, and as a result it fails to be as neutral as I am advertising it to be. In particular, there is a recently popular family of views about responsibility, sometimes called *actual-sequence* views, according to which, roughly, the responsibility of agents is exclusively grounded in actually explanatory sequences of events.[4] These are views that are inspired by Harry Frankfurt's arguments against the traditional model of responsibility in terms of alternative possibilities and by his suggestion that responsibility is just a function of the actually explanatory factors (Frankfurt 1969). So, aren't these actual-sequence views of responsibility at odds with the suggestion that the responsibility of agents by omission can be grounded in causal connections that take place in alternative possible worlds?

I have discussed this in more detail elsewhere,[5] but the short answer is 'no.' Actual-sequence views are in fact compatible with the claim that some counterfactual scenarios (and, in particular, causal processes obtaining in counterfactual scenarios) are relevant to our responsibility. What those views typically deny is the relevance of *alternative possibilities of action* for responsibility. However, alternative possibilities are not just any old possibilities, since they are supposed to involve robust abilities to engage in alternative behaviors (roughly, having the right kind of "access" to the relevant possibilities). Thus, actual-sequence views needn't differ from alternative possibilities views as far as the relevance of some counterfactual facts. The essence of the disagreement concerns, instead, the relevance of *robust alternatives*.

Imagine, for example, that Beebee's view is right and that omissions can be part of causal explanations but they can never themselves be causes. In that case, an actual-sequence view will have to account for the responsibility of agents by omission by appeal to factors that are actually explanatory, in Beebee's sense. And recall that, on Beebee's view, this carries with it a commitment to the relevance of some counterfactual causal processes. But, again, the mere existence of those counterfactual causal processes is not the same thing as the agent's having robust alternatives.

Another way to see that actual-sequence views can consistently accept the relevance of some counterfactual facts is this. The most natural way (although,

4 See e.g. Fischer and Ravizza (1998) and Sartorio (2016).

5 See Sartorio (2016, ch. 1) and Sartorio (forthcoming).

again, not the only possible way) to understand actual sequences is as causal sequences. But notice that many accounts of causation are *counterfactual* accounts—that is, they analyze or ground causation itself in terms of, among other things, some counterfactual facts.[6] This suggests that the very idea that responsibility is grounded in actual sequences, when understood as causal sequences, can potentially carry with it a commitment to the relevance of some counterfactual facts (those facts that help ground the relevant causal facts). So, again, there is no inconsistency in embracing an actual-sequence view of responsibility and accepting the relevance of some counterfactual facts.

So far, I have argued that the metaphysics of omissions doesn't directly bear on our responsibility in a way that would justify excusing us of responsibility in omission cases if it turned out that omissions are causally inefficacious. If the metaphysics of omissions is relevant to our responsibility, it is not in this most obvious kind of way. In the sections below I look at other possibilities.

3. Symmetric Overdetermination

The next natural step is to think about cases with a more complex structure. If the metaphysics of causation involving omissions is to have a bearing on the moral responsibility of agents, it is natural to expect to see this in cases with puzzling causal structures such as those involving the contributions of multiple factors, or multiple agents, and some form of overdetermination or redundancy. Imagine, for example, that two agents have to do their part in order to save the victim's life; say, two swimmers are needed to bring the drowning child back to the shore. Imagine that the swimmers independently and simultaneously decide to stay put and not do their part, in each case for purely selfish reasons, and the child dies as a result. This is a case of what is commonly known as overdetermination: what each swimmer does (or fails to do) is on its own sufficient for the child's drowning. Overdetermination cases are notably trickier because their causal or metaphysical structure is less than fully clear: even if one accepts that omissions can be causes, overdetermination cases raise special challenges and there is much debate about how to make sense of them. In particular, on some views overdeterminers are causes but on other views they are not.[7] So, the question I will focus on next is: Could one's metaphysical views on overdetermination, and in particular on overdetermination by omission, have any consequences for the responsibility of agents in those cases?

[6] For examples of views in this tradition, see the essays collected in Collins, Hall, and Paul (2004).

[7] For a defense of the idea that overdeterminers are causes, see e.g. Schaffer (2003). For a discussion of a view according to which overdeterminers are not causes, see e.g. Lewis (1986: 212).

Although things are considerably less clear in this case, I think that here, too, there are certain inferences that one could *not* plausibly make. Here is one such inference:

From lack of causal efficacy and difference-making to lack of responsibility:

1. The true metaphysics of omissions and causation is one according to which omissions are never causes.

2. There is no other way to ground the responsibility of agents for outcomes in overdetermination cases involving omissions. In particular, in those cases there is a lack of difference-making: each omission makes no difference to the outcome, given the other omission. (In terms of counterfactual dependence: the outcome doesn't counterfactually depend on either of the omissions, for it would still have occurred if only one of the agents had acted.)

3. Therefore, agents are not morally responsible for the outcome in overdetermination cases involving omissions.[8]

This argument would get the two selfish swimmers off the hook for the child's death in the case where their omissions overdetermine his death. But this seems clearly wrong. Each of the swimmers failed to do what they should have done and, had they both done what they should, the child would have been saved, and in the expected way. Moreover, they had no excuse for what they did (or failed to do). So, again, the responsibility of the swimmers seems to be non-negotiable: it is clearer that the conclusion of the argument is false than that the conjunction of the premises is true. As a result, even if it might not be clear which premise is false, the argument still doesn't get off the ground.

In the previous section I noted that the link between responsibility and causation would survive the discovery or realization that omissions cannot be causes—for, even if responsibility by omission couldn't then be grounded in actual causation, it could still be grounded in causation that takes place in the relevant possible worlds. An extension of the same reasoning could be offered here too. Imagine that we couldn't ground the responsibility of the two swimmers in actual causation—either because there is no causation by omission in general, or because overdetermining omissions cannot be causes. We could still partly ground their responsibility in causal connections that obtain in other possible worlds. In this case, given that it is an overdetermination case, the relevant possible worlds would arguably have to be worlds where not just one but *both* of the swimmers act (in whichever way is required to save the child). In those possible worlds, the

[8] Michael Moore argues in this way in his 2009, chapter 18. Moore thinks omissions cannot be causes because only positive events can be causes. Although he thinks that counterfactual dependence is an alternative ground for responsibility that applies to many omission cases, overdetermination cases exhibit a lack of counterfactual dependence. As a result, Moore thinks that agents cannot be responsible in overdetermination omission cases.

swimmers start a causal process that ends in the child's being saved. Even if the details would have to be worked out, it seems clear that, at least in principle, one could potentially ground the swimmers' responsibility for the child's death partly in those counterfactual causal processes.

Let me clarify what I think it is, exactly, that is non-negotiable in these cases. It is the basic fact that the swimmers bear *some* responsibility for the child's death: it is sufficiently clear that they are each responsible, *to some degree*, for the child's death. "Two wrongs don't make a right," we think, and for good reason. Under the proper understanding, the slogan seems clearly true, and it explains why we think that the swimmers cannot be off the hook for the child's death.[9] On the other hand, what I think may be negotiable is *how* responsible each of the swimmers is, or the extent of their responsibility—assuming responsibility is a concept that comes in degrees. (Although, given that the case is intended to be perfectly symmetrical, it is clear that each of the swimmers must bear the same amount of responsibility, whatever that amount may be.) If responsibility comes in degrees, it is an open question whether the existence of overdetermination lessens the responsibility that each agent bears for the outcome. It might be, for example, that when two agents overdetermine an outcome, each agent only bears half of the responsibility (or, in any case, less responsibility than if they had been the only agent involved).[10]

Even then, I think that it would probably be misguided to expect the metaphysics of omission-involving causation *itself* to have any bearing on the degree of the responsibility of agents in overdetermination cases. For whether we think that an agent is less than fully responsible in these cases will depend on more general considerations having to do with the right way to apportion responsibility when more than one agent is responsible (and especially in cases where each agent's contribution is independently sufficient for the outcome), instead of on considerations concerning the metaphysical status of omissions or of causation by omission.

In many cases, it would also be a mistake to expect that the extent of the agents' responsibility in scenarios of this kind will hinge on the right views on *overdetermination* and causation. For the issue of whether overdeterminers are causes is usually seen as depending on more general theoretical considerations that seem to have nothing to do with the grounds for responsibility. To illustrate, consider David Lewis's views on this. Lewis famously discussed the difficulties involved in accepting that overdeterminers can be causes given the framework of a

[9] I have discussed the proper understanding of the slogan in Sartorio (2012). Briefly, the main thought is that the only way in which two wrongs could make a right would be if the two agents' contributions ended up interfering with each other in a way that neutralized them both. (Picture two evil assassins whose bullets intercept and cancel each other in mid-flight.) The two swimmers case does not seem to be like this, and this is why we think that the swimmers cannot be off the hook for the child's death.

[10] For discussion of this issue, see e.g. Cohen (1981), Zimmermann (1985), Bernstein (2016; 2017), and Sartorio (2020).

counterfactual approach to causation (his preferred account).[11] Counterfactual views have trouble with overdetermination cases precisely because of the lack of counterfactual dependence between the individual overdetermining events and the outcome: How can we account for the causal structure of these cases in counterfactual terms if the outcome's occurrence does not counterfactually depend on the overdetermining events? Such lack of counterfactual dependence would appear to leave us with unexplained outcomes: outcomes that don't have any causes, which seems contrary to what we want to say about those cases. Part of Lewis's solution was to suggest that, even if counterfactual views don't seem to have the resources to analyze individual overdeterminers as causes due to the lack of counterfactual dependence (and in this sense they may have to be left as "spoils to the victor"), the event that is the *mereological sum* of the overdetermining events would still come out as a cause (because the outcome does counterfactually depend on that event, Lewis thought). This, he argued, would be enough to fill any causal "gaps" that might need filling. Presumably, Lewis was thinking that this way of filling in the causal gaps would also be enough to ground the moral responsibility of any agents involved.[12]

As this example suggests, it would be implausible to argue that the very same kinds of considerations that motivate the idea that overdeterminers aren't causes *also* motivate a reduction of the responsibility of agents whose behaviors overdetermine the effect. For they are considerations of a quite different type, in that they track theoretical issues about the best way to accommodate certain causal judgments within one's general theory of causation, about the best way to fill in the causally explanatory gaps, etc. And these kinds of issues don't seem to have any obvious implications for the responsibility of agents involved.

There is one way in which the metaphysics of causation could potentially ground the claim that the responsibility of agents is reduced in overdetermination cases: this is if one held the view that causation comes in degrees, and that the existence of overdetermination in general reduces the extent of one's causal contribution.[13] In that case one could argue that overdetermination makes agents less responsible because agents in overdetermination scenarios make a lesser contribution. But, again, note that this is a quite general proposal that doesn't rely on any specific metaphysical feature of *omissions*, or of omission-involving causation. For similar reasons, then, I don't think this shows that the metaphysics of omissions has any bearing on the responsibility of agents.[14]

[11] Lewis (1986), postscript E.

[12] Lewis (1986: 212). Note that this is another way in which responsibility could be grounded in causation without being grounded in actual individual causation; in this case, it would be grounded in actual "collective" causation. I make a similar point about cases involving overdetermination by omissions in Sartorio (2004; 2017). See also the discussion in section 4 below.

[13] See, for example, Chockler and Halpern (2004).

[14] What if the lesser degree of causal contribution were due to the fact that a behavior is an omission (instead of a positive action)? This view combines the idea that causal contributions come in degrees with

4. Asymmetric Overdetermination

So far, we haven't been able to find any clear examples where the metaphysics of omissions bears on our responsibility as agents. But the main reason for this, I think, is that we have focused exclusively on cases where our responsibility judgments (at least the most basic or fundamental judgments concerning our responsibility) are already quite clear. In those cases, as we have seen, the responsibility of agents seems to be resistant to metaphysical discoveries concerning what causes what, or what kinds of things can and cannot be causes. This suggests that we should look, instead, at cases where our responsibility judgments are not initially clear.

What could those cases be? I think the most promising candidates are *asymmetric* cases of redundancy involving omissions—that is, omission-involving cases that include overdetermination of a certain kind but where the overdeterminers are not otherwise "on a par", and in this respect they differ from those scenarios discussed in the preceding section. The reason these examples are better candidates is that it is unclear, in at least some of those cases, which of the agents is responsible for the outcome, and why. Moreover, this is due to the intriguing and unique nature of omissions. As a result, this is an issue that could potentially be decided by metaphysical truths about omissions or about omission-involving causation.

Probably the most famous example of this kind is the "desert traveler" puzzle discussed, for instance, in the classical work on causation in the law by Hart and Honoré (1985).[15] On one variant of the case, a traveler is planning to take a trip into the desert with his water canteen. He has two enemies who want him dead. The two enemies don't know about each other, and so they each independently come up with an evil plan for the traveler to die in the desert. At time T1, the first enemy secretly empties the water from the canteen, and refills it with sand so that the man won't notice the difference in weight. Next, at T2, the second enemy steals the canteen from the man (thinking that it contains water). The man dies from thirst in the desert, at some later time, T3. It is clear that the traveler died, in some sense, because of what the two enemies did. Thus, it is clear that we want to blame someone for the traveler's death—either the first enemy, or the second enemy, or perhaps both. But it is hard to say who killed him, for it is not clear how either one

the claim that omissions make less significant contributions than actions. I don't think this view is very plausible, for two reasons. One is that I don't think it's plausible in general to think that causal contributions come in degrees (I argue for this in Sartorio 2020). But the other is that, to the extent that we can make sense of the idea of causal contributions coming in degrees, it also doesn't seem plausible to argue that *generally* omissions make less significant contributions than actions. Consider, for example, someone's omission to feed their own child and the contribution that this would make to the child's death.

[15] The puzzle is originally from McLaughlin (1925).

could have made a contribution, given what the other one did. How can draining the water from the canteen be relevant to the man's death, if that canteen was going to be miles away from the man when he needed it (because the second enemy was going to steal it)? And, how can stealing a canteen that wasn't filled with water (but with sand, as a result of the first enemy's intervention) be relevant to the man's death from thirst? Hence the puzzle.

At least on the face of it, this is a case where the metaphysics of omissions (and, more generally, of non-occurrences or absences) could potentially be relevant. This isn't because the wrongful behaviors by the two enemies are themselves omissions (they aren't; they are positive actions), but, rather, because their behaviors are only potentially relevant to the outcome of the traveler's death given that they result in non-occurrences or absences that are, themselves, potentially relevant to the traveler's death. Thus, the first enemy's action of draining the water from the canteen results in the *absence* of water in the canteen at T1, and the second enemy's action of stealing the canteen results in the *absence* of the canteen from the traveler's location at T2. If either enemy can be said to have contributed to the traveler's death, it must be because of the contribution that one of these non-occurrences or absences made to it.

Thus, the question "Which of those absences made a contribution to the traveler's death?" becomes at least potentially relevant. Finding out what the right answer to that question is could then help us solve the puzzle about responsibility. And, in looking for an answer to that question, one could hope to get help from the metaphysics of omissions, or of omission-involving causation.

The desert traveler case is different from the *symmetric* overdetermination cases discussed in the previous section in some important respects. As noted above, in the symmetric cases it seems clear that somebody is to blame for the outcome, but, given the symmetry, it also seems clear that both agents are to blame, and to the same extent. The desert traveler case is different in that one agent acts *before* the other, and everything else that that entails. This means, for example, that the first enemy is the *first* to guarantee that the traveler will die in the desert, which could be seen as potentially significant in a way that emphasizes the contribution of the first enemy. On the other hand, it also means that the second enemy is the *last* to act before the traveler dies, and this could be seen as potentially significant in a way that emphasizes the contribution of the second enemy. As a result, in this case it is not at all obvious that both agents are responsible for the outcome, or that they are both responsible to the same extent. All that I think is clear is that *someone* is to blame (either the first enemy or the second or both), but nothing more than that. This is what opens the door to the metaphysical truths being relevant in these cases.

Imagine, for example, that the right metaphysical view entails that one enemy is a cause of the traveler's death but the other is not (in other words, it is a case of what is commonly known as "preemption": one enemy causally preempts the other). In that case, we would have good reason for thinking that one enemy is responsible for the traveler's death and the other is not. Or imagine that the right metaphysical view entails that both enemies are causes. In that case we would have good reason for thinking that both enemies are responsible for the man's death.

In retrospect, I think this is probably why the desert traveler case, as well as other similar examples,[16] have been discussed by responsibility theorists and metaphysicians alike. For in those cases it is natural to think that the answer to the metaphysical question (the question of who makes an actual contribution to the outcome) is key, in that the question of who is responsible for that outcome hinges on it. As a result, if we had reason to believe that one of those metaphysical views is true, this could help us solve the puzzle about responsibility.

Now, I think that neither of those metaphysical views is, *in fact*, right: the desert traveler scenario is not a case of preemption (or a case where one of the agents is a cause and the other one isn't), and it is also not a case where both of the agents are causes. The right thing to say, I believe, is that *neither* agent makes an individual causal contribution.

Let me add that my reason for thinking this is not that I believe omissions or non-occurrences cannot be causes in general. I actually remain neutral on this issue. Also, this wouldn't get to the heart of the puzzle because the same kinds of questions would arise if the right way to think about omissions and other non-occurrences were not in terms of causation but in terms of, say, explanation. For one could reformulate the desert traveler puzzle as a puzzle about explanation: What explains why the traveler died? Is it what the first enemy did, or what the second enemy did, or a combination of both? These questions are just as puzzling as the questions about causation.

My reason for thinking that neither of the views described above is the right metaphysics is, rather, that I think that the desert traveler case is a case of *mutual causal cancellation*, one where the causal powers (or explanatory powers, if omissions cannot be causes) of the two agents' contributions cancel each other out and this results in neither agent being a cause. I have defended this view elsewhere[17] and I won't rehash it here. But, roughly, the thought is that the mutual causal cancellation takes place at the level of individual causes (thus neither enemy is individually a cause of the traveler's death); still, there is causation at the collective level (a collective cause that combines the contributions of the two agents). I also argued that it follows from this that the responsibility question remains an open question, for in order to figure out who is responsible we would

[16] Examples with a similar structure include the two catchers example discussed in Collins (2000) and the planted sharks examples discussed in Sartorio (2017) and Clarke (2014, ch. 6).

[17] Sartorio (2015).

first have to figure out who is responsible for the collective cause, and this question lacks an obvious answer.[18]

At any rate, note that what we were doing at that point (before discussing the implications for responsibility) was straight metaphysics. We were trying to figure out what the best way to model the structure of these cases is, in metaphysical terms, in order to then examine the potential implications for the responsibility of agents. In other words, we clearly saw the metaphysical structure of these cases as at least *potentially* mattering. Even if my own view ends up being that the metaphysical structure of these cases is not enough to settle the responsibility question, I recognize that there are other (sensible) metaphysical views that deny this.[19]

In section 1 above I mentioned the example of causal determinism as a metaphysical thesis that is widely regarded as potentially bearing on our responsibility. As it happens, my take on determinism is quite similar: I am a compatibilist, so I don't actually think that the truth of determinism ultimately matters for our responsibility. But, of course, I recognize that there are sensible views (some incompatibilist views of freedom) on which it does. And until we figure out which view of freedom is true, we won't be able to agree on whether the truth of determinism is relevant to our responsibility. Based just on what we agree on, all we can say is that it is *potentially* relevant. The same goes, I think, for the metaphysics of omissions and of omission-involving causation in the kinds of scenarios discussed in this section.

5. Conclusion

The question of whether and how the metaphysics of omissions is relevant to responsibility is, as we have seen, a difficult one to answer. Responsibility judgments are quite resilient in that they would resist the discovery of most metaphysical truths concerning omissions. I have argued, in particular, that the central metaphysical question of whether omissions have causal powers is of this kind: our responsibility does not hinge, in any clear way, on the right answer to that question. As we have seen, this is partly because our judgments concerning the potential grounds for responsibility are, in comparison, quite flexible.

On the other hand, we have seen that there are other metaphysical truths concerning the causal (or otherwise quasi-causal or explanatory) powers of

[18] Although I did provide a tentative answer to that question (one according to which only the first enemy ends up being responsible). But, again, I cannot get into it here.

[19] By a "sensible" view I mean a view that has at least some plausibility, even if it may be false. In contrast, as suggested above, I don't think it's sensible to argue that the right answer to the question of whether omissions have *causal powers* can help with the responsibility question. For similar questions arise if omissions have causal powers or if they don't.

omissions in more complex kinds of situations that can, at least potentially, bear on our responsibility as agents. These are situations where our responsibility judgments are less clear to start with, and where the metaphysical and responsibility questions are as a result more closely tied together. In those cases, it is more natural to expect the metaphysical truths to bear on our responsibility in significant ways.

I conclude that, as with other kinds of theorizing, theorizing about responsibility should draw on the metaphysics to the extent that it is recommended by an exercise of a sensible process of reflective equilibrium. In cases where our responsibility judgments are very resilient, the responsibility questions will predictably "float free" from the metaphysics. However, in other cases where our responsibility judgments are more uncertain, the metaphysics can be expected to play a more significant role.

Acknowledgments

Thanks to Sara Bernstein, Tyron Goldschmidt, and the participants at a symposium on causation and ethics at the 2020 Central APA.

References

Beebee, Helen (2004). "Causing and Nothingness", in *Causation and Counterfactuals*, ed. John Collins, Ned Hall, and L. A. Paul (Cambridge, MA: MIT Press), pp. 291–308.

Bernstein, Sara (2014). "Omissions as Possibilities", *Philosophical Studies* **167** (1): 1–23.

Bernstein, Sara (2015). "The Metaphysics of Omissions", *Philosophy Compass* **10** (3): 208–18.

Bernstein, Sara (2016). "Causal and Moral Indeterminacy", *Ratio* **29** (4): 434–47.

Bernstein, Sara (2017). "Causal Proportions and Moral Responsibility", in *Oxford Studies in Agency and Responsibility*, vol. 4, ed. David Shoemaker (Oxford: Oxford University Press), pp. 165–82.

Chockler, Hana and Joseph Y. Halpern (2004). "Responsibility and Blame: A Structural-Model Approach", *Journal of Artificial Intelligence Research* **22**: 93–115.

Clarke, Randolph (2014). *Omissions: Agency, Metaphysics, and Responsibility* (New York: Oxford University Press).

Cohen, L. Jonathan (1981). "Who Is Starving Whom?", *Theoria* **47** (2): 65–81.

Collins, John (2000). "Preemptive Prevention", *Journal of Philosophy* **97** (4) (Special Issue: Causation): 223–34.

Collins, John, Ned Hall, and L. A. Paul (2004). *Causation and Counterfactuals* (Cambridge, MA: MIT Press).

Dowe, Phil (2001). "A Counterfactual Theory of Prevention and 'Causation' by Omission", *Australasian Journal of Philosophy* **79** (2): 216–26.

Fischer, John Martin and Mark Ravizza (1998). *Responsibility and Control: A Theory of Moral Responsibility* (Cambridge: Cambridge University Press).

Frankfurt, Harry G. (1969). "Alternate Possibilities and Moral Responsibility", *Journal of Philosophy* **66** (23): 829–39.

Hart, H. L. A. and Tony Honoré (1985). *Causation in the Law*, 2nd edn (Oxford: Oxford University Press).

Lewis, David (1986). "Causation", in *Philosophical Papers*, vol. **II** (New York: Oxford University Press), pp. 159–213.

Lewis, David (2004). "Void and Object", in *Causation and Counterfactuals*, ed. John Collins, Ned Hall, and L. A. Paul (Cambridge, MA: MIT Press), pp. 277–90.

McGrath, Sarah (2005). "Causation by Omission: A Dilemma", *Philosophical Studies* **123** (1/2): 125–48.

McLaughlin, James Angell (1925). "Proximate Cause", *Harvard Law Review* **39** (2): 149–99.

Mellor, D. H. (1995). *The Facts of Causation* (London: Routledge).

Moore, Michael S. (2009). *Causation and Responsibility: An Essay in Law, Morals, and Metaphysics* (Oxford: Oxford University Press).

Sartorio, Carolina (2004). "How to be Responsible for Something without Causing It", *Philosophical Perspectives* **18** (1): 315–36.

Sartorio, Carolina (2012). "Two Wrongs Don't Make a Right: Responsibility and Overdetermination", *Legal Theory* **18**: 473–90.

Sartorio, Carolina (2015). "Resultant Luck and the Thirsty Traveler", *Methode* **4** (6): 153–71.

Sartorio, Carolina (2016). *Causation and Free Will* (Oxford: Oxford University Press).

Sartorio, Carolina (2017). "The Puzzle(s) of Frankfurt-Style Omission Cases", in *The Ethics and Law of Omissions*, ed. Dana Kay Nelkin and Samuel C. Rickless (New York: Oxford University Press), pp. 133–47.

Sartorio, Carolina (2020). "More of a Cause?", *Journal of Applied Philosophy* **37** (3) (Special Issue: Symposium on Causation in War): 346–63.

Sartorio, Carolina (forthcoming). "The Grounds of Our Freedom", *Inquiry* (special issue on Harry Frankfurt's "Alternate Possibilities and Moral Responsibility").

Schaffer, Jonathan (2003). "Overdetermining Causes", *Philosophical Studies* **114** (1/2): 23–45.

Schaffer, Jonathan (2004). "Causes Need Not Be Physically Connected to Their Effects: The Case for Negative Causation", in *Contemporary Debates in Philosophy of Science*, ed. Christopher Hitchcock (Malden, MA: Blackwell), pp. 197–216.

Thomson, Judith Jarvis (2003). "Causation: Omissions", *Philosophy and Phenomenological Research* **66** (1): 81–103.

Varzi, Achille C. (2007). "Omissions and Causal Explanations", in *Agency and Causation in the Human Sciences*, ed. Francesca Castellani and Josef Quitterer (Paderborn: Mentis Verlag), pp. 155-67.

Weinryb, Elazar (1980). "Omissions and Responsibility", *Philosophical Quarterly* **30** (118): 1–18.

Zimmermann, Michael J. (1985). "Sharing Responsibility", *American Philosophical Quarterly* **22** (2): 115–22.

17

Death's Shadow Lightened

Daniel Rubio

1. Introduction

Epicurus famously argued that "death is nothing to us." In claiming this, he meant to exhort his reader to abandon our typical responses to dying—fear, dread, angst, perhaps even the rituals that often accompany death. His argument? Death does no harm, because "when we are here, death is not; and when death is here, we are not." Thus, there are two components to the Epicurean perspective on death. First, death is nothing to us. We should not fear or dread it. Second, death's harm is nothing special. The reason that death is nothing to us is that its harm is mundane or non-existent.

My objective is to provide a defense of the Epicurean position through a kind of argument by cases. Following Korsgaard (1983), we can divide the kinds of value properties a thing or event might instantiate along two axes. The properties might be intrinsic or extrinsic, the value might be final or instrumental. This gives us four categories to explore in search of an adequate harm for death: extrinsic instrumentalist views, extrinsic final views, intrinsic instrumentalist views, and intrinsic final views. After looking in each of these corners, I will conclude that we have yet to identify something that can play the role the harm of death should play.

An important clarification before we begin. I will assume that death leads to non-existence. This is historically a minority position, and there is an interesting philosophical literature on immortality and/or post-mortem life that we will not have time to delve into here.[1] If death is just another part of life, then a whole different suite of moral and metaphysical considerations become relevant.

Once we regard death as the final end of life, it becomes clear that the Epicurean position runs counter to intuition. Non-existence doesn't sound very fun. It precludes all experience, emotion, enjoyment. It is therefore plausible that the event that leads from existence to non-existence, death, is a harm. Nevertheless, I will argue that once we think about what kinds of value-properties an event like death might have, we will see that it cannot be a serious harm to the one who dies.

We will begin with some preliminaries to clarify what is at stake in this investigation, then move on to the most common explanation for the harm of

[1] Williams (1973), Rosati (2013), Gorman (forthcoming) discuss the issue from the perspective of risk and agency, Climenhaga (2018) and Rubio (2020) from the perspective of axiology.

Daniel Rubio, *Death's Shadow Lightened* In: *Non-Being: New Essays on the Metaphysics of Non-Existence.* Edited by: Sara Bernstein and Tyron Goldschmidt, Oxford University Press (2021). © Daniel Rubio.
DOI: 10.1093/oso/9780198846222.003.0017

death in the literature, a view where death's harm is extrinsic and instrumental, known as deprivationism. Deprivationists argue that death is bad for us because we miss out on good things we would have had by living. After addressing deprivationism, I will argue that its flaws extend to all extrinsic instrumentalist views. We will then turn our attention to intrinsicalist views. Once we have identified the flaw in views where the harm of death is intrinsic, we will move on to views where it is extrinsic but final. We will conclude that none of these types of view is adequate.

2. Preliminaries

Before we launch fully into our exploration, it is worth taking a moment to clarify what is at stake in the debate over the badness of death. Epicurus's challenge is to the way we treat death in our emotional lives and moral reasoning. To meet the challenge, the anti-Epicurean must provide an account of the badness of death that is up to the task. Just finding something bad and associated with dying isn't enough.

We can see this by recalling that 'harm,' like 'tall,' is a gradable adjective. That means that it renders, in the context of utterance, a binary judgment on a phenomenon that comes in degrees. Facts about who or what is tall depend on facts about height combined with context. In the context of a 4th grade basketball league, a guard who stands six feet is tall. Very tall. In the context of the National Basketball Association, a guard who stands six feet is not only not tall, but downright short.

In fact, given the right context, any height can count as 'tall.' Even heights, such as twelve inches, that are in most normal contexts not remotely tall. As with height so with harm. By correctly massaging context, we could make an event that in normal contexts does not count as a harm into one that does, by finding some bad thing associated with it. What matters is not whether, when context has been sufficiently shifted, death can count as a harm. What matters is that it count as the right kind of harm in order for it to play the role that it does in our emotional and moral lives.

And what role is that? One of the worst events that can befall a person. We go to efforts large and small to avoid dying, from altering our course as we walk across town to avoid being hit by cars to allowing surgeons to cut into our bodies and remove our organs (inflicting considerable pain, despite their best efforts) in order to postpone death, even only for a few years. The penalty of death is assigned only to the worst crimes, and is routinely called the ultimate punishment. The harm of death must typically be very great to warrant its role.[2]

[2] This is not uncontroversial. McMahan (2002) argues that some of the role death occupies in our moral reasoning, such as the wrongness of killing, it occupies because of other factors such as killing showing a lack of respect for persons. But he does grant (2002: 95–6) that it occupies some of its roles in virtue of the harm it causes, and this will be sufficient.

3. Argument

The central argument is a kind of argument by cases. Using the twin axes of intrinsic/extrinsic and final/instrumental, we can map out the space of possible value-properties. There are four possible types: extrinsic instrumental value, extrinsic final value, instrinsic instrumental value, and intrinsic final value. Because the most common view falls under the extrinsicalist instrumentalist banner, we will start there. Next we will discuss intrinsicalist views, because in so doing we will learn some things that will be useful in discussing extrinsicalist final views, the discussion of which is reserved for the end. Since these exhaust the possible types of value property, an argument that the harm of death (more precisely: a harm adequate to the role of death in our typical theorizing) is not in any of them is an argument that death is not harmful. That is our conclusion.

3.1 Against Extrinsicalist Instrumentalist Views

First I will argue that extrinsicalist instrumentalist views don't deliver an adequate harm for death. I will begin by attacking deprivationism, the premier account of the harm of death. Once I have argued that it is inadequate to play the role the harm of death must play for death to occupy the role it does, I will argue that the flaw in deprivationism generalizes to views where the badness of death is extrinsic and instrumental.

3.1.1 Deprivationism

What is now the majority response to the Epicurean challenge came to contemporary attention in Nagel's classic (1970) "Death." There, Nagel argued that death is bad and to be feared because dying deprives us of potential future goods. Thus death's harm is a kind of opportunity cost; there are so many better things we could be doing while we are busy being dead. Later authors have developed Nagel's suggestion into the view now known as deprivationism. Deprivationists locate the harm of death in the loss of the goods that the deceased would have gained in failing to die.

I think the deprivationist response is mistaken. Loss of goods one would have had, even modulated by any of the various epicycles that have been built into deprivationist views, may in some contexts amount to harm. But simply finding something that is in some contexts a harm and associated with death is at best a pyrrhic response to the Epicurean challenge. Death has a distinctive role in our emotional life and in our moral thinking. The Epicurean challenge is, in its essence, a challenge to whether death really should occupy that role. A response to the Epicurean that finds something harmful and associated with death but not fit to fill death's distinctive moral role does not successfully meet the challenge.

Deprivationism comes in various flavors. Feldman's Total Life View asks us to compare the total value of the life actually lived with the value of the life the person who died would have lived had they not died, with the harm of the death being the difference between the lives.[3] Jeff McMahan's Interest Relative View takes into account Parfit's argument that as the R-relations tying our present to our future selves thin, our present interest in future benefits diminishes. Thus, unlike Feldman's version of the view, McMahan's modulates the value of the goods one would have received but for dying by psychological connectedness to the potential future self who would have enjoyed them.[4] These are not the only flavors, but they are representative.

What they have in common is that they regard the harm of death as a kind of opportunity cost. Opportunity costs are those costs we incur by refraining to do things that would be beneficial. For example, I might have two opportunities for spending my afternoon: I could pay to go to the concert, or I could visit the art exhibit for free. Even though visiting the art exhibit has no direct cost, it does involve an opportunity cost. The concert is a one-off event, and if I skip it I won't be able to experience it. By going to the exhibit I forfeit my opportunity to go to the concert. I incur an opportunity cost.

Almost everything we do carries opportunity cost. Time and resources are limited; by choosing to do some things, we exclude ourselves from doing others. We are deprived of them. If, as deprivationists contend, this is a kind of *prima facie* harm,[5] the badness associated with dying is no different in kind from a very mundane sort of badness. A badness associated with many events in our lives, since all choices involve opportunity costs. The only difference may be in degree. Of course, if the things we give up as opportunity costs are not as good as the things we give them up for then the *prima facie* harms are not always all-things-considered harms. They may be compensated or canceled out by the goods we only gained by incurring them.

This means that death only harms an individual if the life they would have lived had they not died is an improvement on their actual life. Depending on the details of the deprivationist view, anyone whose total life value, or interest-relative life value, or remaining life value, would be worsened by surviving is benefited by their death. Deprivationists claim this as a feature of the view:

Deprivationism implies that when death takes a good life from its victim, that person's death is bad. Conversely, it entails that when death keeps a person from living a bad life, it is good for that person. It also gives us an extent to which death is good or bad: the more of a good life it takes away, the worse it is; the more of a bad life it takes away, the better it is. These are intuitively the right answers to the

[3] Feldman (1992). [4] McMahan (2002).

[5] See Broome (2012), Timmerman (2019), and Bradley (2009) for explicit endorsement.

questions: under what circumstances is death bad, and when it is bad, how bad is it?[6]

However, this gives rise to some problem cases for the deprivationist that it will be good to examine before we deliver our objection, since it will allow us to put the various nuances and refinements deprivationists have used to supplement their core view on the table, which we state here:

CORE DEPRIVATIONISM: an individual's death harms them by (and to the extent it does) depriving them of the goods that they would have obtained had they not died.

The first problem comes from the context-sensitivity inherent in conditional claims. Consider the following case adapted from from McMahan (2002):

YOUNG PEDESTRIAN: Phillip is an unfortunate 20 year old. At around noon, he crossed the street. Unfortunately, halfway across the street, he is hit by a careless driver and instantly killed. Unbeknownst to Phillip or anyone else, he also had an aneurysm building that would have burst and killed him a week later. The presence of the aneurysm was causally irrelevant to the way Phillip actually died.

In YOUNG PEDESTRIAN, what should a deprivationist say about the harm Phillip's death inflicted on him? For simplicity, we will assume that Phillip has been enjoying an average-valued life, and nothing noteworthy (good or ill) would have happened to him in the final week.

A natural thing to do here is to consult theories of the semantics of counterfactuals. Robert Stalnaker's (1968) theory bids us look at the nearest possible world where Phillip is not hit by a car at noon and see what happens.[7] However, nearness is generally reckoned by something like objective similarity, and in the most objectively similar worlds to actuality when Phillip is not killed by a car at noon, he's killed by that same car—moving a bit faster or slower—right around noon. Thus, his death will be of almost no consequence to him. A few moments makes little difference.

Of course, this isn't what we were after. The goal is to remove any car accident during the street crossing, not just one that happens precisely at noon. But even so, this buys Phillip an extra week at most. And yet his death seems more tragic than that.

[6] Bradley (2009: 51–2).

[7] David Lewis's (1973) view is similar, but allows for both ties for nearest world and cases where there is no nearest world, a complication we wish to ignore for now.

At this point deprivationists point to the variability and context-sensitivity of both event-individuation[8] and counterfactuals.[9] Perhaps the relevant counterfactual ignores the specific way in which Phillip dies and focuses instead on his dying young. Perhaps a friendly context selects a custom similarity relation between worlds so that a world missing both the accident and the aneurysm counts as closest. There are a variety of moves to make here, without a clear best option.[10] The exact details won't be relevant, but it will be important to remember that we must do something to get around this problem beyond simply appealing to the best natural-language semantics for 'would.'

A second problem case was introduced by Kai Draper (1999). Draper noted that certain absences, certain non-events, that seem to carry the same kind of harm as death strike us as trivial. This is an important observation. Draper asks us to consider a case like the following:

> THE LAMP: Alice is traveling on a plane to her tour of Europe. During her flight, she is served a mundane snack. Perhaps pretzels. With her snack, there is no lamp containing a benevolent genie who will grant her wishes. As a result of not being given a magic lamp on her flight, Alice goes on to have a perfectly acceptable but mundane life, far less good than she would have had had she been given a magic lamp.

On a permissive reading of what counts as a deprivation, Alice has suffered a grave deprivationist harm at snacktime. Comparing the total value of her actual life to that of her lamp-having life, even modulated by interest-relativity, yields a decisive judgment in favor of the latter. By not getting the lamp, she misses out on all sorts of goods that she would have had were she to have gotten it. Of course, there is no such lamp, but that's a contingent and empirical matter.

This is completely counter-intuitive. Alice was not harmed by not receiving a magic lamp. But she did miss out on ever so many goods by not receiving it. It looks like it takes more than missing out on goods you would have obtained, had things gone differently, to be actually harmed.

Deprivationists have had different reactions to this example. Some embrace it as a harm and explain our intuition that it is not by appeal to a distinction between what is harmful to us and what it is rational to feel in response to an event. Perhaps some even very grave harms don't merit concern.[11] Others attempt to offer adjustments to CORE DEPRIVATIONISM so that not all deprivations count toward making an event harmful, but only a special class. This is Draper's proposed solution, restricting the goods deprived by dying to those goods the deceased had a reasonable expectation of enjoying.[12] Alternatives to this might include a

[8] Feldman (1991), McMahan (2002). [9] Bradley (2009). [10] McMahan (2002).
[11] Bradley (2009). [12] Draper (1999).

restriction to goods the deceased had or would have had a right to, or goods that it was really possible for the deceased to receive (assuming that magical lamps are not, in fact, consistent with what science and history tell us about our world). Still others embrace it as a harm but without any sort of explanation for the intuition.[13]

Some of these responses seem reasonable. After all, it would be absurd to count each moment a grave misfortune simply because we are not receiving a genie's favor at that moment, a favor that would likely be much better for us than whatever it is we happen to be doing. There is something fantastical about the idea of missing the lamp as a deprivation that makes responses like Draper's and Bradley's seem apt. But the lamp problem points us to a structural flaw in deprivationism. Not all deprivations are equal, and we react much more severely to the deprivations of death than we do other deprivations, even those that are more severe by the lights of a total life or even an interest-relative view.

However, we do not need to countenance magical lamps to find mundane deprivations that match or exceed those of many deaths, yet which as humans we do not regard as nearly as terrible and as philosophers we do not regard as nearly as bad. We merely need consider the following cases:

> REMOTE TRAGEDY: An epidemic strikes a remote village. All the villagers die, losing out on many years of life of middling quality (this is not a rich village). The village is sufficiently remote that there is no danger of the disease reaching others, and no one misses the villagers or their village.
>
> A GOOD DAY FOR THE HOUSE: A group of card sharps detect a flaw in one of the casino's proprietary game. With 90 % probability, their card counting and betting strategy will net them millions. Only a few draws would ruin the plan. Unfortunately, the house gets lucky and the plan fails. That night, one of the casino employees realizes the flaw and the opportunity for an easy fortune passes. The sharps go on to live middling lives, but if they had won they would have lived high quality lives full of comfort, good relationships, and meaningful uses for their money.
>
> A BAD CALL ON THE MARKET: The manager of a pension fund thinks he sees an opportunity to play the markets for an easy gain. He is wrong; some unlikely geopolitical developments in Central Asia spook the market and his easy gain turns into a major loss. As a result, the fund has to decrease the monthly payment to the retirees it serves by one fifteenth in order to stay solvent. The retirees all experience a smallish diminishment in the quality of their golden years.
>
> TOO MANY LULZ: The dating website Online Match has a fairly good track record, connecting people so that on average it plants the seeds of many happy

[13] Timmerman (2019) does not explicitly discuss this case, but it follows from his version of deprivationism that not finding the lamp is bad for Alice in the specific sense of 'bad for' that he is interested in, and that he argues is the most joint-carving one on offer.

relationships every day. Bored one day, some hackers decide to take it out with a Distributed Denial of Service attack. The website is off for 24 hours, and as a result lots people miss out on what would have been a happy relationship. None of them go on to find anyone.

We may stipulate, by filling in the appropriate numbers, that each of these cases results in deprivations of the same total disvalue. I have not provided numbers of my own in order to remain neutral on such difficult questions as how to aggregate welfare and how to trade various goods off against each other. The starting case, of course, is the first one where people actually die. Each of the following cases provides some sort of contrast. The casino case involves a drop from what would have been a high level of well-being to something more middling; the retiree case involves a small drop in well-being spread over a large population; the dating case involves a loss not primarily in money (which, of course, is an instrumental good that may be used to gain or pursue final goods) but in relationships, plausibly a final good.

Even when we have filled in the cases so that the victims suffer deprivations of the same total disvalue, however that is reckoned, it should be clear that one of these is not like the others. Remote Tragedy is considerably worse. If we could prevent only one of these, that would be the one. And yet by deprivationist reckoning, they should be equivalent.

It will do no good to appeal to knock-on effects of REMOTE TRAGEDY as an explanation for why it is worse. There are none. Although usually the destruction of a village will prompt sadness among those who hear of it and mourning amongst those who knew the victims, there are no such considerations at play here. That is why remote tragedy is remote. Appeal to harms to people other than the victims has no purchase. The badness here is the badness of death alone.

Although these cases press on the same weak point in the deprivationist theory that the lamp case does, they have advantages that the lamp case lacks. Rather than being fantastical events, they are all quite mundane. Casinos exist; stock markets crash dating websites go down. Events like these happen all the time, and even though they inflict as much or more deprivationist harm upon their victims as death does, we do not count them as nearly as bad. Moreover, the goods that the people on the wrong end of those events are deprived of are ordinary goods: money, relationships, achievements, meaning.

I think this shows the inadequacy of deprivationist responses to the problem. It would be odd, as Bradley suggests, to think that it is fitting to respond in one way to a deprivation of some suite of goods due to death but in another to the loss of an equally valuable suite of goods (perhaps, if we contrived the case carefully, the exact same suite of goods) due to a stock market crash, a bad run at a casino, or a website going down.

Furthermore, in some of the cases I've given, Draper's requirement that the harmful deprivations be those that the victim reasonably expected to receive is fulfilled. The card sharps reasonably expected to beat the house; the stock manager reasonably expected to do well, the pensioners reasonably expected that their benefits would not decrease, and the lovers-that-weren't reasonably expected at least that the website would not be down. Since we are not relying on the fantastical or implausible, responses that trade on the fantasticality or implausibility of finding a magic lamp are no longer available.

3.1.2 Generalization

This flaw is not unique to deprivationism. It infects any extrinsicalist instrumentalist account of death's harm. How? Since the disvalue in an extrinsicalist instrumentalist view comes about because of a death's relation to other things—such as potential goods, plans, desires, and so on—we should be able to find examples of events other than deaths that stand in a similar relation to those same things. We can then compare cases where those events occur but no one dies, and compare them to cases where there are deaths. That is essentially what the series of cases in section 3.1.1 does for deprivationism. If what we have seen is representative, then we can predict that as above the death events will seem worst.

In fact, there is a body of research in psychology that confirms this prediction. Life is what psychologists call a "sacred value," and one of the characteristics of a sacred value is resistance to equivalences and trade-offs.[14] Thus we can expect that examples of events where lives are lost will consistently be viewed as worse than examples of events where other instrumental goods are lost, even when—by deprivationist lights—the other goods lost are of greater value than the lives.

3.2 Against Intrinsicalist Views

With the most common kind of extrinsicalist view out of the way, we will take a detour through intrinsicalist views before discussing views where the harm of death is extrinsic but final. We will do so because a discussion of death's intrinsic properties and why its harm cannot be among them will teach us some lessons that will become important in discussing the extrinsicalist final view. It will turn out that, if death is harmful, it is harmful partially in virtue of the arrow of time, and this will be significant throughout the rest of the discussion.

3.2.1 The Case for an Intrinsicalist View

At first glance, there's a strong case for regarding the harm of death as intrinsic. This comes from comparing the behavioral-motivational profile of our approach to death with that of the classic intrinsic harm: pain. We regard pain as harmful in

[14] Tetlock et al. (2000).

itself. We do not avoid pain in order to avoid some other harmful thing that accompanies pain. We expend other general-purpose intrumental goods like money in order to avoid pain. We do not tend to trade pain off against extrinsic harms, unless doing so serves to gain us intrinsic goods or helps us to avoid other intrinsic harms.

Likewise with death. We do not typically avoid death in order to avoid other unpleasant things; we typically avoid death because we do not want to die. In fact, we make considerable sacrifices to avoid death. We are not as a general rule willing to trade dying off to obtain other self-regarding goods (although we may for other-regarding goods); it's difficult to pay someone to die, unless the money goes to some other person like a family member or friend. In fact, death is one of the few harms that we will trade pain, even considerable pain, to avoid. We behave as if death itself is bad, not as if there is something that tends to accompany death that we wish to avoid by avoiding dying.

This suggests, but does not demand, that death's harm is inherent to dying. The natural first step in looking for something bad about dying is to look at the intrinsic properties of death itself. To this investigation we now turn.

3.2.2 Intrinsicality

It is notoriously tricky to say what makes a property intrinsic. The contemporary discussion of intrinsicality begins with Langton and Lewis (1998). In an earlier exchange, Jaegwon Kim (1982) had suggested a definition of intrinsicality in terms of accompaniment: a property is intrinsic just in case it can be instantiated by an object that is all alone, i.e. the only thing that exists. Lewis (1983) replied that 'being alone' is not intrinsic, but it obviously fits the definition.

Langton and Lewis argued that with some work, Kim's definition could be modified into something satisfactory. In order to do so, they introduce the idea of a property being independent of accompaniment. A property is independent of accompaniment in one of four ways: it can be instantiated alone, it can be instantiated while accompanied, a lonely thing can fail to instantiate it, and an accompanied thing can fail to instantiate it.

This does well on the easy cases. Intuitively intrinsic properties, like shape properties, are rightly classified as intrinsic. And intuitively extrinsic properties, like 'being alone,' are rightly classified as extrinsic. However, independence from accompaniment alone is not good enough. As Langton and Lewis noted, certain disjunctive properties will still cause trouble. For example, 'being cubical and alone or non-cubical and accompanied' is independent of accompaniment, but intuitively not intrinsic. In order to deal with cases like it they invoked the notion of naturalness, in an effort that is widely considered a failure.

However, recent work on intrinsicality has still found the idea of independence from accompaniment useful. In particular, Witmer, Butchard, and Trogdon (2005) have made it the core of an analysis of intrinsicality that seems well suited to our context. They begin, not with intrinsic properties, but with having a property in an intrinsic way. Since we are dealing with a property (value) that can be both intrinsic and extrinsic, this is a useful notion to have on hand. They define having a property in an intrinsic way as follows:[15]

INTRINSIC WAY: something has a property in an intrinsic way just in case (i) that property is independent of accompaniment, and (ii) any other property in virtue of which it has that property is also independent of accompaniment.

They then define an intrinsic property as one that such that all possible instances of it are had in an intrinsic way.

This may be adequate for the project of identifying the purely intrinsic properties. But in our case, we are interested in what it is to have a property like value that can be either intrinsic or extrinsic. The idea of having a property in an intrinsic fashion is useful, but we will need to relativize all of the definitions to instances of that property rather than to properties *simpliciter*.

This begins with independence of accompaniment. Langton and Lewis defined what it is for a property to be independent of accompaniment; how would this relativize to some object having a property independent of accompaniment? A first pass simply takes the Langton and Lewis definition and inserts relativizations:

INSTANCE INDEPENDENCE A: some object having a property is independent of accompaniment just in case: (i) the object has the property, (ii) it can have that property while lonely, (iii) it can have that property while accompanied, (iv) it can lack that property while lonely, and (v) it can lack that property while accompanied.

This definition will work for some paradigm cases. When I am lying down, my shape is more-or-less straight. If I were the only thing in the world, it would still be more-or-less straight. But I can sit up, and doing so would make my shape bent. If I were the only thing in the world, it would still be bent. So it looks my instantiating my shape-properties fits INSTANCE INDEPENDENCE A.

However, as it stands, this is not an adequate adaptation of the Langton and Lewis definition. I might have some of my intrinsic properties independent of accompaniment (and in fact this is likely the case with value properties), in which case they will fail clauses (iv) and (v) of the definition on account of my being unable to lack them. This suggests that we might be better off without those clauses, leading us to a second pass:

[15] Witmer et al. (2005: 333).

INSTANCE INDEPENDENCE B: some object having a property is independent of accompaniment just in case: (i) the object has that property (ii) it can have that property while lonely, and (iii) it can have that property while accompanied.

This definition now does not exclude essential properties from being had intrinsically. But it errs in the opposite direction: it now counts every essential property as being had independent of accompaniment, since a thing must have its essential properties whether or not it is accompanied. The original definition avoided this problem by allowing all possible instances of a property to count when testing for Langton and Lewis's clauses (iii) and (iv). Perhaps by borrowing this strategy, we can avoid the issue.

INSTANCE INDEPENDENCE C: some object having a property is independent of accompaniment just in case: (i) the object has that property, (ii) it can have that property while lonely, (iii) it can have that property while accompanied, (iv) something can lack that property while lonely, and (v) something can lack that property while accompanied.

Now we have a definition that looks well suited to handle essential properties. Unlike our second pass, it does not reckon every essential property as had independent of accompaniment. But it leaves the possibility open.

With a notion of independence of accompaniment that applies not to just to properties in general but to particular instances of property-having, we can now adapt the second step of the Witmer et al analysis of intrinsicality.

INSTANCE INTRINSICALITY: an instance of object having a property is in an intrinsic way just in case (i) the object having that property is independent of accompaniment, and (ii) for any other property such that the object has the property in virtue of it, the object has that property independent of accompaniment.

Like the Witmer et al. definition, this will fit both the paradigm cases and the problem cases. And it will allow us to test for when something has a property intrinsically, which will allow us to explore what (if any) properties a death has intrinsically.

3.2.3 The Intrinsic Value-Properties of a Death

The central question for an intrinsicalist view of death is to find an answer to the question: "What are the intrinsic properties of a death, and is it reasonable to count 'harm' among them?" Now that we have a working definition for when something has a property intrinsically, we can begin to answer this question. And it begins by thinking about what a death is.

A death is the ending of a life. That tells us several things. First, a death is an event. The ontology of events is contentious, and we do not need to take a stand here, but it means that to think sensibly about what properties an event would have alone we must assume that we can think sensibly about lonely events. Second, a death sits at one temporal boundary of a sequence of events, a life. We can think of a life as a four-dimensional complex event that traces through space-time. Crucially, a death sits at the later boundary of a life. The early boundary of a life, if one there be, is a birth. On the assumption that there is some harm in dying, this asymmetry is axiologically significant. Whatever else it may be, one's birth is not itself a harm. This means that the harm of death partly depends on the arrow of time.

The point here is related to what is known as the Lucretian "symmetry" argument. Lucretius argued that death is not harmful because birth is not harmful and there is an axiological symmetry between death and birth. I am going to proceed on the assumption that this premise is false, and that there is an axiological asymmetry between birth and death.[16] Since the arrow of time is the difference between the earlier and later temporal boundaries of a life (it determines which is earlier and which is later), the arrow of time must partially ground this asymmetry.

What is the arrow of time? We can see the arrow of time in many things. At a physical level, things tend toward equilibrium. Gasses expand to fill open spaces; they do not bunch up in a corner. Entropy increases. Causes precede their effects. We have reliable memories but not reliable foresight, departments of history but not departments of prophecy. Time, unlike space, has a directionality to it.

Why does time have this feature? Nothing in our standard fundamental physical laws requires it. An intriguing proposal reduces Time's Arrow to the probabilities of statistical mechanics, plus the hypothesis that our universe began in a low entropy state.[17] The details of this proposal aren't important; what matters is that the arrow of time itself depends on facts about the entropy of the boundary conditions of the universe and the physical grounds of the probabilities of statistical mechanics. When we place a death alone in a world, these things do not plausibly come with it. But without them, there is no arrow of time. Without an arrow of time, there is no death. There are merely axiologically symmetric endpoints to a life.

It is difficult to imagine what it would be like to live a life absent a direction to time. Much of how we conceive of ourselves involves before-and-after thinking. But we might be able to draw some inspiration from how we think of ourselves in space. Space, unlike time, has no inherent directionality. There is no absolute up or down, left or right. Our perception of these directions is relative. From the

[16] Kamm (1988). [17] Albert (2000), Loewer (2012), Chen (2020).

perspective of someone living in the United States, people living on the other side of the world are upside down, and vice versa. When Neil Armstrong stood on the Moon, 'up' for him was very different than when he was on Earth. More importantly, we do not attach axiological significance to any of the boundaries of our life in spacelike dimensions. 'The high point of a life' may occasionally serve as a metaphor for its best period, but it is not, except by accident, the literal boundary of the life in any of the spacelike dimensions.

If the way we perceive the spatial boundaries of our existence is a guide to how we would perceive its temporal boundaries if time, like space, had no inherent direction, then it looks like we cannot divorce the badness of death from the arrow of time. Consequently, death does not have its value-properties intrinsically because it does not have them independent of accompaniment. It must be accompanied by the physical grounds for the arrow of time. While the exact nature of those grounds is disputed, it is plausible that they involve the physical grounds of statistical mechanics and/or the entropy-properties of the boundary conditions of the universe. Not things that we will find when we consider a death alone. So death's harm cannot be intrinsic; the arrow of time is required to create the birth/death asymmetry.

Before we depart this discussion, it's worth anticipating an objection. One might be tempted to object that I have assumed a controversial four-dimensionalism about persons. According to the four dimensionalists, persons are 4D space-time worms (or series of stages, but we can ignore that wrinkle) that exist at specific times by having temporal parts at them. This is indeed controversial metaphysics, but I have not assumed it. I am assuming that lives are four-dimensional events spread out in time. But I have said nothing about the relationship between persons and lives. For all I have said, persons might be 4D worms spread through the extended events we call lives, or they might be 3D objects that sweep through their lives without having parts at different times. On that I am neutral.[18]

3.3 Against Extrinsicalist Final Views

Now that we have explored the relationship between the harm of death (if any there be) and the arrow of time, we are in a position to evaluate proposals where the harm of death is extrinsic but final. This view incorporates many of the insights we have already encountered. Unlike instrumentalist views, it is not vulnerable to parody cases where it predicts that a number of deaths are both

[18] I have spoken as if presentism about time is false, by characterizing lives as complex 4D events spread out in space-time. But that is just for ease of exposition. Taking the time we could translate everything into presentism-friendly terminology.

quantitatively and qualitatively equally as bad as other losses of instrumental value as might be found in a casino or a stock market. Unlike intrinsicalist views, it concedes that death does not have its harm independent of accompaniment. A defender of the extrinsicalist final view can acknowledge that a death has its value properties partially in virtue of the arrow of time. This makes it the most challenging account of the badness of death for an Epicurean to deal with.

How might we object to an extrinsicalist final view? One line of objection might co-opt a common argument for extrinsicalist instrumentalist views: the fact that not all deaths are equally bad for the ones who die. The death of someone who has lived a full life, seen their great projects to successful conclusion, and can look forward only to diminished capacities may still be bad for them, but the death of a young person in the prime of life with many hopes and dreams unpursued and unfulfilled seems worse for them. If death is a final bad, this kind of asymmetry is a mystery. Unlike pain, deaths do not come in duration or intensity, so a view according to which not all deaths are equal but death is a final harm has some explaining to do. It already must dismantle many of our intuitive judgments about death's harm. Why not one further, the judgment that death is a harm?

Another line of argument might point to some of the things that turn out to ground the harm of death as providing a bad explanation. We might begin this objection by noting that if the extrinsicalist final view is true, we get some odd counterfactuals, such as: if the initial entropy of the universe had been much higher, my death would not be harmful. This is because the arrow of time is partially explained by the initial entropy of the universe, and the harm of my death is partially explained by the arrow of time. Odd counterfactuals are often the canary in the coal mine for bad explanations. It is part of the role of explanations to explain not only what does happen, but what would happen under plausible suppositions. When these start going wrong, it's a clue that we haven't quite gotten the explanation of the actual phenomenon right. And yet if an extrinsicalist final view is right, the badness of death is explained by the arrow of time. Might we then instead reject the proposal which ties the arrow of time to statistical mechanics and the entropy at the boundary conditions of the universe? It is controversial. But other substitutes are not much better. One rival, for instance, simply takes the arrow of time as a brute fact about the universe.[19] This adds mystery to mystery. Another takes it as a feature of how similarity works in evaluating counterfactuals.[20] This will support similarly support strange counterfactuals about how, were the behavior of 'would' in English somewhat different, my death would not be harmful. Another wrong explanation.

Alone, neither line of argumentation is conclusive. Advocates of the view could find ways to explain or explain away the intuition that some deaths are worse than

[19] Maudlin (2007). [20] Lewis (1979).

others. They could also appeal to the controversial nature of attempts to account for the arrow of time, holding out for future research to provide a solution that does not give rise to the odd counterfactuals that point to it as a bad explanation for the harm of death. But these moves together seem a significant cost. They leave the view holding that death is an extrinsic final harm, but that we neither have a good grasp on which properties in the world are those in virtue of which it is a harm nor on what about death, beyond its seeming harmful, our account of death's harm is in a good position to explain. We have at best a pyrrhic victory.

4. Conclusion

Thus concludes our exploration. We have mapped out the possible kinds of value property into four categories: extrinsic instrumentalist, extrinsic final, intrinsic instrumentalist, and intrinsic final. In each category we have searched for a harm associated with death that is fit to play the role death plays in our moral theorizing and emotional lives. We found none. We are not exactly in a position to say, with the strongest form of Epicureanism, that death has no harm whatsoever associated with it. But the harms we have found are mundane, unconcerning in contexts that do not involve death. Perhaps they prevent us from saying that death is nothing to us. But if they are all that's bad about dying, we should find death's shadow considerably lightened.

Acknowledgments

Thanks are due to the following for helpful comments and conversations: Jimmy Goodrich, Michael Rabenberg, Ewan Kingston, Ryan Darr, the editors of this volume: Sara Bernstein and Tyron Goldschmidt, and audiences at the Princeton University Center for Human Values and the Princeton Values Forum.

References

Albert, David Z. (2000). *Time and Chance* (Cambridge, MA: Harvard University Press).

Bradley, Ben (2009). *Well-Being and Death* (Oxford: Oxford University Press).

Broome, John (2012). "The Badness of Death and the Goodness of Life", in *The Oxford Handbook of Philosophy of Death*, ed. Ben Bradley, Fred Feldman, and Jens Johansson (New York: Oxford University Press), pp. 218–33.

Chen, Eddy Keming (2020). "Time's Arrow in a Quantum Universe: On the Status of Statistical Mechanical Probabilities", in *Statistical Mechanics and Scientific*

Explanation: Determinism, Indeterminism and Laws of Nature, ed. Valia Allori (Singapore: World Scientific), pp. xxx–xx.

Climenhaga, Nevin (2018). "Infinite Value and the Best of All Possible Worlds", *Philosophy and Phenomenological Research* **97** (2): 367–92.

Draper, Kai (1999). "Disappointment, Sadness, and Death", *Philosophical Review* **108** (3): 387–414.

Feldman, Fred (1991). "Some Puzzles About the Evil of Death", *Philosophical Review* **100** (2): 205–27.

Feldman, Fred (1992). *Confrontations with the Reaper: A Philosophical Study of the Nature and Value of Death* (New York: Oxford University Press).

Gorman, August (forthcoming). "Taking Stock of the Risks of Life Without Death", in *Exploring The Philosophy of Death and Dying: Classical and Contemporary Perspectives*, ed. Michael Cholbi and Travis Timmerman (New York: Routledge).

Kamm, F. M. (1988). "Why is Death Bad and Worse than Prenatal Non-Existence?", *Pacific Philosophical Quarterly* **69** (2): 161–4.

Kim, Jaegwon (1982). "Psychophysical Supervenience", *Philosophical Studies* **41** (1): 51–70.

Korsgaard, Christine (1983). "Two Distinctions in Goodness", *Philosophical Review* **92** (2): 169–95.

Langton, Rae and David Lewis (1998). "Defining 'Intrinsic'", *Philosophy and Phenomenological Research* **58** (2): 333–45.

Lewis, David (1973). *Counterfactuals* (Oxford: Basil Blackwell).

Lewis, David (1979). "Counterfactual Dependence and Time's Arrow", *Noûs* **47** (3): 453–66.

Lewis, David (1983). "Extrinsic Properties", *Philosophical Studies* **44** (2): 197–200.

Loewer, Barry (2012). "Two Accounts of Laws and Time", *Philosophical Studies* **160** (1): 115–37.

Maudlin, Tim (2007). *The Metaphysics Within Physics* (Oxford: Oxford University Press).

McMahan, Jeff (2002). *The Ethics of Killing: Problems at the Margins of Life* (New York: Oxford University Press).

Nagel, Thomas (1970). "Death", *Noûs* **4** (1): 73–80.

Rosati, Connie S. (2013). "The Makropulos Case Revisted: Reflections on Immortality and Agency", in *The Oxford Handbook of Philosophy of Death*, ed. Ben Bradley, Fred Feldman, and Jens Johansson (New York: Oxford University Press), pp. 355–90.

Rubio, Daniel (2020), "In Defense of No Best World," *Australasian Journal of Philosophy* 98 (4): 811–825.

Stalnaker, Robert (1968). "A Theory of Conditionals", in *Studies in Logical Theory*, American Philosophical Quarterly Monograph Series 2, ed. Nicholas Rescher (Oxford: Blackwell), pp. 98–112.

Tetlock, Philip E., Orie V. Kristel, S. Beth Elson, Melanie C. Green, and Jennifer S. Lerner (2000). "The Psychology of the Unthinkable: Taboo Trade-Offs, Forbidden Base Rates, and Heretical Counterfactuals", *Journal of Personality and Social Psychology* **78** (5): 853–70.

Timmerman, Travis (2019). "A Dilemma for Epicureanism", *Philosophical Studies* **176**: 241–57.

Williams, Bernard (1973). "The Makropulos Case: Reflections on the Tedium of Immortality", in *Problems of the Self* (Cambridge: Cambridge University Press), pp. 82–100.

Witmer, D. Gene, William Butchard, and Kelly Trogdon (2005). "Intrinsicality Without Naturalness", *Philosophy and Phenomenological Research* **70** (2): 326–50.

Index